T0336897

NUMERICAL SOLUTION OF DIFFERENTIAL EQUATIONS

This introduction to finite difference and finite element methods is aimed at advanced undergraduate and graduate students who need to solve differential equations. The prerequisites are few (basic calculus, linear algebra, and ordinary and partial differential equations) and so the book will be accessible and useful to readers from a range of disciplines across science and engineering.

Part I begins with finite difference methods. Finite element methods are then introduced in Part II. In each part, the authors begin with a comprehensive discussion of one-dimensional problems, before proceeding to consider two or higher dimensions. An emphasis is placed on numerical algorithms, related mathematical theory, and essential details in the implementation, while some useful packages are also introduced. The authors also provide well-tested Matlab® codes, all available online.

ZHILIN LI is a tenured full professor at the Center for Scientific Computation & Department of Mathematics at North Carolina State University. His research area is in applied mathematics in general, particularly in numerical analysis for partial differential equations, moving interface/free boundary problems, irregular domain problems, computational mathematical biology, and scientific computing and simulations for interdisciplinary applications. Li has authored one monograph, *The Immersed Interface Method*, and also edited several books and proceedings.

ZHONGHUA QIAO is an associate professor in the Department of Applied Mathematics at the Hong Kong Polytechnic University.

TAO TANG is a professor in the Department of Mathematics at Southern University of Science and Technology, China.

NUMERICAL SOLUTION OF DIFFERENTIAL EQUATIONS

Introduction to Finite Difference and Finite Element Methods

ZHILIN LI
North Carolina State University, USA

ZHONGHUA QIAO
Hong Kong Polytechnic University, China

TAO TANG
Southern University of Science and Technology, China

CAMBRIDGE
UNIVERSITY PRESS

University Printing House, Cambridge CB2 8BS, United Kingdom

One Liberty Plaza, 20th Floor, New York, NY 10006, USA

477 Williamstown Road, Port Melbourne, VIC 3207, Australia

314-321, 3rd Floor, Plot 3, Splendor Forum, Jasola District Centre, New Delhi - 110025, India

103 Penang Road, #05-06/07, Visioncrest Commercial, Singapore 238467

Cambridge University Press is part of the University of Cambridge.

It furthers the University's mission by disseminating knowledge in the pursuit of education, learning and research at the highest international levels of excellence.

www.cambridge.org
Information on this title: www.cambridge.org/9781107163225
DOI: 10.1017/9781316678725

© Zhilin Li, Zhonghua Qiao, and Tao Tang 2018

This publication is in copyright. Subject to statutory exception and to the provisions of relevant collective licensing agreements, no reproduction of any part may take place without the written permission of Cambridge University Press.

First published 2018

A catalogue record for this publication is available from the British Library

ISBN 978-1-107-16322-5 Hardback
ISBN 978-1-316-61510-2 Paperback

Cambridge University Press has no responsibility for the persistence or accuracy of URLs for external or third-party internet websites referred to in this publication, and does not guarantee that any content on such websites is, or will remain, accurate or appropriate.

Table of Contents

Preface *page* ix

1 Introduction 1
- 1.1 Boundary Value Problems of Differential Equations 1
- 1.2 Further Reading 5

PART I FINITE DIFFERENCE METHODS 7

2 Finite Difference Methods for 1D Boundary Value Problems 9
- 2.1 A Simple Example of a Finite Difference Method 9
- 2.2 Fundamentals of Finite Difference Methods 14
- 2.3 Deriving FD Formulas Using the Method of Undetermined Coefficients 19
- 2.4 Consistency, Stability, Convergence, and Error Estimates of FD Methods 21
- 2.5 FD Methods for 1D Self-adjoint BVPs 27
- 2.6 FD Methods for General 1D BVPs 29
- 2.7 The Ghost Point Method for Boundary Conditions Involving Derivatives 30
- 2.8 An Example of a Nonlinear BVP 34
- 2.9 The Grid Refinement Analysis Technique 37
- 2.10 * 1D IIM for Discontinuous Coefficients 39
- Exercises 44

3 Finite Difference Methods for 2D Elliptic PDEs 47
- 3.1 Boundary and Compatibility Conditions 49
- 3.2 The Central Finite Difference Method for Poisson Equations 51
- 3.3 The Maximum Principle and Error Analysis 55

3.4 Finite Difference Methods for General Second-order Elliptic PDEs 60
3.5 Solving the Resulting Linear System of Algebraic Equations 61
3.6 A Fourth-order Compact FD Scheme for Poisson Equations 67
3.7 A Finite Difference Method for Poisson Equations in Polar Coordinates 69
3.8 Programming of 2D Finite Difference Methods 72
Exercises 75

4 FD Methods for Parabolic PDEs 78
4.1 The Euler Methods 80
4.2 The Method of Lines 85
4.3 The Crank–Nicolson scheme 87
4.4 Stability Analysis for Time-dependent Problems 89
4.5 FD Methods and Analysis for 2D Parabolic Equations 97
4.6 The ADI Method 99
4.7 An Implicit–explicit Method for Diffusion and Advection Equations 104
4.8 Solving Elliptic PDEs using Numerical Methods for Parabolic PDEs 105
Exercises 105

5 Finite Difference Methods for Hyperbolic PDEs 108
5.1 Characteristics and Boundary Conditions 109
5.2 Finite Difference Schemes 110
5.3 The Modified PDE and Numerical Diffusion/Dispersion 115
5.4 The Lax–Wendroff Scheme and Other FD methods 117
5.5 Numerical Boundary Conditions 120
5.6 Finite Difference Methods for Second-order Linear Hyperbolic PDEs 121
5.7 Some Commonly Used FD Methods for Linear System of Hyperbolic PDEs 127
5.8 Finite Difference Methods for Conservation Laws 127
Exercises 131

PART II FINITE ELEMENT METHODS 133

6 Finite Element Methods for 1D Boundary Value Problems 135
6.1 The Galerkin FE Method for the 1D Model 135
6.2 Different Mathematical Formulations for the 1D Model 138
6.3 Key Components of the FE Method for the 1D Model 143

6.4 Matlab Programming of the FE Method for the 1D Model Problem 152
Exercises 156

7 Theoretical Foundations of the Finite Element Method 158
7.1 Functional Spaces 158
7.2 Spaces for Integral Forms, $L^2(\Omega)$ and $L^p(\Omega)$ 160
7.3 Sobolev Spaces and Weak Derivatives 164
7.4 FE Analysis for 1D BVPs 168
7.5 Error Analysis of the FE Method 173
Exercises 178

8 Issues of the FE Method in One Space Dimension 181
8.1 Boundary Conditions 181
8.2 The FE Method for Sturm–Liouville Problems 185
8.3 High-order Elements 189
8.4 A 1D Matlab FE Package 195
8.5 The FE Method for Fourth-order BVPs in 1D 208
8.6 The Lax–Milgram Lemma and the Existence of FE Solutions 214
8.7 *1D IFEM for Discontinuous Coefficients 221
Exercises 223

9 The Finite Element Method for 2D Elliptic PDEs 228
9.1 The Second Green's Theorem and Integration by Parts in 2D 228
9.2 Weak Form of Second-order Self-adjoint Elliptic PDEs 231
9.3 Triangulation and Basis Functions 233
9.4 Transforms, Shape Functions, and Quadrature Formulas 246
9.5 Some Implementation Details 248
9.6 Simplification of the FE Method for Poisson Equations 251
9.7 Some FE Spaces in $H^1(\Omega)$ and $H^2(\Omega)$ 257
9.8 The FE Method for Parabolic Problems 272
Exercises 275

Appendix: Numerical Solutions of Initial Value Problems 279
A.1 System of First-order ODEs of IVPs 279
A.2 Well-posedness of an IVP 280
A.3 Some Finite Difference Methods for Solving IVPs 281
A.4 Solving IVPs Using Matlab ODE Suite 284
Exercises 288

References 289

Index 291

Preface

The purpose of this book is to provide an introduction to finite difference and finite element methods for solving ordinary and partial differential equations of boundary value problems. The book is designed for beginning graduate students, upper level undergraduate students, and students from interdisciplinary areas including engineers and others who need to obtain such numerical solutions. The prerequisite is a basic knowledge of calculus, linear algebra, and ordinary differential equations. Some knowledge of numerical analysis and partial differential equations would also be helpful but not essential.

The emphasis is on the understanding of finite difference and finite element methods and essential details in their implementation with reasonably mathematical theory. Part I considers finite difference methods, and Part II is about finite element methods. In each part, we start with a comprehensive discussion of one-dimensional problems before proceeding to consider two or higher dimensions. We also list some useful references for those who wish to know more in related areas.

The materials of this textbook in general can be covered in an academic year. The two parts of the book are essentially independent. Thus it is possible to use only one part for a class.

This is a textbook based on materials that the authors have used, and some are from Dr. Randall J. LeVeque's notes, in teaching graduate courses on the numerical solution of differential equations. Most sample computer programming is written in Matlab®. Some advantages of Matlab are its simplicity, a wide range of library subroutines, double precision accuracy, and many existing and emerging tool-boxes.

A website www4.ncsu.edu/~zhilin/`FD_FEM_Book` has been set up, to post or link computer codes accompanying this textbook.

We would like to thank Dr. Roger Hoskin, Lilian Wang, Peiqi Huang, and Hongru Chen for proofreading the book, or for providing Matlab code.

1
Introduction

1.1 Boundary Value Problems of Differential Equations

We discuss *numerical* solutions of problems involving ordinary differential equations (ODEs) or partial differential equations (PDEs), especially linear first- and second-order ODEs and PDEs, and problems involving systems of first-order differential equations.

A *differential equation* involves derivatives of an unknown function of one independent variable (say $u(x)$), or the partial derivatives of an unknown function of more than one independent variable (say $u(x,y)$, or $u(t,x)$, or $u(t,x,y,z)$, *etc.*). Differential equations have been used extensively to model many problems in daily life, such as pendulums, Newton's law of cooling, resistor and inductor circuits, population growth or decay, fluid and solid mechanics, biology, material sciences, economics, ecology, kinetics, thermodynamics, sports and computer sciences.[1] Examples include the Laplace equation for potentials, the Navier–Stokes equations in fluid dynamics, biharmonic equations for stresses in solid mechanics, and Maxwell equations in electromagnetics. For more examples and for the mathematical theory of PDEs, we refer the reader to Evans (1998) and references therein.

However, although differential equations have such wide applications, too few can be solved exactly in terms of elementary functions such as polynomials, $\log x$, e^x, trigonometric functions ($\sin x$, $\cos x$, ...), *etc.* and their combinations. Even if a differential equation can be solved analytically, considerable effort and sound mathematical theory are often needed, and the closed form of the solution may even turn out to be too messy to be useful. If the analytic solution of the differential equation is unavailable or too difficult to obtain, or

[1] There are other models in practice, for example, statistical models.

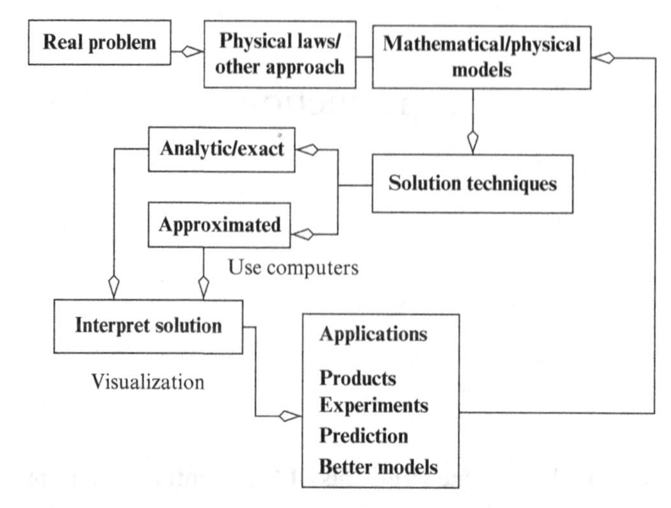

Figure 1.1. A flowchart of a problem-solving process.

takes some complicated form that is unhelpful to use, we may try to find an approximate solution. There are two traditional approaches:

1. Semi-analytic methods. Sometimes we can use series, integral equations, perturbation techniques, or asymptotic methods to obtain an approximate solution expressed in terms of simpler functions.
2. Numerical solutions. Discrete numerical values may represent the solution to a certain accuracy. Nowadays, these number arrays (and associated tables or plots) are obtained using computers, to provide effective solutions of many problems that were impossible to obtain before.

In this book, we mainly adopt the second approach and focus on numerical solutions using computers, especially the use of finite difference (FD) or finite element (FE) methods for differential equations. In Figure 1.1, we show a flowchart of the problem-solving process.

Some examples of ODE/PDEs are as follows.

1. Initial value problems (IVP). A canonical first-order system is

$$\frac{d\mathbf{y}}{dt} = \mathbf{f}(t, \mathbf{y}), \quad \mathbf{y}(t_0) = \mathbf{y}_0\,; \tag{1.1}$$

and a single higher-order differential equation may be rewritten as a first-order system. For example, a second-order ODE

$$\begin{aligned} &u''(t) + a(t)u'(t) + b(t)u(t) = f(t), \\ &u(0) = u_0, \quad \boxed{u'(0) = v_0}. \end{aligned} \tag{1.2}$$

can be converted into a first-order system by setting $y_1(t) = u$ and $y_2(t) = u'(t)$.

An ODE IVP can often be solved using Runge–Kutta methods, with adaptive time steps. In Matlab, there is the ODE-Suite which includes ode45, ode23, ode23s, ode15s, *etc.* For a stiff ODE system, either ode23s or ode15s is recommended; see Appendix for more details.

2. Boundary value problems (BVP). An example of an ODE BVP is

$$\begin{aligned} &u''(x) + a(x)u'(x) + b(x)u(x) = f(x), \quad 0 < x < 1, \\ &u(0) = u_0, \quad \boxed{u(1) = u_1}; \end{aligned} \tag{1.3}$$

and a PDE BVP example is

$$\begin{aligned} u_{xx} + u_{yy} &= f(x,y), \quad (x,y) \in \Omega, \\ u(x,y) &= u_0(x,y), \quad (x,y) \in \partial\Omega, \end{aligned} \tag{1.4}$$

where $u_{xx} = \frac{\partial^2 u}{\partial x^2}$ and $u_{yy} = \frac{\partial^2 u}{\partial y^2}$, in a domain Ω with boundary $\partial\Omega$. The above PDE is linear and classified as *elliptic*, and there are two other classifications for linear PDE, namely, *parabolic* and *hyperbolic*, as briefly discussed below.

3. BVP and IVP, *e.g.*,

$$\begin{aligned} &u_t = a u_{xx} + f(x,t), \\ &u(0,t) = g_1(t), \quad u(1,t) = g_2(t), \qquad \text{BC} \\ &u(x,0) = u_0(x), \qquad \text{IC}, \end{aligned} \tag{1.5}$$

where BC and IC stand for boundary condition(s) and initial condition, respectively, where $u_t = \frac{\partial u}{\partial t}$.

4. Eigenvalue problems, *e.g.*,

$$\begin{aligned} &u''(x) = \lambda u(x), \\ &u(0) = 0, \quad u(1) = 0. \end{aligned} \tag{1.6}$$

In this example, both the function $u(x)$ (the *eigenfunction*) and the scalar λ (the *eigenvalue*) are unknowns.

5. Diffusion and reaction equations, *e.g.*,

$$\frac{\partial u}{\partial t} = \nabla \cdot (\beta \nabla u) + \mathbf{a} \cdot \nabla u + f(u) \tag{1.7}$$

where $\mathbf{a}$ is a vector, $\nabla \cdot (\beta \nabla u)$ is a diffusion term, $\mathbf{a} \cdot \nabla u$ is called an advection term, and $f(u)$ a reaction term.

6. Systems of PDE. The incompressible Navier–Stokes model is an important nonlinear example:

$$\begin{aligned} &\rho\,(\mathbf{u}_t + (\mathbf{u}\cdot\nabla)\mathbf{u}) = \nabla p + \mu\Delta\mathbf{u} + \mathbf{F}, \\ &\nabla\cdot\mathbf{u} = 0. \end{aligned} \tag{1.8}$$

In this book, we will consider BVPs of differential equations in one dimension (1D) or two dimensions (2D). A linear second-order PDE has the following general form:

$$\begin{aligned} &a(x,y)u_{xx} + 2b(x,y)u_{xy} + c(x,y)u_{yy} \\ &\qquad + d(x,y)u_x + e(x,y)u_y + g(x,y)u(x,y) = f(x,y) \end{aligned} \tag{1.9}$$

where the coefficients are independent of $u(x,y)$ so the equation is linear in u and its partial derivatives. The solution of the 2D linear PDE is sought in some bounded domain Ω; and the classification of the PDE form (1.9) is:

- Elliptic if $b^2 - ac < 0$ for all $(x,y) \in \Omega$,
- Parabolic if $b^2 - ac = 0$ for all $(x,y) \in \Omega$, and
- Hyperbolic if $b^2 - ac > 0$ for all $(x,y) \in \Omega$.

The appropriate solution method typically depends on the equation class. For the first-order system

$$\frac{\partial\mathbf{u}}{\partial t} = A(\mathbf{x})\frac{\partial\mathbf{u}}{\partial\mathbf{x}}, \tag{1.10}$$

the classification is determined from the eigenvalues of the coefficient matrix $A(\mathbf{x})$.

Finite difference and finite element methods are suitable techniques to solve differential equations (ODEs and PDEs) numerically. There are other methods as well, for example, finite volume methods, collocation methods, spectral methods, *etc.*

1.1.1 Some Features of Finite Difference and Finite Element Methods

Many problems can be solved numerically by some finite difference or finite element methods. We strongly believe that any numerical analyst should be familiar with both methods and some important features listed below.

Finite difference methods:

- Often relatively simple to use, and quite easy to understand.
- Easy to implement for regular domains, *e.g.*, rectangular domains in Cartesian coordinates, and circular or annular domains in polar coordinates.
- Their discretization and approximate solutions are pointwise, and the fundamental mathematical tool is the Taylor expansion.
- There are many fast solvers and packages for regular domains, *e.g.*, the Poisson solvers Fishpack (Adams et al.) and Clawpack (LeVeque, 1998).
- Difficult to implement for complicated geometries.
- Have strong regularity requirements (the existence of high-order derivatives).

Finite element methods:

- Very successful for structural (elliptic type) problems.
- Suitable approach for problems with complicated boundaries.
- Sound theoretical foundation, at least for elliptic PDE, using Sobolev space theory.
- *Weaker* regularity requirements.
- Many commercial packages, *e.g.*, Ansys, Matlab PDE Tool-Box, Triangle, and PLTMG.
- Usually coupled with multigrid solvers.
- Mesh generation can be difficult, but there are now many packages that do this, *e.g.*, Matlab, Triangle, Pltmg, Fidap, Gmsh, and Ansys.

1.2 Further Reading

This textbook provides an introduction to finite difference and finite element methods. There are many other books for readers who wish to become expert in finite difference and finite element methods.

For FD methods, we recommend Iserles (2008); LeVeque (2007); Morton and Mayers (1995); Strikwerda (1989) and Thomas (1995). The textbooks by Strikwerda (1989) and Thomas (1995) are classical, while Iserles (2008); LeVeque (2007) and Morton and Mayers (1995) are relatively new. With LeVeque (2007), the readers can find the accompanying Matlab code from the author's website.

A classic book on FE methods is Ciarlet (2002), while Johnson (1987) and Strang and Fix (1973) have been widely used as graduate textbooks. The series by Carey and Oden (1983) not only presents the mathematical background of FE methods, but also gives some details on FE method programming in Fortran. Newer textbooks include Braess (2007) and Brenner and Scott (2002).

Part I

Finite Difference Methods

2

Finite Difference Methods for 1D Boundary Value Problems

2.1 A Simple Example of a Finite Difference Method

Let us consider a model problem

$$u''(x) = f(x), \quad 0 < x < 1, \quad u(0) = u_a, \quad u(1) = u_b,$$

to illustrate the general procedure using a finite difference method as follows.

1. *Generate a grid.* A grid is a finite set of points on which we seek the function values that represent an approximate solution to the differential equation. For example, given an integer parameter $n > 0$, we can use a *uniform* Cartesian grid

$$x_i = ih, \quad i = 0, 1, \ldots, n, \quad h = \frac{1}{n}.$$

   The parameter n can be chosen according to accuracy requirement. If we wish that the approximate solution has four significant digits, then we can take $n = 100$ or larger, for example.

2. *Represent the derivative by some finite difference formula* at every grid point where the solution is unknown, to get an algebraic system of equations. Note that for a twice differentiable function $\phi(x)$, we have

$$\phi''(x) = \lim_{\Delta x \to 0} \frac{\phi(x - \Delta x) - 2\phi(x) + \phi(x + \Delta x)}{(\Delta x)^2}.$$

   Thus at a grid point x_i, we can approximate $u''(x_i)$ using nearby function values to get a finite difference formula for the second-order derivative

$$u''(x_i) \approx \frac{u(x_i - h) - 2u(x_i) + u(x_i + h)}{h^2},$$

with some error in the approximation. In the finite difference method, we replace the differential equation at each grid point x_i by

$$\frac{u(x_i - h) - 2u(x_i) + u(x_i + h)}{h^2} = f(x_i) + error,$$

where the error is called the local *truncation error* and will be reconsidered later. Thus we *define* the finite difference (FD) solution (an approximation) for $u(x)$ at all x_i as the solution U_i (if it exists) of the following linear system of algebraic equations:

$$\begin{aligned}
\frac{u_a - 2U_1 + U_2}{h^2} &= f(x_1) \\
\frac{U_1 - 2U_2 + U_3}{h^2} &= f(x_2) \\
\frac{U_2 - 2U_3 + U_4}{h^2} &= f(x_3) \\
\cdots &= \cdots \\
\frac{U_{i-1} - 2U_i + U_{i+1}}{h^2} &= f(x_i) \\
\cdots &= \cdots \\
\frac{U_{n-3} - 2U_{n-2} + U_{n-1}}{h^2} &= f(x_{n-2}) \\
\frac{U_{n-2} - 2U_{n-1} + u_b}{h^2} &= f(x_{n-1}).
\end{aligned}$$

Note that the finite difference equation at each grid point involves solution values at three grid points, *i.e.*, at x_{i-1}, x_i, and x_{i+1}. The set of these three grid points is called the *finite difference stencil*.

3. *Solve the system of algebraic equations*, to get an approximate solution at each grid point. The system of algebraic equations can be written in the matrix and vector form

$$\begin{bmatrix}
-\frac{2}{h^2} & \frac{1}{h^2} & & & & \\
\frac{1}{h^2} & -\frac{2}{h^2} & \frac{1}{h^2} & & & \\
& \frac{1}{h^2} & -\frac{2}{h^2} & \frac{1}{h^2} & & \\
& & \ddots & \ddots & \ddots & \\
& & & \frac{1}{h^2} & -\frac{2}{h^2} & \frac{1}{h^2} \\
& & & & \frac{1}{h^2} & -\frac{2}{h^2}
\end{bmatrix}
\begin{bmatrix} U_1 \\ U_2 \\ U_3 \\ \vdots \\ U_{n-2} \\ U_{n-1} \end{bmatrix}
=
\begin{bmatrix} f(x_1) - u_a/h^2 \\ f(x_2) \\ f(x_3) \\ \vdots \\ f(x_{n-2}) \\ f(x_{n-1}) - u_b/h^2 \end{bmatrix}
\tag{2.1}$$

The tridiagonal system of linear equations above can be solved efficiently in $O(Cn)$ operations by the Crout or Cholesky algorithm, see for example, Burden and Faires (2010), where C is a constant, typically $C=5$ in this case.

4. *Implement and debug the computer code.* Run the program to get the output. Analyze and visualize the results (tables, plots, *etc.*).
5. *Error analysis.* Algorithmic consistency and stability implies convergence of the finite difference method, which will be discussed later. The convergence is pointwise, *i.e.*, $\lim_{h\to 0} \|u(x_i) - U_i\|_\infty = 0$. The finite difference method requires the solution $u(x)$ to have up to *second-order* continuous derivatives.

2.1.1 A Matlab Code for the Model Problem

Below we show a Matlab function called *two_point.m*, for the model problem, and use this Matlab function to illustrate how to convert the algorithm to a computer code.

```
function [x,U] = two_point(a,b,ua,ub,f,n)

%%%%%%%%%%%%%%%%%%%%%%%%%%%%%%%%%%%%%%%%%%%%%%%%%%%%%%%%%%%%%%
%   This matlab function two_point solves the following      %
%   two-point boundary value problem: u''(x) = f(x)          %
%   using the centered finite difference scheme.             %
%   Input:                                                   %
%    a, b: Two end points.                                   %
%    ua, ub: Dirichlet boundary conditions at a and b        %
%    f: external function f(x).                              %
%    n: number of grid points.                               %
%   Output:                                                  %
%    x: x(1),x(2),...x(n-1) are grid points                  %
%    U: U(1),U(2),...U(n-1) are approximate solution at      %
%    grid points                                             %
%%%%%%%%%%%%%%%%%%%%%%%%%%%%%%%%%%%%%%%%%%%%%%%%%%%%%%%%%%%%%%

h = (b-a)/n; h1=h*h;

A = sparse(n-1,n-1);
F = zeros(n-1,1);

for i=1:n-2,
  A(i,i) = -2/h1; A(i+1,i) = 1/h1; A(i,i+1)= 1/h1;
end
  A(n-1,n-1) = -2/h1;

for i=1:n-1,
  x(i) = a+i*h;
  F(i) = feval(f,x(i));
```

```
end
  F(1) = F(1) - ua/h1;
  F(n-1) = F(n-1) - ub/h1;

U = A\F;

return
%%%%%---------   End of the program -------------------------
```

We can call the Matlab function two_point directly in a Matlab command window, but a better way is to put all Matlab commands in a Matlab file (called an M-file), referred to here as *main.m*. The advantage of this is to keep a record, and we can also revisit or modify the file whenever we want.

To illustrate, suppose the differential equation is defined in the interval of (0, 1), with $f(x) = -\pi^2 \cos(\pi x)$, $u(0) = 0$, and $u(1) = -1$. A sample Matlab M-file is then as follows.

```
%%%%%%%% Clear all unwanted variables and graphs.
  clear;  close all
%%%%%%% Input

a=0; b=1; n=40;
ua=1; ub=-1;

%%%%%% Call the solver: U is the FD solution at the grid
       points.

[x,U] = two_point(a,b,ua,ub,'f',n);

%%%%%%%%%%%%%%%%% Plot and show the error %%%%%%%%%%%%%%%%%%

plot(x,U,'o'); hold   % Plot the computed solution

u=zeros(n-1,1);
for i=1:n-1,
  u(i) = cos(pi*x(i));
end
plot(x,u)               %%% Plot the true solution at the grid
                        %%% points on the same plot.
%%%%%%% Plot the error

figure(2); plot(x,U-u)

norm(U-u,inf)              %%% Print out the maximum error.
```

It is easy to check that the exact solution of the BVP is $\cos(\pi x)$. If we plot the computed solution, the finite difference approximation to the true solution at the grid points (use plot(x, u, 'o'), and the exact solution represented by the

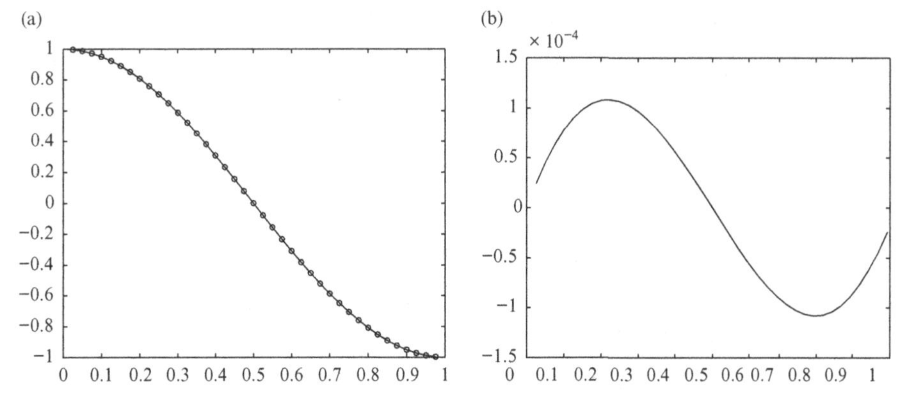

Figure 2.1. (a) A plot of the computed solution (little 'o's) with $n = 40$, and the exact solution (solid line). (b) The plot of the error.

solid line in Figure 2.1(a), the difference at the grid points is not too evident. However, if we plot the difference of the computed solution and the exact solution, which we call the error, we see that there is indeed a small difference of $O(10^{-3})$, *cf.* Figure 2.1(b), but in practice we may nevertheless be content with the accuracy of the computed numerical solution.

Questions One May Ask from This Example:

- Are there other finite difference formulas to approximate derivatives? If so, how do we derive them? The reader may have already encountered other formulas in an elementary numerical analysis textbook.
- How do we know whether a finite difference method works or not? If it works, how accurate is it? Specifically, what is the error of the computed solution?
- Do round-off errors affect the computed solution? If so, by how much?
- How do we deal with boundary conditions other than Dirichlet conditions (involving only function values) as above, notably Neumann conditions (involving derivatives) or mixed boundary conditions?
- Do we need different finite difference methods for different problems? If so, are the procedures similar?
- How do we know that we are using the most efficient method? What are the criteria, in order to implement finite difference methods efficiently?

We will address these questions in the next few chapters.

2.2 Fundamentals of Finite Difference Methods

The Taylor expansion is the most important tool in the analysis of finite difference methods. It may be written as an infinite series

$$u(x+h) = u(x) + hu'(x) + \frac{h^2}{2}u''(x) + \cdots + \frac{h^k}{k!}u^{(k)}(x) + \cdots \tag{2.2}$$

if $u(x)$ is "analytic" (differentiable to any order), or as a finite sum

$$u(x+h) = u(x) + hu'(x) + \frac{h^2}{2}u''(x) + \cdots + \frac{h^k}{k!}u^{(k)}(\xi), \tag{2.3}$$

where $x < \xi < x+h$ (or $x+h < \xi < x$ if $h < 0$), if $u(x)$ is differentiable up to k-th order. The second form of the Taylor expansion is sometimes called the extended mean value theorem. As indicated earlier, we may represent derivatives of a differential equation by finite difference formulas at grid points to get a linear or nonlinear algebraic system. There are several kinds of finite difference formulas to consider, but in general their accuracy is directly related to the magnitude of h (typically small).

2.2.1 *Forward, Backward, and Central Finite Difference Formulas for $u'(x)$*

Let us first consider the first derivative $u'(x)$ of $u(x)$ at a point $\bar{x}$ using the nearby function values $u(\bar{x} \pm h)$, where h is called the step size. There are three commonly used formulas:

$$\text{Forward FD:} \quad \Delta_+ u(\bar{x}) = \frac{u(\bar{x}+h) - u(\bar{x})}{h} \sim u'(\bar{x}), \tag{2.4}$$

$$\text{Backward FD:} \quad \Delta_- u(\bar{x}) = \frac{u(\bar{x}) - u(\bar{x}-h)}{h} \sim u'(\bar{x}), \tag{2.5}$$

$$\text{Central FD:} \quad \delta u(\bar{x}) = \frac{u(\bar{x}+h) - u(\bar{x}-h)}{2h} \sim u'(\bar{x}). \tag{2.6}$$

Below we derive these finite difference formulas from geometric intuitions and calculus.

From calculus, we know that

$$u'(\bar{x}) = \lim_{h\to 0} \frac{u(\bar{x}+h) - u(\bar{x})}{h}.$$

Assume $|h|$ is small and $u'(x)$ is continuous, then we expect that $\frac{u(\bar{x}+h)-u(\bar{x})}{h}$ is close to but usually not exactly $u'(\bar{x})$. Thus an approximation to the first derivative at $\bar{x}$ is the *forward finite difference* denoted and defined by

$$\Delta_+ u(\bar{x}) = \frac{u(\bar{x}+h) - u(\bar{x})}{h} \sim u'(\bar{x}), \tag{2.7}$$

where an error is introduced and $h > 0$ is called the step size, the distance between two points. Geometrically, $\Delta_+ u(\bar{x})$ is the slope of the secant line that connects the two points $(\bar{x}, u(\bar{x}))$ and $(\bar{x}+h, u(\bar{x}+h))$, and in calculus we recognize it tends to the slope of the tangent line at $\bar{x}$ in the limit $h \to 0$.

To determine how closely $\Delta_+ u(\bar{x})$ represents $u'(\bar{x})$, if $u(x)$ has second-order continuous derivatives we can invoke the extended mean value theorem (Taylor series) such that

$$u(\bar{x}+h) = u(x) + u'(\bar{x})h + \frac{1}{2} u''(\xi)\, h^2, \tag{2.8}$$

where $0 < \xi < h$. Thus we obtain the error estimate

$$E_f(h) = \frac{u(\bar{x}+h) - u(\bar{x})}{h} - u'(\bar{x}) = \frac{1}{2} u''(\xi) h = O(h), \tag{2.9}$$

so the error, defined as the difference of the approximate value and the exact one, is proportional to h and the discretization (2.7) is called *first-order accurate.* In general, if the error has the form

$$E(h) = Ch^p, \qquad p > 0, \tag{2.10}$$

then the method is called p-th order accurate.

Similarly, we can analyze the *backward finite difference* formula

$$\Delta_- u(\bar{x}) = \frac{u(\bar{x}) - u(\bar{x}-h)}{h}, \qquad h > 0, \tag{2.11}$$

for approximating $u'(\bar{x})$, where the error estimate is

$$E_b(h) = \frac{u(\bar{x}) - u(\bar{x}-h)}{h} - u'(\bar{x}) = -\frac{1}{2} u''(\xi)\, h = O(h), \tag{2.12}$$

so this formula is also first-order accurate.

Geometrically (see Figure 2.2), one may expect the slope of the secant line that passes through $(\bar{x}+h, u(\bar{x}+h))$ and $(\bar{x}-h, u(\bar{x}-h))$ is a better approximation to the slope of the tangent line of $u(\bar{x})$ at $(\bar{x}, u(\bar{x}))$, suggesting that the corresponding *central finite difference* formula

$$\delta u(\bar{x}) = \frac{u(\bar{x}+h) - u(\bar{x}-h)}{2h}, \qquad h > 0, \tag{2.13}$$

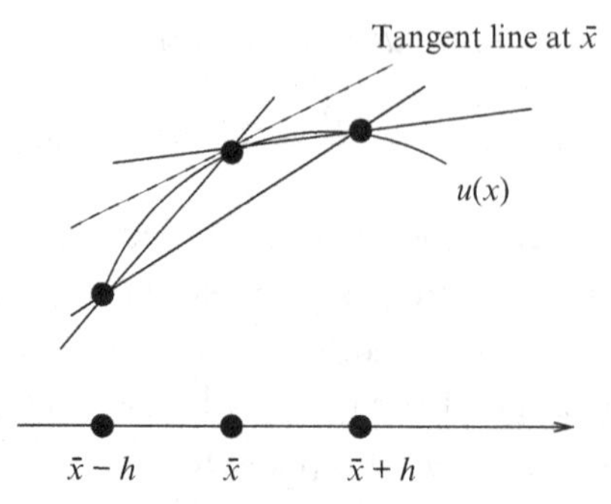

Figure 2.2. Geometric illustration of the forward, backward, and central finite difference formulas for approximating $u'(\bar{x})$.

for approximating the first-order derivative may be more accurate. In order to get the relevant error estimate, we need to retain more terms in the Taylor expansion:

$$u(x+h) = u(x) + hu'(x) + \frac{1}{2}u''(x)h^2 + \frac{1}{6}u'''(x)h^3 + \frac{1}{24}u^{(4)}(x)h^4 + \cdots,$$

$$u(x-h) = u(x) - hu'(x) + \frac{1}{2}u''(x)h^2 - \frac{1}{6}u'''(x)h^3 + \frac{1}{24}u^{(4)}(x)h^4 + \cdots,$$

which leads to

$$E_c(h) = \frac{u(\bar{x}+h) - u(\bar{x}-h)}{2h} - u'(\bar{x}) = \frac{1}{6}u'''(\bar{x})h^2 + \cdots = O(h^2) \quad (2.14)$$

where $\cdots$ stands for higher-order terms, so the central finite difference formula is second-order accurate. It is easy to show that (2.13) can be rewritten as

$$\delta u(\bar{x}) = \frac{u(\bar{x}+h) - u(\bar{x}-h)}{2h} = \frac{1}{2}\left(\Delta_+ + \Delta_-\right)u(\bar{x}).$$

There are other higher-order accurate formulas too, *e.g.*, the third-order accurate finite difference formula

$$\delta_3 u(\bar{x}) = \frac{2u(\bar{x}+h) + 3u(\bar{x}) - 6u(\bar{x}-h) + u(\bar{x}-2h)}{6h}. \quad (2.15)$$

2.2.2 Verification and Grid Refinement Analysis

Suppose now that we have learned or developed a numerical method and associated analysis. If we proceed to write a computer code to implement the method, how do we know that our code is bug-free and our analysis is correct? One way is by a *grid refinement analysis*.

Grid refinement analysis can be illustrated by a case where we know the exact solution.[1] Starting with a fixed h, say $h = 0.1$, we decrease h by half to

[1] Of course, we usually do not know it and there are other techniques to validate a computed solution to be discussed later.

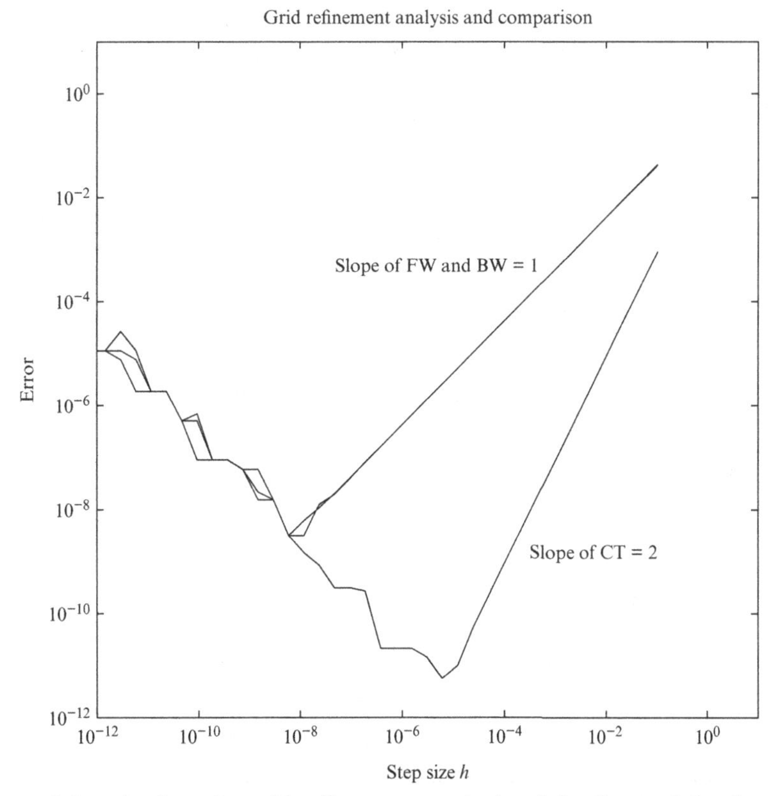

Figure 2.3. A plot of a grid refinement analysis of the forward, backward, and central finite difference formulas for $u'(x)$ using the log–log plot. The curves for forward and backward finite difference are almost identical and have slope one. The central formula is second-order accurate and the slope of the plot is two. As h gets smaller, round-off errors become evident and eventually dominant.

see how the error changes. For a first-order method, the error should decrease by a factor of two, *cf.* (2.9), and for a second-order method the error should be decrease by a factor of four, *cf.* (2.14), *etc.* We can plot the errors versus h in a log–log scale, where the slope is the order of convergence if the scales are identical on both axes. The forward, backward, and central finite difference formula rendered in a Matlab script file compare.m are shown below. For example, consider the function $u(x) = \sin x$ at $x = 1$, where the exact derivative is of course $\cos 1$. We plot the errors versus h in log–log scale in Figure 2.3, where we see that the slopes do indeed correctly produce the convergence order. As h decreases further, the round-off errors become dominant, which affects the actual errors. Thus for finite difference methods, we cannot take h arbitrarily small hoping to improve the accuracy. For this example, the best h that would provide the smaller error is $h \sim \sqrt{\epsilon} = \sqrt{10^{-16}} \sim 10^{-8}$ for the forward and backward formulas while it is $h \sim \sqrt[3]{\epsilon} \sim 10^{-5}$ for the central formula, where ϵ is the machine precision which is around 10^{-16} in Matlab for most computers. The best h can be estimated by balancing the formula error and the round-off errors. For the central formula, they are $O(h^2)$ and ϵ/h, respectively. The best h then is estimated by $h^2 = \epsilon/h$ or $h = \sqrt[3]{\epsilon}$. The following is a Matlab script file called *compare.m* that generates Figure 2.3.

```
% Compare truncation errors of the forward, backward,
% and central scheme for approximating u'(x). Plot the
% error and estimate the convergence order.
%    u(x) = sin(x) at x=1.   Exact derivative: u'(1) =
%    cos(1).

clear; close all
  h = 0.1;
  for i=1:5,
    a(i,1) = h;
    a(i,2) = (sin(1+h)-sin(1))/h - cos(1);
    a(i,3) = (sin(1) - sin(1-h))/h - cos(1);
    a(i,4) = (sin(1+h)-sin(1-h))/(2*h)- cos(1);
    h = h/2;
  end

  format short e  % Use this option to see the first
                  % a few significant digits.

  a = abs(a);      % Take absolute values of the matrix.
  h1 = a(:,1);     % Extract the first column which is h.
  e1 = a(:,2); e2 = a(:,3); e3 = a(:,4);

  loglog(h1,e1,h1,e2,h1,e3)
  axis('equal'); axis('square')
  axis([1e-6 1e1 1e-6 1e1])
  gtext('Slope of FW and BW = 1')
  gtext('Slope of CD =2')

%%%%%%%%%%%%%%%%% End Of Matlab Program %%%%%%%%%%%%%%%

%Computed Results:

%      h           forward        backward       central

%   1.0000e-01   -4.2939e-02    4.1138e-02   -9.0005e-04
%   5.0000e-02   -2.1257e-02    2.0807e-02   -2.2510e-04
%   2.5000e-02   -1.0574e-02    1.0462e-02   -5.6280e-05
%   1.2500e-02   -5.2732e-03    5.2451e-03   -1.4070e-05
%   6.2500e-03   -2.6331e-03    2.6261e-03   -3.5176e-06
```

2.3 Deriving FD Formulas Using the Method of Undetermined Coefficients

Sometimes we need a "one-sided" finite difference, for example to approximate a first derivative at some boundary value $\bar{x} = b$, and such approximations may also be used more generally. Thus to approximate a first derivative to second-order accuracy, we may anticipate a formula involving the values $u(\bar{x})$, $u(\bar{x} - h)$, and $u(\bar{x} - 2h)$ in using the *method of undetermined coefficients* in which we write

$$u'(\bar{x}) \sim \gamma_1 u(\bar{x}) + \gamma_2 u(\bar{x} - h) + \gamma_3 u(\bar{x} - 2h).$$

Invoking the Taylor expansion at $\bar{x}$ yields

$$\begin{aligned}
&\gamma_1 u(\bar{x}) + \gamma_2 u(\bar{x} - h) + \gamma_3 u(\bar{x} - 2h) \\
&\quad = \gamma_1 u(\bar{x}) + \gamma_2 \left((u(\bar{x}) - hu'(\bar{x}) + \frac{h^2}{2} u''(\bar{x}) - \frac{h^3}{6} u'''(\bar{x}) \right) \\
&\qquad + \gamma_3 \left((u(\bar{x}) - 2hu'(\bar{x}) + \frac{4h^2}{2} u''(\bar{x}) - \frac{8h^3}{6} u'''(\bar{x}) \right) + O(\max |\gamma_k| h^4),
\end{aligned}$$

which should approximate $u'(\bar{x})$ if we ignore the high-order term. So we set

$$\begin{aligned}
\gamma_1 + \gamma_2 + \gamma_3 &= 0 \\
-h\gamma_2 - 2h\gamma_3 &= 1 \\
h^2\gamma_2 + 4h^2\gamma_3 &= 0 .
\end{aligned}$$

It is easy to show that the solution to this linear system is

$$\gamma_1 = \frac{3}{2h}, \qquad \gamma_2 = -\frac{2}{h}, \qquad \gamma_3 = \frac{1}{2h},$$

and hence we obtain the one-sided finite difference scheme

$$u'(\bar{x}) = \frac{3}{2h} u(\bar{x}) - \frac{2}{h} u(\bar{x} - h) + \frac{1}{2h} u(\bar{x} - 2h) + O(h^2). \tag{2.16}$$

Another one-sided finite difference formula is immediately obtained by setting $-h$ for h, namely,

$$u'(\bar{x}) = -\frac{3}{2h} u(\bar{x}) + \frac{2}{h} u(\bar{x} + h) - \frac{1}{2h} u(\bar{x} + 2h) + O(h^2). \tag{2.17}$$

One may also differentiate a polynomial interpolation formula to get a finite difference scheme. For example, given a sequence of points $(x_i, u(x_i))$, $i = 0, 1, 2, \dots, n$, the Lagrange interpolating polynomial is

$$p_n(x) = \sum_{i=0}^{n} l_i(x) u(x_i), \quad \text{where} \quad l_i(x) = \prod_{j=0, j\neq i}^{n} \frac{(x - x_j)}{(x_i - x_j)},$$

suggesting that $u'(\bar{x})$ can be approximated by

$$u'(\bar{x}) \sim p_n'(\bar{x}) = \sum_{i=0}^{n} l_i'(\bar{x}) u(x_i).$$

2.3.1 FD Formulas for Second-order Derivatives

We can apply finite difference operators twice to get finite difference formulas to approximate the second-order derivative $u''(\bar{x})$, *e.g.*, the central finite difference formula

$$\begin{aligned} \Delta_+\Delta_- u(\bar{x}) &= \Delta_+ \frac{u(\bar{x}) - u(\bar{x} - h)}{h} \\ &= \frac{1}{h}\left(\frac{u(\bar{x}+h) - u(\bar{x})}{h} - \frac{u(\bar{x}) - u(\bar{x}-h)}{h}\right) \\ &= \frac{u(\bar{x}-h) - 2u(\bar{x}) + u(\bar{x}+h)}{h^2} \\ &= \Delta_-\Delta_+ u(\bar{x}) = \delta^2 u(\bar{x}) \end{aligned} \tag{2.18}$$

approximates $u''(\bar{x})$ to $O(h^2)$.

Using the same finite difference operator twice produces a one-sided finite difference formula, *e.g.*,

$$\begin{aligned} \Delta_+\Delta_+ u(\bar{x}) &= (\Delta_+)^2 u(\bar{x}) = \Delta_+ \frac{u(\bar{x}+h) - u(\bar{x})}{h} \\ &= \frac{1}{h}\left(\frac{u(\bar{x}+2h) - u(\bar{x}+h)}{h} - \frac{u(\bar{x}+h) - u(\bar{x})}{h}\right) \\ &= \frac{u(\bar{x}) - 2u(\bar{x}+h) + u(\bar{x}+2h)}{h^2}, \end{aligned} \tag{2.19}$$

also approximates $u''(\bar{x})$, but only to first-order accuracy $O(h)$.

In a similar way, finite difference operators can be used to derive approximations for partial derivatives. We obtain similar forms not only for partial derivatives u_x, u_{xx}, *etc.*, but also for mixed partial derivatives, *e.g.*,

$$\begin{aligned}
&\delta_x\delta_y u(\bar{x},\bar{y}) \\
&\quad = \frac{u(\bar{x}+h,\bar{y}+h)+u(\bar{x}-h,\bar{y}-h)-u(\bar{x}+h,\bar{y}-h)-u(\bar{x}-h,\bar{y}+h)}{4h^2} \\
&\quad \approx \frac{\partial^2 u}{\partial x \partial y}(\bar{x},\bar{y})\,, \qquad (2.20)
\end{aligned}$$

if we adopt a uniform step size in both x and y directions. Here we use the x subscript on δ_x to denote the central finite difference operator in the x direction, and so on.

2.3.2 FD Formulas for Higher-order Derivatives

We can likewise apply either lower-order finite difference formulas or the method of undetermined coefficients to obtain finite difference formulas for approximating third-order derivatives. For example,

$$\begin{aligned}
\Delta_+\delta^2 u(\bar{x}) &= \Delta_+\frac{u(\bar{x}-h)-2u(\bar{x})+u(\bar{x}+h)}{h^2} \\
&= \frac{-u(\bar{x}-h)+3u(\bar{x})-3u(\bar{x}+h)+u(\bar{x}+2h)}{h^3} \\
&= u'''(\bar{x})+\frac{h}{2}u^{(4)}(\bar{x})+\cdots
\end{aligned}$$

is first-order accurate. If we use the central formula

$$\frac{-u(\bar{x}-2h)+2u(\bar{x}-h)-2u(\bar{x}+h)+u(\bar{x}+2h)}{2h^3} = u'''(\bar{x})+\frac{h^2}{4}u^{(5)}(\bar{x})+\cdots\,,$$

then we can have a second-order accurate scheme. In practice, we seldom need more than fourth-order derivatives. For higher-order differential equations, we usually convert them to first- or second-order systems.

2.4 Consistency, Stability, Convergence, and Error Estimates of FD Methods

When a finite difference method is used to solve a differential equation, it is important to know how accurate the resulting approximate solution is compared to the true solution.

2.4.1 Global Error

If $\mathbf{U}=[U_1, U_2, \ldots, U_n]^T$ denotes the approximate solution generated by a finite difference scheme with no round-off errors and $\mathbf{u}=[u(x_1), u(x_2), \ldots, u(x_n)]$ is the exact solution at the grid points $x_1, x_2, \ldots, x_n$, then the *global error* vector is defined as $\mathbf{E}=\mathbf{U}-\mathbf{u}$. Naturally, we seek a smallest upper bound for the error vector, which is commonly measured using one of the following norms:

- The *maximum* or *infinity* norm $\|\mathbf{E}\|_\infty=\max_i\{|e_i|\}$. If the error is large at even one grid point then the maximum norm is also large, so this norm is regarded as the strongest measurement.
- The *1-norm*, an average norm defined as $\|\mathbf{E}\|_1=\sum_i h_i|e_i|$, analogous to the L^1 norm $\int |e(x)|\,dx$, where $h_i=x_{i+1}-x_i$.
- The *2-norm*, another average norm defined as $\|\mathbf{E}\|_2=(\sum_i h_i|e_i|^2)^{1/2}$, analogous to the L^2 norm $(\int |e(x)|^2\,dx)^{1/2}$.

If $\|E\|\le Ch^p$, $p>0$, we call the finite difference method *p-th order accurate*. We prefer to use a reasonably high-order accurate method while keeping the computational cost low.

Definition 2.1. A finite difference method is called *convergent* if $\lim_{h\to 0}\|\mathbf{E}\|=0$.

2.4.2 Local Truncation Errors

Local truncation errors refer to the differences between the original differential equation and its finite difference approximations at grid points. Local truncation errors measure how well a finite difference discretization approximates the differential equation.

For example, for the two-point BVP

$$u''(x)=f(x), \quad 0<x<1, \quad u(0)=u_a, \quad u(1)=u_b,$$

the local truncation error of the finite difference scheme

$$\frac{U_{i-1}-2U_i+U_{i+1}}{h^2}=f(x_i)$$

at x_i is

$$T_i=\frac{u(x_i-h)-2u(x_i)+u(x_i+h)}{h^2}-f(x_i), \qquad i=1,2,\ldots,n-1.$$

Thus on moving the right-hand side to the left-hand side, we obtain the local truncation error by rearranging or rewriting the finite difference equation to resemble the original differential equation, and then substituting the true solution $u(x_i)$ for U_i.

We define the local truncation error as follows. Let $P(d/dx)$ denote a differential operator on u in a linear differential equation, *e.g.*,

- $Pu=f$ represents $u''(x)=f(x)$ if $P\left(\frac{d}{dx}\right)=\frac{d^2}{dx^2}$; and
- $Pu=f$ represents $u'''+au''+bu'+cu=f(x)$ if

$$P\left(\frac{d}{dx}\right)=\frac{d^3}{dx^3}+a(x)\frac{d^2}{dx^2}+b(x)\frac{d}{dx}+c(x).$$

Let P_h be a corresponding finite difference operator, *e.g.*, for the second-order differential equation $u''(x)=f(x)$, a possible finite difference operator is

$$P_hu(x)=\frac{u(x-h)-2u(x)+u(x+h)}{h^2}.$$

More examples will be considered later. In general, the local truncation error is then defined as

$$T(x)=P_hu-Pu, \tag{2.21}$$

where it is noted that u is the exact solution. For example, for the differential equation $u''(x)=f(x)$ and the three-point central difference scheme (2.18) the local truncation error is

$$\begin{aligned} T(x)=P_hu-Pu&=\frac{u(x-h)-2u(x)+u(x+h)}{h^2}-u''(x)\\ &=\frac{u(x-h)-2u(x)+u(x+h)}{h^2}-f(x). \end{aligned} \tag{2.22}$$

Note that local truncation errors depend on the solution in the finite difference stencil (three-point in this example) but not on the solution globally (far away), hence the *local* tag.

Definition 2.2. A finite difference scheme is called *consistent* if

$$\lim_{h\to 0}T(x)=\lim_{h\to 0}(P_hu-Pu)=0. \tag{2.23}$$

Usually we should use consistent finite difference schemes.

If $|T(x)|\le Ch^p$, $p>0$, then we say that the discretization is p-th order accurate, where $C=O(1)$ is the *error constant* dependent on the solution $u(x)$. To check whether or not a finite difference scheme is consistent, we Taylor expand all the terms in the local truncation error at a master grid point x_i. For example, the three-point central finite difference scheme for $u''(x)=f(x)$ produces

$$T(x)=\frac{u(x-h)-2u(x)+u(x+h)}{h^2}-u''(x)=\frac{h^2}{12}u^{(4)}(x)+\cdots=O(h^2)$$

such that $|T(x)| \le Ch^2$, where $C = \max_{0\le x\le 1} |\frac{1}{12}u^{(4)}(x)|$ — *i.e.*, the finite difference scheme is consistent and the discretization is second-order accurate.

Now let us examine another finite difference scheme for $u''(x) = f(x)$, namely,

$$\frac{U_i - 2U_{i+1} + U_{i+2}}{h^2} = f(x_i), \quad i = 1, 2, \ldots, n-2,$$

$$\frac{U_{n-2} - 2U_{n-1} + u(b)}{h^2} = f(x_{n-1}).$$

The discretization at x_{n-1} is second-order accurate since $T(x_{n-1}) = O(h^2)$, but the local truncation error at all other grid points is

$$T(x_i) = \frac{u(x_i) - 2u(x_{i+1}) + u(x_{i+2})}{h^2} - f(x_i) = O(h),$$

i.e., at all grid points where the solution is unknown. We have $\lim_{h\to 0} T(x_i) = 0$, so the finite difference scheme is consistent. However, if we implement this finite difference scheme we may get weird results, because it does not use the boundary condition at $x = a$, which is obviously wrong. Thus consistency cannot guarantee the convergence of a scheme, and we need to satisfy another condition, namely, its *stability*.

Consider the representation

$$A\mathbf{u} = \mathbf{F} + \mathbf{T}, \qquad A\mathbf{U} = \mathbf{F} \qquad \Longrightarrow \qquad A(\mathbf{u} - \mathbf{U}) = \mathbf{T} = -A\mathbf{E}, \qquad (2.24)$$

where $\mathbf{E} = \mathbf{U} - \mathbf{u}$, A is the coefficient matrix of the finite difference equations, $\mathbf{F}$ is the modified source term that takes the boundary condition into account, and $\mathbf{T}$ is the local truncation error vector at the grid points where the solution is unknown. Thus, if A is nonsingular, then $\|\mathbf{E}\| = \|A^{-1}\mathbf{T}\| \le \|A^{-1}\|\|\mathbf{T}\|$. However, if A is singular, then $\|E\|$ may become arbitrarily large so the finite difference method may not converge. This is the case in the example above, whereas for the central finite difference scheme (2.22) we have $\|\mathbf{E}\| \le \|A^{-1}\|\, h^2$ and we can prove that $\|A^{-1}\|$ is bounded by a constant. Note that the global error depends on both $\|A^{-1}\|$ and the local truncation error vector $\mathbf{T}$.

Definition 2.3. A finite difference method for the BVPs is stable if A is invertible and

$$\|A^{-1}\| \le C, \quad \text{for all} \quad 0 < h < h_0, \qquad (2.25)$$

where C and h_0 are two constants that are independent of h.

From the definitions of consistency and stability, and the discussion above, we reach the following theorem:

Theorem 2.4. *A consistent and stable finite difference method is convergent.*

Usually it is easy to prove consistency but more difficult to prove stability.

To prove the convergence of the central finite difference scheme (2.22) for $u''(x)=f(x)$, we can apply the following lemma:

Lemma 2.5. *Consider a symmetric tridiagonal matrix $A\in R^{n\times n}$ whose main diagonals and off-diagonals are two constants, d and α, respectively. Then the eigenvalues of A are*

$$\lambda_j=d+2\alpha\cos\left(\frac{\pi j}{n+1}\right),\quad j=1,2,\ldots,n, \tag{2.26}$$

and the corresponding eigenvectors are

$$x_k^j=\sin\left(\frac{\pi kj}{n+1}\right),\quad k=1,2,\ldots,n. \tag{2.27}$$

The lemma can be proved by direct verification (from $A\mathbf{x}^j=\lambda_j\mathbf{x}^j$). We also note that the eigenvectors $\mathbf{x}^j$ are mutually orthogonal in the R^n vector space.

Theorem 2.6. *The central finite difference method for $u''(x)=f(x)$ and a Dirichlet boundary condition is convergent, with $\|E\|_\infty\le\|E\|_2\le Ch^{3/2}$.*

Proof From the finite difference method, we know that the finite difference coefficient matrix $A\in R^{(n-1)\times(n-1)}$ and it is tridiagonal with $d=-2/h^2$ and $\alpha=1/h^2$, so the eigenvalues of A are

$$\lambda_j=-\frac{2}{h^2}+\frac{2}{h^2}\cos\left(\frac{\pi j}{n}\right)=\frac{2}{h^2}\left(\cos(\pi jh)-1\right).$$

Noting that the eigenvalues of A^{-1} are $1/\lambda_j$ and A^{-1} is also symmetric, we have[2]

$$\begin{aligned}\|A^{-1}\|_2&=\frac{1}{\min|\lambda_j|}\\&=\frac{h^2}{2(1-\cos(\pi h))}=\frac{h^2}{2(1-(1-(\pi h)^2/2+(\pi h)^4/4!+\cdots)}<\frac{1}{\pi^2}.\end{aligned}$$

Using the inequality $\|A^{-1}\|_\infty\le\sqrt{n-1}\,\|A^{-1}\|_2$, therefore, we have

$$\|\mathbf{E}\|_\infty\le\|A^{-1}\|_\infty\,\|\mathbf{T}\|_\infty\le\sqrt{n-1}\,\|A^{-1}\|_2\,\|\mathbf{T}\|_\infty\le\frac{\sqrt{n-1}}{\pi^2}\,Ch^2\le\bar{C}h^{3/2},$$

[2] We can also use the identity $1-\cos(\pi h)=2\sin^2\frac{\pi h}{2}$ to get $\|A^{-1}\|_2=\frac{h^2}{2(1-\cos(\pi h))}=\frac{h^2}{4\sin^2\frac{\pi h}{2}}\approx\frac{1}{\pi^2}$.

since $\sqrt{n-1} \sim O(1/\sqrt{h})$. The error bound is *overestimated* since we can also prove that the infinity norm is also proportional to h^2 using the maximum principle or the Green function approach (*cf.* LeVeque, 2007).

Remark 2.7. The eigenvectors and eigenvalues of the coefficient matrix in (2.1) can be obtained by considering the Sturm–Liouville eigenvalue problem

$$u''(x) - \lambda u = 0, \quad u(0) = u(1) = 0. \tag{2.28}$$

It is easy to check that the eigenvalues are

$$\lambda_k = -(k\pi)^2, \quad k = 1, 2, \dots, \tag{2.29}$$

and the corresponding eigenvectors are

$$u_k(x) = \sin(k\pi x). \tag{2.30}$$

The discrete form at a grid point is

$$u_k(x_i) = \sin(k\pi ih), \quad i = 1, 2, \dots, n-1, \tag{2.31}$$

one of the eigenvectors of the coefficient matrix in (2.1). The corresponding eigenvalue can be found using the definition $A\mathbf{x} = \lambda\mathbf{x}$.

2.4.3 The Effect of Round-off Errors

From knowledge of numerical linear algebra, we know that

- $\|A\|_2 = \max|\lambda_j| = \frac{2}{h^2}(1 - \cos(\pi(n-1)h)) \sim \frac{4}{h^2} = 4n^2$, therefore the condition number of A satisfies $\kappa(A) = \|A\|_2\|A^{-1}\|_2 \sim n^2$.
- The relative error of the computed solution $\mathbf{U}$ for a stable scheme satisfies

$$\begin{aligned}\frac{\|\mathbf{U}-\mathbf{u}\|}{\|\mathbf{u}\|} &\le \text{local truncation error} + \text{round-off error in solving } A\mathbf{U} = \mathbf{F}\\ &\le \|A^{-1}\|\|\mathbf{T}\| + \bar{C}g(n)\|A\|\|A^{-1}\|\,\epsilon\\ &\le Ch^2 + \bar{C}g(n)\frac{1}{h^2}\,\epsilon,\end{aligned}$$

where $g(n)$ is the growth factor of the algorithm for solving the linear system of equations and ϵ is the machine precision. For most computers, $\epsilon \sim 10^{-8}$ when we use the single precision, and $\epsilon \sim 10^{-16}$ for the double precision.

Usually, the global error decreases as h decreases, but in the presence of round-off errors it may actually increase if h is too small! We can roughly estimate such a critical h. To keep the discussion simple, assume that $C \sim O(1)$

and $g(n) \sim O(1)$, roughly the critical h occurs when the local truncation error is about the same as the round-off error (in magnitute), *i.e.*,

$$h^2 \sim \frac{1}{h^2}\epsilon, \qquad \Longrightarrow \qquad n \sim \frac{1}{h} = \frac{1}{\epsilon^{1/4}}$$

which is about 100 for the single precision with the machine precision 10^{-8} and 10,000 for the double precision with the machine precision 10^{-16}. Consequently, if we use the single precision there is no point in taking more than 100 grid points and we get roughly four significant digits at best, so we usually use the double precision to solve BVPs. Note that Matlab uses double precision by default.

2.5 FD Methods for 1D Self-adjoint BVPs

Consider 1D self-adjoint BVPs of the form

$$\left(p(x)u'(x)\right)' - q(x)u(x) = f(x), \quad a < x < b, \tag{2.32}$$

$$u(a) = u_a, \quad u(b) = u_b, \quad \text{or other BC.} \tag{2.33}$$

This is also called a Sturm–Liouville problem. The existence and uniqueness of the solution is assured by the following theorem.

Theorem 2.8. *If $p(x) \in C^1(a,b)$, $q(x) \in C^0(a,b)$, $f(x) \in C^0(a,b)$, $q(x) \geq 0$ and there is a positive constant such that $p(x) \geq p_0 > 0$, then there is unique solution $u(x) \in C^2(a,b)$.*

Here $C^0(a,b)$ is the space of all continuous functions in $[a, b]$, $C^1(a,b)$ is the space of all functions that have continuous first-order derivative in $[a, b]$, and so on. We will see later that there are weaker conditions for finite element methods, where integral forms are used. The proof of this theorem is usually given in advanced differential equations courses.

Let us assume the solution exists and focus on the finite difference method for such a BVP, which involves the following steps.

Step 1: *Generate a grid.* For simplicity, consider the uniform Cartesian grid

$$x_i = a + ih, \quad h = \frac{b-a}{n}, \quad i = 0, 1, \ldots, n,$$

where in particular $x_0 = a$, $x_n = b$. Sometimes an adaptive grid may be preferred, but not for the central finite difference scheme.

Step 2: *Substitute derivatives with finite difference formulas at each grid point* where the solution is unknown. This step is also called the *discretization*.

Define $x_{i+\frac{1}{2}} = x_i + h/2$, so $x_{i+\frac{1}{2}} - x_{i-\frac{1}{2}} = h$. Thus using the central finite difference formula at a typical grid point x_i with half grid size, we obtain

$$\frac{p_{i+\frac{1}{2}} u'(x_{i+\frac{1}{2}}) - p_{i-\frac{1}{2}} u'(x_{i-\frac{1}{2}})}{h} - q_i u(x_i) = f(x_i) + E_i^1,$$

where $p_{i+\frac{1}{2}} = p(x_{i+\frac{1}{2}})$, $q_i = q(x_i)$, $f_i = f(x_i)$, and $E_i^1 = Ch^2$. Applying the central finite difference scheme for the first-order derivative then gives

$$\frac{p_{i+\frac{1}{2}} \frac{u(x_{i+1}) - u(x_i)}{h} - p_{i-\frac{1}{2}} \frac{u(x_i) - u(x_{i-1})}{h}}{h} - q_i u(x_i) = f(x_i) + E_i^1 + E_i^2,$$

for $i = 1, 2, \ldots, n-1$.

The consequent finite difference solution $U_i \approx u(x_i)$ is then defined as the solution of the linear system of equations

$$\frac{p_{i+\frac{1}{2}} U_{i+1} - \left(p_{i+\frac{1}{2}} + p_{i-\frac{1}{2}}\right) U_i + p_{i-\frac{1}{2}} U_{i-1}}{h^2} - q_i U_i = f_i, \tag{2.34}$$

for $i = 1, 2, \ldots, n-1$. In a matrix-vector form, this linear system can be written as $A\mathbf{U} = \mathbf{F}$, where

$$A = \begin{bmatrix} -\frac{p_{1/2}+p_{3/2}}{h^2} - q_1 & \frac{p_{3/2}}{h^2} & & & \\ \frac{p_{3/2}}{h^2} & -\frac{p_{3/2}+p_{5/2}}{h^2} - q_2 & \frac{p_{5/2}}{h^2} & & \\ & \ddots & \ddots & \ddots & \\ & & & \frac{p_{n-3/2}}{h^2} & -\frac{p_{n-3/2}+p_{n-1/2}}{h^2} - q_{n-1} \end{bmatrix},$$

$$\mathbf{U} = \begin{bmatrix} U_1 \\ U_2 \\ U_3 \\ \vdots \\ U_{n-2} \\ U_{n-1} \end{bmatrix}, \qquad \mathbf{F} = \begin{bmatrix} f(x_1) - \dfrac{p_{1/2} u_a}{h^2} \\ f(x_2) \\ f(x_3) \\ \vdots \\ f(x_{n-2}) \\ f(x_{n-1}) - \dfrac{p_{n-1/2} u_b}{h^2} \end{bmatrix}.$$

It is important to note that A is symmetric, negative definite, weakly diagonally dominant, and an M-matrix. Those properties guarantee that A is nonsingular.

The differential equation may also be written in the nonconservative form

$$p(x)u'' + p'(x)u' - qu = f(x),$$

where second-order finite difference formulas can be applied. However,

- the derivative of $p(x)$ or its finite difference approximation is needed and
- the coefficient matrix of the corresponding finite difference equations is no longer symmetric, nor negative positive definite, nor diagonally dominant.

Consequently, we tend to avoid using the nonconservative form if possible.

The local truncation error of the conservative finite difference scheme is

$$T_i = \frac{p_{i+\frac{1}{2}}u(x_{i+1}) - (p_{i+\frac{1}{2}} + p_{i-\frac{1}{2}})u(x_i) + p_{i-\frac{1}{2}}u(x_{i-1})}{h^2} - q_i u(x_i) - f_i. \quad (2.35)$$

Note that $P\,(d/dx) = (d/dx)\,(p\ d/dx) - q$ is the differential operator. It is easy to show that $|T_i| \le Ch^2$, but it is more difficult to show that $\|A^{-1}\| \le C$. However, we can use the maximum principle to prove second-order convergence of the finite difference scheme, as explained later.

2.6 FD Methods for General 1D BVPs

Consider the problem

$$p(x)u''(x) + r(x)u'(x) - q(x)u(x) = f(x), \quad a < x < b, \quad (2.36)$$

$$u(a) = u_a, \quad u(b) = u_b, \quad \text{or other BC.} \quad (2.37)$$

There are two different discretization techniques that we can use depending on the magnitude of $r(x)$.

1. Central finite difference discretization for all derivatives:
$$p_i \frac{U_{i-1} - 2U_i + U_{i+1}}{h^2} + r_i \frac{U_{i+1} - U_{i-1}}{2h} - q_i U_i = f_i, \quad (2.38)$$
for $i = 1, 2, \ldots, n-1$, where $p_i = p(x_i)$ and so on. An advantage of this discretization is that the method is second-order accurate, but a disadvantage is that the coefficient matrix may not be diagonally dominant even if $q(x) \ge 0$ and $p(x) > 0$. If u denotes the velocity in some applications, then $r(x)u'(x)$ is often called an advection term. When the advection is strong (*i.e.*, $|r(x)|$ is large), the central finite difference approximation is likely to have nonphysical oscillations, *e.g.*, when $r_i \sim 1/h$.
2. The upwinding discretization for the first-order derivative and the central finite difference scheme for the diffusion term:
$$p_i \frac{U_{i-1} - 2U_i + U_{i+1}}{h^2} + r_i \frac{U_{i+1} - U_i}{h} - q_i U_i = f_i, \text{ if } r_i \ge 0,$$
$$p_i \frac{U_{i-1} - 2U_i + U_{i+1}}{h^2} + r_i \frac{U_i - U_{i-1}}{h} - q_i U_i = f_i, \text{ if } r_i < 0.$$

This scheme increases the diagonal dominance of the finite difference coefficient matrix if $q(x) \geq 0$, but it is only first-order accurate. It is often easier and more accurate to solve a linear system of equations with diagonally dominant matrices. If $|r(x)|$ is very large (say $|r(x)| \sim 1/h$), the finite difference solution using the upwinding scheme will not have nonphysical oscillations compared with the central finite difference scheme.

Note that if $p(x)=1$, $r(x)=0$, and $q(x) \leq 0$, then the BVP is a 1D Helmholtz equation that may be difficult to solve if $|q(x)|$ is large, say $q(x) \sim 1/h^2$.

2.7 The Ghost Point Method for Boundary Conditions Involving Derivatives

In this section, we discuss how to treat Neumann and mixed (Robin) boundary conditions. Let us first consider the problem

$$u''(x)=f(x), \quad a<x<b,$$

$$u'(a)=\alpha, \qquad u(b)=u_b\,,$$

where the solution at $x=a$ is unknown. If we use a uniform Cartesian grid $x_i = a+ih$, then U_0 is one component of the solution. We can still use the central finite difference discretization at interior grid points

$$\frac{U_{i-1}-2U_i+U_{i+1}}{h^2}=f_i, \quad i=1,2,\ldots,n-1,$$

but we need an additional equation at $x_0=a$ given the Neumann boundary condition at a. One approach is to take

$$\frac{U_1-U_0}{h}=\alpha \quad \text{or} \quad \frac{-U_0+U_1}{h^2}=\frac{\alpha}{h}, \tag{2.39}$$

and the resulting linear system of equations is again tridiagonal and symmetric negative definite:

$$\begin{bmatrix} -\frac{1}{h^2} & \frac{1}{h^2} & & & & \\ \frac{1}{h^2} & -\frac{2}{h^2} & \frac{1}{h^2} & & & \\ & \frac{1}{h^2} & -\frac{2}{h^2} & \frac{1}{h^2} & & \\ & & \ddots & \ddots & \ddots & \\ & & & \frac{1}{h^2} & -\frac{2}{h^2} & \frac{1}{h^2} \\ & & & & \frac{1}{h^2} & -\frac{2}{h^2} \end{bmatrix} \begin{bmatrix} U_0 \\ U_1 \\ U_2 \\ \vdots \\ U_{n-2} \\ U_{n-1} \end{bmatrix} = \begin{bmatrix} \frac{\alpha}{h} \\ f(x_1) \\ f(x_2) \\ \vdots \\ f(x_{n-2}) \\ f(x_{n-1}) - \frac{u_b}{h^2} \end{bmatrix}. \tag{2.40}$$

However, this approach is only first-order accurate if $\alpha \neq 0$.

To maintain second-order accuracy, the *ghost point method* is recommended, where a ghost grid point $x_{-1}=x_0-h=a-h$ is added and the solution is extended to the interval $[a-h,a]$. Then the central finite difference scheme can be used at all grid points where the solution is unknown, *i.e.*, for $i=0,1,\ldots,n-1$. However, we now have n equations and $n+1$ unknowns including U_{-1}, so one more equation is needed to close the system. The additional equation is the central finite difference equation for the Neumann boundary condition

$$\frac{U_1-U_{-1}}{2h}=\alpha, \tag{2.41}$$

which yields $U_{-1}=U_1-2h\alpha$. Inserting this into the central finite difference equation at $x=a$, *i.e.*, at x_0, now treated as an "interior" grid point, we have

$$\frac{U_{-1}-2U_0+U_1}{h^2}=f_0,$$

$$\frac{U_1-2h\alpha-2U_0+U_1}{h^2}=f_0,$$

$$\frac{-U_0+U_1}{h^2}=\frac{f_0}{2}+\frac{\alpha}{h},$$

where the coefficient matrix is precisely the same as for (2.40) and the only difference in this second-order method is the component $f_0/2+\alpha/h$ in the vector on the right-hand side, rather than α/h in the previous first-order method.

To discuss the stability, we can use the eigenvalues of the coefficient matrix, so the 2-norm of the inverse of the coefficient matrix, $\|A^{-1}\|_2$. The eigenvalues can again be associated with the related continuous problem

$$u''(x)-\lambda u=0, \quad u'(0)=0, \quad u(1)=0. \tag{2.42}$$

It is easy to show that the eigenvectors are

$$u_k(x)=\cos\left(\frac{\pi x}{2}+k\pi x\right) \tag{2.43}$$

corresponding to the eigenvalues $\lambda_k=-(\pi/2+k\pi)^2$, from which we can conclude that the ghost point approach is stable. The convergence follows by combining the stability and the consistency.

We compare the two finite difference methods in Figure 2.4, where the differential equation $u''(x)=f(x)$ is subject to a Dirichlet boundary condition at $x=0$ and a Neumann boundary condition at $x=0.5$. When $f(x)=-\pi^2\cos\pi x$, $u(0)=1$, $u'(0.5)=-\pi$, the exact solution is $u(x)=\cos\pi x$. Figure 2.4(a) shows the grid refinement analysis using both the backward finite

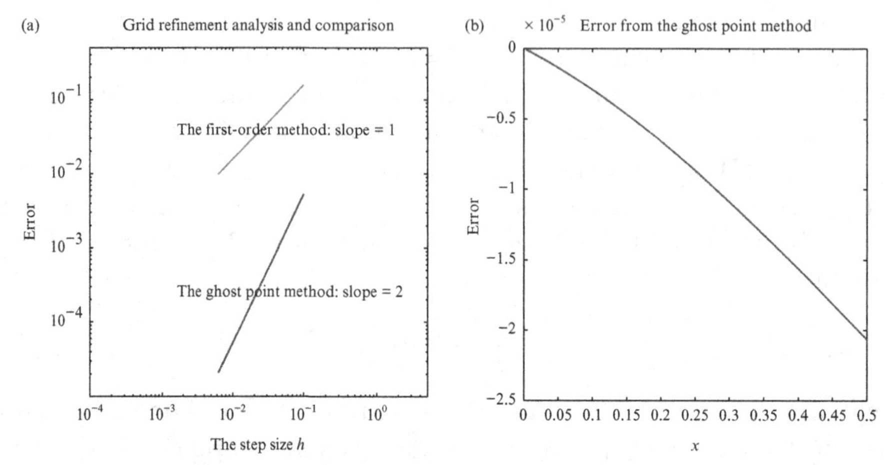

Figure 2.4. (a) A grid refinement analysis of the ghost point method and the first-order method. The slopes of the curves are the order of convergence. (b) The error plot of the computed solution from the ghost point method.

difference method (*bw_at_b.m*) and the ghost point method (*ghost_at_b.m*). The error in the second-order method is evidently much smaller than that in the first-order method. In Figure 2.4(b), we show the error plot of the ghost point method, and note the error at $x=b$ is no longer zero.

2.7.1 A Matlab Code of the Ghost Point Method

```
function [x,U] = ghost_at_b(a,b,ua,uxb,f,n)

%%%%%%%%%%%%%%%%%%%%%%%%%%%%%%%%%%%%%%%%%%%%%%%%%%%%%%%%%%%%%
%   This matlab function two_point solves the following     %
%   two-point boundary value problem: u''(x) = f(x)         %
%   using the center finite  difference scheme.             %
%   Input:                                                  %
%    a, b: Two end points.                                  %
%    ua, uxb: Dirichlet and Neumann boundary conditions     %
%    at a and b                                             %
%    f: external function f(x).                             %
%    n: number of grid points.                              %
%   Output:                                                 %
%    x: x(1),x(2),...x(n) are grid points                   %
%    U: U(1),U(2),...U(n) are approximate solution at       %
%    grid points.                                           %
%%%%%%%%%%%%%%%%%%%%%%%%%%%%%%%%%%%%%%%%%%%%%%%%%%%%%%%%%%%%%
```

```
h = (b-a)/n; h1=h*h;

A = sparse(n,n);
F = zeros(n,1);

for i=1:n-1,
  A(i,i) = -2/h1; A(i+1,i) = 1/h1; A(i,i+1)= 1/h1;
end
  A(n,n) = -2/h1;
  A(n,n-1) = 2/h1;

for i=1:n,
  x(i) = a+i*h;
  F(i) = feval(f,x(i));
end
  F(1) = F(1) - ua/h1;
  F(n) = F(n) - 2*uxb/h;

U = A\F;

return
```

2.7.1.1 The Matlab Driver Program

Below is a Matlab driver code to solve the two-point BVP

$$u''(x) = f(x), \quad a < x < b,$$

$$u(a) = ua, \qquad u'(b) = uxb.$$

```
%%%%%%%% Clear all unwanted variable and graphs.

clear;  close all

%%%%%%% Input

a =0; b=1/2;
ua = 1; uxb = -pi;

%%%%%% Call solver: U is the FD solution

n=10;
k=1;

for k=1:5
    [x,U] = ghost_at_b(a,b,ua,uxb,'f',n);
    %ghost-point method.
  u=zeros(n,1);
  for i=1:n,
    u(i) = cos(pi*x(i));
  end
```

```
  h(k) = 1/n;
  e(k) = norm(U-u,inf);      %%% Print out the maximum error.
  k = k+1; n=2*n;
end

log-log(h,e,h,e,'o'); axis('equal'); axis('square'),
title('The error plot in log-log scale, the slope = 2');
figure(2); plot(x,U-u); title('Error')
```

2.7.2 Dealing with Mixed Boundary Conditions

The ghost point method can be used to discretize a mixed boundary condition. Suppose that $\alpha u'(a) + \beta u(a) = \gamma$ at $x = a$, where $\alpha \neq 0$. Then we can discretize the boundary condition by

$$\alpha \frac{U_1 - U_{-1}}{2h} + \beta U_0 = \gamma,$$

$$\text{or} \qquad U_{-1} = U_1 + \frac{2\beta h}{\alpha} U_0 - \frac{2h\gamma}{\alpha},$$

and substitute this into the central finite difference equation at $x = x_0$ to get

$$\left(-\frac{2}{h^2} + \frac{2\beta}{\alpha h}\right) U_0 + \frac{2}{h^2} U_1 = f_0 + \frac{2\gamma}{\alpha h}, \tag{2.44}$$

$$\text{or} \qquad \left(-\frac{1}{h^2} + \frac{\beta}{\alpha h}\right) U_0 + \frac{1}{h^2} U_1 = \frac{f_0}{2} + \frac{\gamma}{\alpha h}, \tag{2.45}$$

yielding a symmetric coefficient matrix.

2.8 An Example of a Nonlinear BVP

Discretrizing a nonlinear differential equation generally produces a nonlinear algebraic system. Furthermore, if we can solve the nonlinear system then we can get an approximate solution. We present an example in this section to illustrate the procedure.

Consider the following nonlinear (a quasilinear) BVP:

$$\begin{gathered} \frac{d^2 u}{dx^2} - u^2 = f(x), \qquad 0 < x < \pi, \\ u(0) = 0, \qquad u(\pi) = 0. \end{gathered} \tag{2.46}$$

If we apply the central finite difference scheme, then we obtain the system of nonlinear equations

$$\frac{U_{i-1} - 2U_i + U_{i+1}}{h^2} - U_i^2 = f(x_i), \quad i = 1, 2, \ldots, n-1. \tag{2.47}$$

The nonlinear system of equations above can be solved in several ways:

- Approximate the nonlinear ODE using a linearization process. Unfortunately, not all linearization processes will work.
- Substitution method, where the nonlinear term is approximated upon an iteration using the previous approximation. For an example, given an initial guess $U^{(0)}(x)$, we get a new approximation using

  $$\frac{U_{i-1}^{k+1} - 2U_i^{k+1} + U_{i+1}^{k+1}}{h^2} - U_i^k U_i^{k+1} = f(x_i), \qquad k = 0, 1, \ldots, \tag{2.48}$$

  involving a two-point BVP at each iteration. The main concerns are then whether or not the method converges, and the rate of the convergence if it does.
- Solve the nonlinear system of equations using advanced methods, *i.e.*, Newton's method as explained below, or its variations.

In general, a nonlinear system of equations $\mathbf{F}(\mathbf{U}) = 0$ is obtained if we discretize a nonlinear ODE or PDE, *i.e.*,

$$\begin{cases} F_1(U_1, U_2, \ldots, U_m) = 0, \\ F_2(U_1, U_2, \ldots, U_m) = 0, \\ \quad \vdots \quad \vdots \quad \vdots \quad \vdots \\ F_m(U_1, U_2, \ldots, U_m) = 0, \end{cases} \tag{2.49}$$

where for the example, we have $m = n - 1$, and

$$F_i(U_1, U_2, \ldots, U_m) = \frac{U_{i-1} - 2U_i + U_{i+1}}{h^2} - U_i^2 - f(x_i), \quad i = 1, 2, \ldots, n-1.$$

The system of the nonlinear equations can generally be solved by Newton's method or some sort of variation. Given an initial guess $\mathbf{U}^{(0)}$, the Newton iteration is

$$\mathbf{U}^{(k+1)} = \mathbf{U}^{(k)} - (J(\mathbf{U}^{(k)}))^{-1}\mathbf{F}(\mathbf{U}^{(k)}) \tag{2.50}$$

or

$$\begin{cases} J(\mathbf{U}^{(k)})\Delta\mathbf{U}^{(k)} = -\mathbf{F}(\mathbf{U}^{(k)}), \\ \mathbf{U}^{(k+1)} = \mathbf{U}^{(k)} + \Delta\mathbf{U}^{(k)}, \end{cases} \quad k = 0, 1, \ldots$$

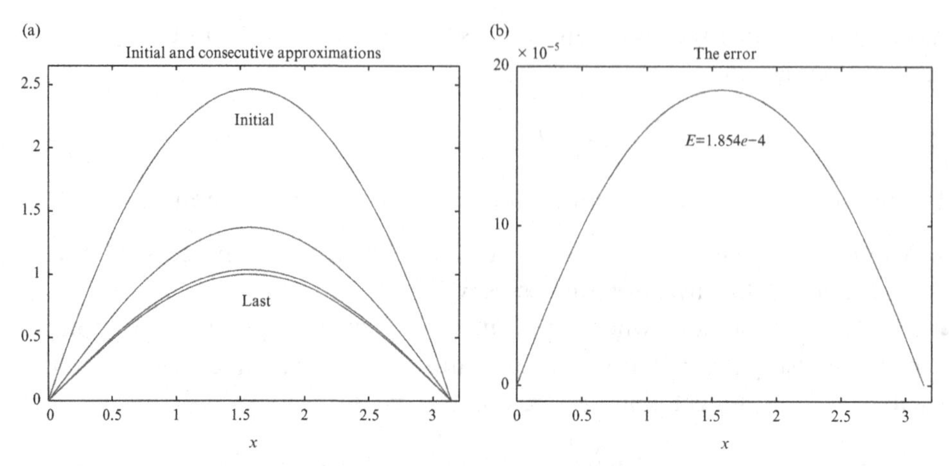

Figure 2.5. (a) Plot of the initial and consecutive approximations to the nonlinear system of equations and (b) the error plot.

where $J(\mathbf{U})$ is the Jacobian matrix defined as

$$\begin{bmatrix} \dfrac{\partial F_1}{\partial U_1} & \dfrac{\partial F_1}{\partial U_2} & \cdots & \dfrac{\partial F_1}{\partial U_m} \\ \dfrac{\partial F_2}{\partial U_1} & \dfrac{\partial F_2}{\partial U_2} & \cdots & \dfrac{\partial F_2}{\partial U_m} \\ \vdots & \vdots & \vdots & \vdots \\ \dfrac{\partial F_m}{\partial U_1} & \dfrac{\partial F_m}{\partial U_2} & \cdots & \dfrac{\partial F_m}{\partial U_m} \end{bmatrix}.$$

For the example problem, we have

$$J(\mathbf{U}) = \frac{1}{h^2}\begin{bmatrix} -2-2h^2U_1 & 1 & & \\ 1 & -2-2h^2U_2 & 1 & \\ & \ddots & \ddots & \ddots \\ & & 1 & -2-2h^2U_{n-1} \end{bmatrix}.$$

We implemented Newton's method in a Matlab code *non_tp.m*. In Figure 2.5(a), we show the initial and consecutive approximations to the nonlinear BVP using Newton's method with $U_i^0 = x_i(\pi - x_i)$. With an $n=40$ mesh and the tolerance $tol = 10^{-8}$, it takes only 6 iterations to converge. The infinity error of the computed solution at the grid points is $\|E\|_\infty = 1.8540 \times 10^{-4}$. In the right plot, the errors at the grid points are plotted.

It is not always easy to find $J(\mathbf{U})$ and it can be computationally expensive. Furthermore, Newton's method is only locally convergent, *i.e.*, it requires a close initial guess to guarantee its convergence. Quasi-Newton methods, such as the Broyden and BFGS rank-one and rank-two update methods, and also conjugate gradient methods, can avoid evaluating the Jacobian matrix. The main issues remain global convergence, the convergence order (Newton's method is quadratically convergent locally), and computational issues (storage, *etc.*). A well-known software package called MINPACK is available through the netlib (*cf.* Dennis and Schnabel, 1996 for more complete discussions).

2.9 The Grid Refinement Analysis Technique

After we have learned or developed a numerical method, together with its convergence analysis (consistency, stability, order of convergence, and computational complexity such as operation counts, storage), we need to validate and confirm the analysis numerically. The algorithmic behavior becomes clearer through the numerical results, and there are several ways to proceed.

- Analyze the output. Examine the boundary conditions and maximum/ minimum values of the numerical solutions, to see whether they agree with the ODE or PDE theory and your intuition.
- Compare the numerical solutions with experiential data, with sufficient parameter variations.
- Do a grid refinement analysis, whether an exact solution is known or not.

Let us now explain the grid refinement analysis when there is an exact solution. Assume a method is p-th order accurate, such that $\|E_h\| \sim Ch^p$ if h is small enough, or

$$\log \|E_h\| \approx \log C + p \log h. \tag{2.51}$$

Thus, if we plot $\log \|E_h\|$ against $\log h$ using the same scale, then the slope p is the convergence order. Furthermore, if we divide h by half to get $\|E_{h/2}\|$, then we have the following relations:

$$\text{ratio} = \frac{\|E_h\|}{\|E_{h/2}\|} \approx \frac{Ch^p}{C(h/2)^p} = 2^p, \tag{2.52}$$

$$p \approx \frac{\log\left(\|E_h\|/\|E_{h/2}\|\right)}{\log 2} = \frac{\log(\text{ratio})}{\log 2}. \tag{2.53}$$

For a first-order method ($p=1$), the ratio approaches number two as h approaches zero. For a second-order method ($p=2$), the ratio approaches number four as h approaches zero, and so on. Incidentally, the method is called superlinear convergent if p is some number between one and two.

For a simple problem, we may come up with an exact solution easily. For example, for most single linear single ODE or PDE, we can simply set an exact solution $u_e(\mathbf{x})$ and hence determine other functions and parameters such as the source term $f(\mathbf{x})$, boundary and initial conditions, *etc.* For more complicated systems of ODE or PDE, the exact solution is often difficult if not impossible to construct, but one can search the literature to see whether there are similar examples, *e.g.*, some may have become benchmark problems. Any new method may then be compared with benchmark problem results.

If we do not have the exact solution, the order of convergence can still be estimated by comparing a numerical solution with one obtained from a finer mesh. Suppose the numerical solution converges and satisfies

$$u_h = u_e + Ch^p + \cdots \tag{2.54}$$

where u_h is the numerical solution and u_e is the true solution, and let u_{h_*} be the solution obtained from the finest mesh

$$u_{h_*} = u_e + C{h_*}^p + \cdots . \tag{2.55}$$

Thus we have

$$u_h - u_{h_*} \approx C\left(h^p - {h_*}^p\right), \tag{2.56}$$

$$u_{h/2} - u_{h_*} \approx C\left((h/2)^p - {h_*}^p\right). \tag{2.57}$$

From the estimates above, we obtain the ratio

$$\frac{u_h - u_{h_*}}{u_{h/2} - u_{h_*}} \approx \frac{h^p - {h_*}^p}{(h/2)^p - {h_*}^p} = \frac{2^p\left(1-(h_*/h)^p\right)}{1-(2h_*/h)^p}, \tag{2.58}$$

from which we can estimate the order of accuracy p. For example, on doubling the number of grid points successively we have

$$\frac{h_*}{h} = 2^{-k}, \qquad k = 2, 3, \ldots, \tag{2.59}$$

then the ratio in (2.58) is

$$\frac{\tilde{u}(h) - \tilde{u}(h^*)}{\tilde{u}(\frac{h}{2}) - \tilde{u}(h^*)} = \frac{2^p\left(1-2^{-kp}\right)}{1-2^{p(1-k)}}. \tag{2.60}$$

In particular, for a first-order method ($p=1$) this becomes

$$\frac{\tilde{u}(h)-\tilde{u}(h^*)}{\tilde{u}(\frac{h}{2})-\tilde{u}(h^*)}=\frac{2\left(1-2^{-k}\right)}{1-2^{1-k}}=\frac{2^k-1}{2^{k-1}-1}.$$

If we take $k=2,3,\ldots$, then the ratios above are

$$3,\quad \frac{7}{3}\simeq 2.333,\quad \frac{15}{7}\simeq 2.1429,\quad \frac{31}{15}\simeq 2.067,\quad \cdots.$$

Similarly, for a second-order method ($p=2$), (2.60) becomes

$$\frac{\tilde{u}(h)-\tilde{u}(h^*)}{\tilde{u}(\frac{h}{2})-\tilde{u}(h^*)}=\frac{4\left(1-4^{-k}\right)}{1-4^{1-k}}=\frac{4^k-1}{4^{k-1}-1},$$

and the ratios are

$$5,\quad \frac{63}{15}=4.2,\quad \frac{255}{63}\simeq 4.0476,\quad \frac{1023}{255}\simeq 4.0118,\quad \cdots$$

when $k=2,3,\ldots$

To do the grid refinement analysis for 1D problems, we can take $n=10,20,40,\ldots,640$, depending on the size of the problem and the computer speed; for 2D problems $(10,10)$, $(20,20)$, $\ldots$, $(640,640)$, or $(16,16)$, $(32,32)$, $\ldots$, $(512,512)$; and for 3D problems $(8,8,8)$, $(16,16,16)$, $\ldots$, $(128,128,128)$, if the computer used has enough memory.

To present the grid refinement analysis, we can tabulate the grid size n and the ratio or order, so the order of convergence can be seen immediately. Another way is to plot the error versus the step size h, in a log–log scale with the same scale on both the horizontal and vertical axes, then the slope of the approximate line is the order of convergence.

2.10 * 1D IIM for Discontinuous Coefficients

In some applications, the coefficient of a differential equation can have a finite discontinuity. Examples include composite materials, two-phase flows such as ice and water, *etc.* Assume that we have a two-point BVP,

$$(pu')'-qu=f(x),\qquad 0<x<1,\qquad u(0)=u_0,\quad u(1)=u_1.$$

Assume that the coefficient $p(x)$ is a piecewise constant

$$p(x)=\begin{cases}\beta^- \text{ if } 0<x<\alpha,\\ \beta^+ \text{ if } \alpha<x<1,\end{cases}\tag{2.61}$$

where $0<\alpha<1$ is called an interface, β^- and β^+ are two positive but different constants. For simplicity, we assume that both q and f are continuous in a domain (0, 1). In a number of applications, u stands for the temperature that should be continuous physically, which means $[u]=0$ across the interface, where

$$[u]=\lim_{x\to\alpha+} u(x) - \lim_{x\to\alpha^-} u(x) = u^+ - u^-, \tag{2.62}$$

denotes the jump of $u(x)$ at the interface α. The quantity

$$[\beta u_x]=\lim_{x\to\alpha+} \beta(x)u'(x) - \lim_{x\to\alpha^-} \beta(x)u'(x) = \beta^+ u_x^+ - \beta^- u_x^- \tag{2.63}$$

is called the jump in the flux. If there is no source at the interface, then the flux should also be continuous which leads to another jump condition $[\beta u_x]=0$. The two jump conditions

$$[u]=0, \qquad [\beta u_x]=0 \tag{2.64}$$

are called the natural jump conditions. Note that since β has a finite jump at α, so does u_x unless $u_x^-=0$ and $u_x^+=0$ which is unlikely.

Using a finite difference method, there are several commonly used methods to deal with the discontinuity in the coefficients.

- Direct discretization if $x_{i-1/2}\neq\alpha$ for $i=1,2,\ldots$ since $p_{i-1/2}$ is well-defined. If the interface $\alpha=x_{j-1/2}$ for some j, then we can define the value of $p(x)$ at $x_{j-1/2}$ as the average, that is, $p_{j-1/2}=(\beta^- + \beta^+)/2$.
- The smoothing method using

$$\beta_\epsilon(x) = \beta^-(x) + \left(\beta^+(x) - \beta^-(x)\right) H_\epsilon(x-\alpha), \tag{2.65}$$

where H_ϵ is a smoothed Heaviside function

$$H_\epsilon(x) = \begin{cases} 0, & \text{if } x<-\epsilon, \\ \dfrac{1}{2}\left(1+\dfrac{x}{\epsilon}+\dfrac{1}{\pi}\sin\dfrac{\pi x}{\epsilon}\right), & \text{if } |x|\le\epsilon, \\ 1, & \text{if } x>\epsilon, \end{cases} \tag{2.66}$$

often ϵ is taken as h or Ch for some constant $C\ge 1$ in a finite difference discretization.

- Harmonic averaging of $p(x)$ defined as

$$p_{i+\frac{1}{2}} = \left[\frac{1}{h}\int_{x_i}^{x_{i+1}} p^{-1}(x)\,dx\right]^{-1}. \tag{2.67}$$

For the natural jump conditions, the harmonic averaging method provides second-order accurate solution in the maximum norm due to error cancellations even though the finite difference discretization may not be consistent.

The methods mentioned above are simple but work only with natural jump conditions. The first two methods are less accurate than the third one. The error analysis for these methods are not straightforward. The second and third approaches cannot be directly generalized to 2D or 3D problems with general interfaces.

We now explain the Immersed Interface Method (IIM) for this problem which can be applied for more general jump conditions $[\beta u_x] = c$ and even with discontinuous solutions ($[u] \neq 0$). We refer the reader to Li and Ito (2006) for more details about the method.

Assume we have a mesh x_i, $i = 0, 1, \ldots, n$. Then there is an integer j such that $x_j \leq \alpha < x_{j+1}$. Except for grid points x_j and x_{j+1}, other grid points are called *regular* since the standard three-point finite stencil does not contain the interface α. The standard finite difference scheme is still used at regular grid points.

At irregular grid points x_j and x_{j+1}, the finite difference approximations need to be modified to take the discontinuity in the coefficient into account. Note that when $f(x)$ is continuous, we also have

$$\beta^+ u_{xx}^+ - q^+ u^+ = \beta^- u_{xx}^- - q^- u^- .$$

Since we assume that $q^+ = q^-$, and $u^+ = u^-$, we can express the limiting quantities from $+$ side in terms of those from the $-$ side to get,

$$u^+ = u^-, \qquad u_x^+ = \frac{\beta^-}{\beta^+} u_x^- + \frac{c}{\beta^+}, \qquad u_{xx}^+ = \frac{\beta^-}{\beta^+} u_{xx}^- . \tag{2.68}$$

The finite difference equations are determined from the method of undetermined coefficients:

$$\begin{aligned} &\gamma_{j,1} u_{j-1} + \gamma_{j,2} u_j + \gamma_{j,3} u_{j+1} - q_j u_j = f_j + C_j, \\ &\gamma_{j+1,1} u_j + \gamma_{j+1,2} u_{j+1} + \gamma_{j+1,3} u_{j+2} - q_{j+1} u_{j+1} = f_{j+1} + C_{j+1}. \end{aligned} \tag{2.69}$$

For the simple model problem, the coefficients of the finite difference scheme have the following closed form:

$$\begin{aligned} &\gamma_{j,1} = \left(\beta^- - [\beta](x_j - \alpha)/h\right)/D_j, && \gamma_{j+1,1} = \beta^-/D_{j+1}, \\ &\gamma_{j,2} = \left(-2\beta^- + [\beta](x_{j-1} - \alpha)/h\right)/D_j, && \gamma_{j+1,2} = \left(-2\beta^+ + [\beta](x_{j+2} - \alpha)/h\right)/D_{j+1}, \\ &\gamma_{j,3} = \beta^+/D_j, && \gamma_{j+1,3} = \left(\beta^+ - [\beta](x_{j+1} - \alpha)/h\right)/D_{j+1}, \end{aligned}$$

where

$$D_j = h^2 + [\beta](x_{j-1} - \alpha)(x_j - \alpha)/2\beta^-,$$
$$D_{j+1} = h^2 - [\beta](x_{j+2} - \alpha)(x_{j+1} - \alpha)/2\beta^+.$$

It has been been shown in Huang and Li (1999) and Li (1994) that $D_j \neq 0$ and $D_{j+1} \neq 0$ if $\beta^- \beta^+ > 0$. The correction terms are:

$$C_j = \gamma_{j,3}\,(x_{j+1} - \alpha)\,\frac{c}{\beta^+}, \qquad C_{j+1} = \gamma_{j+1,1}\,(\alpha - x_{j+1})\,\frac{c}{\beta^-}. \tag{2.70}$$

Remark 2.9. If $\beta^+ = \beta^-$, *i.e.*, the coefficient is continuous, then the coefficients of the finite difference scheme are the same as that from the standard central finite difference scheme as if there was no interface. The correction term is not zero if c is not zero corresponding to a singular source $c\delta(x - \alpha)$. On the other hand, if $c = 0$, then the correction terms are also zero. But the coefficients are changed due to the discontinuity in the coefficient of the ODE BVP.

2.10.1 A Brief Derivation of the Finite Difference Scheme at an Irregular Grid Point

We illustrate the idea of the IIM in determining the finite difference coefficients $\gamma_{j,1}, \gamma_{j,2}$ and $\gamma_{j,3}$ in (2.69). We want to determine the coefficients so that the local truncation error is as small as possible in the magnitude. The main tool is the Taylor expansion in expanding $u(x_{j-1})$, $u(x_j)$, and $u(x_{j+1})$ from each side of the interface α. After the expansions, then we can use the interface relations (2.68) to express the quantities of $u^{\pm}$, $u_x^{\pm}$, and $u_{xx}^{\pm}$ in terms of the quantities from one particular side.

It is reasonable to assume that the $u(x)$ has up to third-order derivatives in $(0, \alpha)$ and $(\alpha, 1)$ *excluding* the interface. Using the Tailor expansion for $u(x_{j+1})$ at α, we have

$$u(x_{j+1}) = u^+(\alpha) + (x_{j+1} - \alpha)\,u_x^+(\alpha) + \frac{1}{2}\,(x_{j+1} - \alpha)^2\,u_{xx}^+(\alpha) + O(h^3).$$

Using the jump relation (2.68), the expression above can be written as

$$u(x_{j+1}) = u^-(\alpha) + (x_{j+1} - \alpha)\left(\frac{\beta^-}{\beta^+}u_x^-(\alpha) + \frac{c}{\beta^+}\right)$$
$$+ \frac{1}{2}\,(x_{j+1} - \alpha)^2\,\frac{\beta^-}{\beta^+}u_{xx}^-(\alpha) + O(h^3).$$

The Taylor expansions of $u(x_{j-1})$ and $u(x_j)$ at α from the left hand side have the following expression

$$u(x_l) = u^-(\alpha) + (x_l - \alpha)u_x^-(\alpha) + \frac{1}{2}(x_l - \alpha)^2 u_{xx}^-(\alpha) + O(h^3), \quad l = j-1, j.$$

Therefore we have the following

$$\begin{aligned}
&\gamma_{j,1}u(x_{j-1}) + \gamma_{j,2}u(x_j) + \gamma_{j,3}u(x_{j+1}) = (\gamma_{j,1} + \gamma_{j,2} + \gamma_{j,3})u^-(\alpha) \\
&+ \left((x_{j-1} - \alpha)\gamma_{j,1} + (x_j - \alpha)\gamma_{j,2} + \frac{\beta^-}{\beta^+}(x_{j+1} - \alpha)\gamma_{j,3} \right) u_x^-(\alpha) \\
&+ \gamma_{j,3}(x_{j+1} - \alpha)\frac{c}{\beta^+} \\
&+ \frac{1}{2}\left((x_{j-1} - \alpha)^2\gamma_{j,1} + (x_j - \alpha)^2\gamma_{j,2} + \frac{\beta^-}{\beta^+}(x_{j+1} - \alpha)^2 \right) u_{xx}^-(\alpha) \\
&+ O(\max_l |\gamma_{j,l}|\, h^3),
\end{aligned}$$

after the Taylor expansions and collecting terms for $u^-(\alpha)$, $u_x^-(\alpha)$ and $u_{xx}^-(\alpha)$.

By matching the finite difference approximation with the differential equation at α from the $-$ side,[3] we get the system of equations for the coefficients γ_j's below:

$$\begin{aligned}
&\gamma_{j,1} + \gamma_{j,2} + \gamma_{j,3} &= 0 \\
&-(\alpha - x_{j-1})\,\gamma_{j,1} - (\alpha - x_j)\,\gamma_{j,2} + \frac{\beta^-}{\beta^+}(x_{j+1} - \alpha)\gamma_{j,3} &= 0 \\
&\frac{1}{2}(\alpha - x_{j-1})^2\,\gamma_{j,1} + \frac{1}{2}(\alpha - x_j)^2\,\gamma_{j,2} + \frac{\beta^-}{2\beta^+}(x_{j+1} - \alpha)^2\,\gamma_{j,3} &= \beta^-.
\end{aligned} \tag{2.71}$$

It is easy to verify that the γ_j's in the left column in the previous page satisfy the system above. Once those γ_j's have been computed, it is easy to set the correction term C_j to match the remaining leading terms of the differential equation.

[3] It is also possible to further expand at $x = x_j$ to match the differential equation at $x = x_j$. The order of convergence will be the same.

Exercises

1. When dealing with irregular boundaries or using adaptive grids, nonuniform grids are needed. Derive the finite difference coefficients for the following:

$$(A): \qquad u'(\bar{x}) \approx \alpha_1 u(\bar{x}-h_1) + \alpha_2 u(\bar{x}) + \alpha_3 u(\bar{x}+h_2),$$

$$(B): \qquad u''(\bar{x}) \approx \alpha_1 u(\bar{x}-h_1) + \alpha_2 u(\bar{x}) + \alpha_3 u(\bar{x}+h_2),$$

$$(C): \qquad u'''(\bar{x}) \approx \alpha_1 u(\bar{x}-h_1) + \alpha_2 u(\bar{x}) + \alpha_3 u(\bar{x}+h_2).$$

   Are they consistent? In other words, as $h = \max\{h_1, h_2\}$ approaches zero, does the error also approach zero? If so, what are the orders of accuracy? Do you see any potential problems with the schemes you have derived?

2. Consider the following finite difference scheme for solving the two-point BVP $u''(x) = f(x)$, $a < x < b$, $u(a) = u_a$ and $u(b) = u_b$:

$$\frac{U_{i-1} - 2U_i + U_{i+1}}{h^2} = f(x_i), \quad i = 2, 3, \ldots, n-1, \tag{2.72}$$

   where $x_i = a + ih$, $i = 0, 1, \ldots, n$, $h = (b-a)/n$. At $i = 1$, the finite difference scheme is

$$\frac{U_1 - 2U_2 + U_3}{h^2} = f(x_1). \tag{2.73}$$

   (a) Find the local truncation errors of the finite difference scheme at x_i, $i = 2, 3, \ldots, n-1$, and x_1. Is this scheme consistent?
   (b) Does this scheme converge? Justify your answer.

3. Program the central finite difference method for the self-adjoint BVP

$$(\beta(x)u')' - \gamma(x)u(x) = f(x), \quad 0 < x < 1,$$

$$u(0) = u_a, \qquad au(1) + bu'(1) = c,$$

   using a uniform grid and the central finite difference scheme

$$\frac{\beta_{i+\frac{1}{2}}(U_{i+1} - U_i)/h - \beta_{i-\frac{1}{2}}(U_i - U_{i-1})/h}{h} - \gamma(x_i)U_i = f(x_i). \tag{2.74}$$

   Test your code for the case where

$$\beta(x) = 1 + x^2, \quad \gamma(x) = x, \quad a = 2, \quad b = -3, \tag{2.75}$$

   and the other functions or parameters are determined from the exact solution

$$u(x) = e^{-x}(x-1)^2. \tag{2.76}$$

   Plot the computed solution and the exact solution, and the error for a particular grid $n = 80$. Do the grid refinement analysis, to determine the order of accuracy of the global solution. Also try to answer the following questions:

   - Can your code handle the case when $a = 0$ or $b = 0$?
   - If the central finite difference scheme is used for the equivalent differential equation

$$\beta u'' + \beta' u' - \gamma u = f, \tag{2.77}$$

     what are the advantages or disadvantages?

4. Consider the finite difference scheme for the 1D steady state *convection–diffusion* equation

$$\epsilon u'' - u' = -1, \quad 0 < x < 1, \tag{2.78}$$

$$u(0) = 1, \quad u(1) = 3. \tag{2.79}$$

   (a) Verify the exact solution is

$$u(x) = 1 + x + \left(\frac{e^{x/\epsilon} - 1}{e^{1/\epsilon} - 1} \right). \tag{2.80}$$

   (b) Compare the following two finite difference methods for $\epsilon = 0.3, 0.1, 0.05$, and 0.0005.
   (1) Central finite difference scheme:

$$\epsilon \frac{U_{i-1} - 2U_i + U_{i+1}}{h^2} - \frac{U_{i+1} - U_{i-1}}{2h} = -1. \tag{2.81}$$

   (2) Central-upwind finite difference scheme:

$$\epsilon \frac{U_{i-1} - 2U_i + U_{i+1}}{h^2} - \frac{U_i - U_{i-1}}{h} = -1. \tag{2.82}$$

   Do the grid refinement analysis for each case to determine the order of accuracy. Plot the computed solution and the exact solution for $h = 0.1$, $h = 1/25$, and $h = 0.01$. You can use the Matlab command *subplot* to put several graphs together.

   (c) From your observations, in your opinion which method is better?

5. (*) For the BVP

$$u'' = f, \quad 0 < x < 1, \tag{2.83}$$

$$u(0) = 0, \quad u'(1) = \sigma, \tag{2.84}$$

   show that the finite difference method using the central formula and the *ghost point* method at $x = 1$ are stable and consistent. Find their convergence order and prove it.

6. For the set of points (x_i, u_i), $i = 0, 1, \ldots, N$, find the Lagrange interpolation polynomial from the formula

$$p(x) = \sum_{i=0}^{N} l_i(x)\, u_i, \qquad l_i(x) = \prod_{j=0, j\neq i}^{N} \frac{x - x_j}{x_i - x_j}. \tag{2.85}$$

   By differentiating $p(x)$ with respect to x, one can get different finite difference formulas for approximating different derivatives. Assuming a uniform mesh

$$x_{i+1} - x_i = x_i - x_{i-1} = \cdots = x_1 - x_0 = h,$$

   derive a central finite difference formula for $u^{(4)}$ and a one-sided finite difference formula for $u^{(3)}$ with $N = 4$.

7. Derive the finite difference method for

$$u''(x) - q(x)u(x) = f(x), \quad a < x < b, \tag{2.86}$$

$$u(a) = u(b), \quad \text{periodic BC}, \tag{2.87}$$

   using the central finite difference scheme and a uniform grid. Write down the system of equations $A_h U = F$. How many unknowns are there (ignoring redundancies)? Is the coefficient matrix A_h tridiagonal?

Hint: Note that $U_0 = U_n$, and set unknowns as $U_1, U_2, \ldots, U_n$.

If $q(x) = 0$, does the solution exist? Derive a compatibility condition for which the solution exists.
If the solution exists, is it unique? How do we modify the finite difference method to make the solution unique?

8. Modify the Matlab code *non_tp.m* to apply the finite difference method and Newton's nonlinear solver to find a numerical solution to the nonlinear pendulum model

$$\begin{aligned} &\frac{d^2\theta}{dt^2} + K\sin\theta = 0, \qquad 0 < \theta < 2\pi, \\ &\theta(0) = \theta_1, \qquad \theta(2\pi) = \theta_2, \end{aligned} \tag{2.88}$$

where K, θ_1 and θ_2 are parameters. Compare the solution with the linearized model $\dfrac{d^2\theta}{dt^2} + K\theta = 0$.

3

Finite Difference Methods for 2D Elliptic PDEs

There are many important applications of elliptic PDEs, see page 4 for the definition of elliptic PDEs. Below we give some examples of linear and nonlinear equations of elliptic PDEs.

- Laplace equations in 2D,

$$u_{xx} + u_{yy} = 0 . \tag{3.1}$$

The solution u is sometimes called a potential function, because a conservative vector field $\mathbf{v}$ (*i.e.*, such that $\nabla \times \mathbf{v} = 0$) is given by $\mathbf{v} = \nabla u$ (or alternatively $\mathbf{v} = -\nabla u$), where ∇ is the gradient operator that acts as a vector. In 2D, the gradient operator is $\nabla = [\frac{\partial}{\partial x}, \frac{\partial}{\partial y}]^T$. If u is a scalar, the $\nabla u = [\frac{\partial u}{\partial x}, \frac{\partial u}{\partial y}]^T$ is the gradient vector of u, and if $\mathbf{v}$ is a vector, then $\nabla \cdot \mathbf{v} = div(\mathbf{v})$ is the divergence of the vector $\mathbf{v}$. The scalar $\nabla \cdot \nabla u = u_{xx} + u_{yy}$ is the Laplacian of u, which is denoted as $\nabla^2 u$ literally from its definition. It is also common to use the notation of $\Delta u = \nabla^2 u$ for the Laplacian of u. If the conservative vector field $\mathbf{v}$ is also divergence free, (*i.e.*, $div(\mathbf{v}) = \nabla \cdot \mathbf{v} = 0$, then we have $\nabla \cdot \mathbf{v} = \nabla \cdot \nabla u = \Delta u = 0$, that is, the potential function is the solution of a Laplace equation.

- Poisson equations in 2D,

$$u_{xx} + u_{yy} = f . \tag{3.2}$$

- Generalized Helmholtz equations,

$$u_{xx} + u_{yy} - \lambda^2 u = f . \tag{3.3}$$

Many incompressible flow solvers are based on solving one or several Poisson or Helmholtz equations, *e.g.*, the projection method for solving incompressible Navier–Stokes equations for flow problems (Chorin, 1968; Li and Lai,

2001; Minion, 1996) at low or modest Reynolds number, or the stream–vorticity formulation method for large Reynolds number (Calhoun, 2002; Li and Wang, 2003). In particular, there are some fast Poisson solvers available for regular domains, *e.g.*, in Fishpack (Adams et al.).

- Helmholtz equations,

$$u_{xx} + u_{yy} + \lambda^2 u = f. \tag{3.4}$$

The Helmholtz equation arises in scattering problems, when λ is a wave number, and the corresponding problem may not have a solution if λ^2 is an eigenvalue of the corresponding BVP. Furthermore, the problem is hard to solve numerically if λ is large.

- General self-adjoint elliptic PDEs,

$$\nabla \cdot (a(x,y)\nabla u(x,y)) - q(x,y)u = f(x,y) \tag{3.5}$$

$$\text{or} \quad (au_x)_x + (au_y)_y - q(x,y)u = f(x,y). \tag{3.6}$$

We assume that $a(x,y)$ does not change sign in the solution domain, *e.g.*, $a(x,y) \geq a_0 > 0$, where a_0 is a constant, and $q(x,y) \geq 0$ to guarantee that the solution exists and it is unique.

- General elliptic PDEs (diffusion and advection equations),

$$\begin{aligned} &a(x,y)u_{xx} + 2b(x,y)u_{xy} + c(x,y)u_{yy} \\ &\qquad + d(x,y)u_x + e(x,y)u_y + g(x,y)u(x,y) = f(x,y), \quad (x,y) \in \Omega, \end{aligned}$$

if $b^2 - ac < 0$ for all $(x,y) \in \Omega$. This equation can be rewritten as

$$\nabla \cdot (a(x,y)\nabla u(x,y)) + \mathbf{w}(x,y) \cdot \nabla u + c(x,y)u = f(x,y) \tag{3.7}$$

after a transformation, where $\mathbf{w}(x,y)$ is a vector.

- Diffusion and reaction equation,

$$\nabla \cdot (a(x,y)\nabla u(x,y)) = f(u). \tag{3.8}$$

Here $\nabla \cdot (a(x,y)\nabla u(x,y))$ is called a diffusion term, the nonlinear term $f(u)$ is called a reaction term, and if $a(x,y) \equiv 1$ the PDE is a nonlinear Poisson equation.

- p-Laplacian equation,

$$\nabla \cdot \left(|\nabla u|^{p-2} \nabla u \right) = 0, \quad p \geq 2, \tag{3.9}$$

where $|\nabla u| = \sqrt{u_x^2 + u_y^2}$ in 2D.

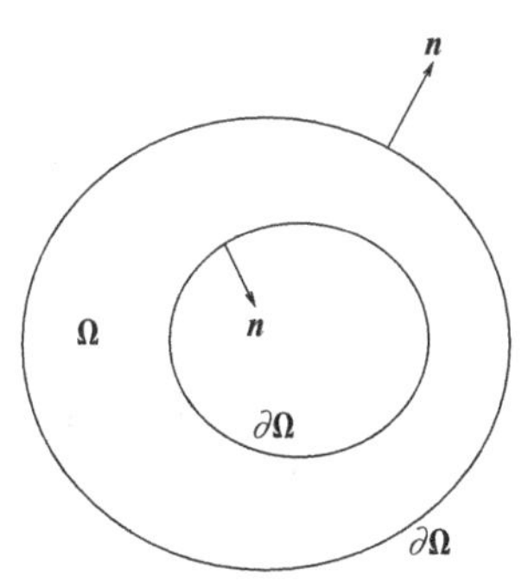

Figure 3.1. A diagram of a 2D domain Ω, its boundary ∂Ω, and its unit normal direction.

- Minimal surface equation,

$$\nabla \cdot \left(\frac{\nabla u}{\sqrt{1 + |\nabla u|^2}} \right) = 0. \tag{3.10}$$

We note that an elliptic PDE ($P(\frac{\partial}{\partial x}, \frac{\partial}{\partial y})u = 0$) can be regarded as the steady state solution of a corresponding parabolic PDE ($u_t = P(\frac{\partial}{\partial x}, \frac{\partial}{\partial y})u$). Furthermore, if a linear PDE is defined on a rectangle domain then a finite difference approximation (in each dimension) can be used for both the equation and the boundary conditions, but the more difficult part is to solve the resulting linear system of algebraic equations efficiently.

3.1 Boundary and Compatibility Conditions

Let us consider a 2D second-order elliptic PDE on a domain Ω, with boundary ∂Ω whose unit normal direction is **n** according to the "right side rule" (*cf.* Figure 3.1). Some common boundary conditions are as follows.

- Dirichlet boundary condition: the solution is known on the boundary,

$$u(x, y)|_{\partial\Omega} = u_0(x, y).$$

- Neumann or flux boundary condition: the normal derivative is given along the boundary,

$$\frac{\partial u}{\partial n} \equiv \mathbf{n} \cdot \nabla u = u_n = u_x n_x + u_y n_y = g(x, y),$$

where $\mathbf{n} = (n_x, n_y)$ ($n_x^2 + n_y^2 = 1$) is the unit normal direction.

- Mixed boundary condition:

$$\left(\alpha(x,y)u(x,y)+\beta(x,y)\frac{\partial u}{\partial n}\right)\Big|_{\partial\Omega}=\gamma(x,y)$$

  is given along the boundary $\partial\Omega$.
- In some cases, a boundary condition is periodic, *e.g.*, for $\Omega=[a,\ b]\times[c,\ d]$, $u(a,y)=u(b,y)$ is periodic in the x-direction, and $u(x,c)=u(x,d)$ is periodic in the y-direction.

There can be different boundary conditions on different parts of the boundary, *e.g.*, for a channel flow in a domain $(a,b)\times(c,d)$, the flux boundary condition may apply at $x=a$, and a no-slip boundary condition $\mathbf{u}=0$ at the boundaries $y=c$ and $y=d$. It is challenging to set up a correct boundary condition at $x=b$ (outflow). One approximate to the outflow boundary condition is to set $\frac{\partial u}{\partial x}=0$.

For a Poisson equation with a purely Neumann boundary condition, there is no solution unless a *compatibility* condition is satisfied. Consider the following problem:

$$\Delta u=f(x,y),\quad (x,y)\in\Omega,\quad \frac{\partial u}{\partial n}\Big|_{\partial\Omega}=g(x,y).$$

On integrating over the domain Ω

$$\iint_{\Omega}\Delta u dxdy=\iint_{\Omega}f(x,y)dxdy,$$

and applying the Green's theorem gives

$$\iint_{\Omega}\Delta u dxdy=\oint_{\partial\Omega}\frac{\partial u}{\partial n}ds,$$

so we have the compatibility condition

$$\iint_{\Omega}\Delta u dxdy=\oint_{\partial\Omega}g\,ds=\iint_{\Omega}f(x,y)dxdy \tag{3.11}$$

for the solution to exist. If the compatibility condition is satisfied and $\partial\Omega$ is smooth, then the solution does exist but it is not unique. Indeed, $u(x,y)+C$ is a solution for arbitrary constant C if $u(x,y)$ is a solution, but we can specify the solution at a particular point (*e.g.*, $u(x_0,y_0)=0$) to render it well-defined.

3.2 The Central Finite Difference Method for Poisson Equations

Let us now consider the following problem, involving a Poisson equation and a Dirichlet BC:

$$u_{xx} + u_{yy} = f(x, y), \quad (x, y) \in \Omega = (a, b) \times (c, d), \tag{3.12}$$

$$u(x, y)|_{\partial\Omega} = u_0(x, y). \tag{3.13}$$

If $f \in C(\Omega)$, then the solution $u(x, y) \in C^2(\Omega)$ exists and it is unique. Later on, we can relax the condition $f \in C(\Omega)$ if the finite element method is used in which we seek a weak solution. An analytic solution is often difficult to obtain, and a finite difference approximation can be obtained through the following procedure.

- Step 1: Generate a grid. For example, a uniform Cartesian grid can be generated with two given parameters m and n:

$$x_i = a + ih_x, \quad i = 0, 1, 2, \ldots, m, \quad h_x = \frac{b - a}{m}, \tag{3.14}$$

$$y_j = c + jh_y, \quad j = 0, 1, 2, \ldots, n, \quad h_y = \frac{d - c}{n}. \tag{3.15}$$

  In seeking an approximate solution U_{ij} at the grid points (x_i, y_j) where $u(x, y)$ is unknown, there are $(m - 1)(n - 1)$ unknowns.
- Step 2: Approximate the partial derivatives at grid points with finite difference formulas involving the function values at nearby grid points. For example, if we adopt the three-point central finite difference formula for second-order partial derivatives in the x- and y-directions, respectively, then

$$\frac{u(x_{i-1}, y_j) - 2u(x_i, y_j) + u(x_{i+1}, y_j)}{(h_x)^2} + \frac{u(x_i, y_{j-1}) - 2u(x_i, y_j) + u(x_i, y_{j+1})}{(h_y)^2}$$
$$= f_{ij} + T_{ij}, \quad i = 1, \ldots, m - 1, \quad j = 1, \ldots, n - 1, \tag{3.16}$$

  where $f_{ij} = f(x_i, y_j)$. The local truncation error satisfies

$$T_{ij} \sim \frac{(h_x)^2}{12} \frac{\partial^4 u}{\partial x^4}(x_i, y_j) + \frac{(h_y)^2}{12} \frac{\partial^4 u}{\partial y^4}(x_i, y_j) + O(h^4), \tag{3.17}$$

  where

$$h = \max\{\, h_x, h_y \,\}. \tag{3.18}$$

We ignore the error term in (3.16) and replace the exact solution values $u(x_i, y_j)$ at the grid points with the approximate solution values U_{ij} obtained from solving the linear system of algebraic equations, *i.e.*,

$$\frac{U_{i-1,j}+U_{i+1,j}}{(h_x)^2}+\frac{U_{i,j-1}+U_{i,j+1}}{(h_y)^2}-\left(\frac{2}{(h_x)^2}+\frac{2}{(h_y)^2}\right)U_{ij}=f_{ij},$$
$$i=1,2,\ldots,m-1, \quad j=1,2,\ldots,n-1\,. \tag{3.19}$$

The finite difference equation at a grid point (x_i, y_j) involves five grid points in a five-point stencil, (x_{i-1}, y_j), (x_{i+1}, y_j), (x_i, y_{j-1}), (x_i, y_{j+1}), and (x_i, y_j). The grid points in the finite difference stencil are sometimes labeled east, north, west, south, and the center in the literature. The center (x_i, y_j) is called the *master grid point*, where the finite difference equation is used to approximate the PDE.

It is obvious that the finite difference discretization is second-order accurate and consistent since

$$\lim_{h\to 0} T_{ij}=0, \quad \text{and} \quad \lim_{h\to 0} \|\mathbf{T}\|_\infty=0, \tag{3.20}$$

where $\mathbf{T}$ is the local truncation error matrix formed by $\{T_{ij}\}$.

- Solve the linear system of algebraic equations (3.19), to get the approximate values for the solution at all of the grid points.
- Error analysis, implementation, visualization, *etc.*

3.2.1 The Matrix–vector Form of the FD Equations

In solving the algebraic system of finite difference equations by a direct method such as Gaussian elimination or some sparse matrix technique, knowledge of the matrix structure is important, although less so for an iterative solver such as the Jacobi, Gauss–Seidel, or SOR(ω) methods. In the matrix-vector form $A\mathbf{U}=\mathbf{F}$, the unknown $\mathbf{U}$ is a 1D array. From 2D Poisson equations the unknowns $\{U_{ij}\}$ are a 2D array, but we can order it to get a 1D array. We also need to order the finite difference equations, and it is a common practice to *use the same ordering for the equations as for the unknown array*. There are two commonly used orderings, namely, the *natural ordering*, a natural choice for sequential computing, and *red–black ordering*, considered to be a good choice for parallel computing.

3.2.1.1 The Natural Row Ordering

In the natural row ordering, we order the unknowns and equations row by row. Thus the k-th finite difference equation corresponding to (i, j) has the following

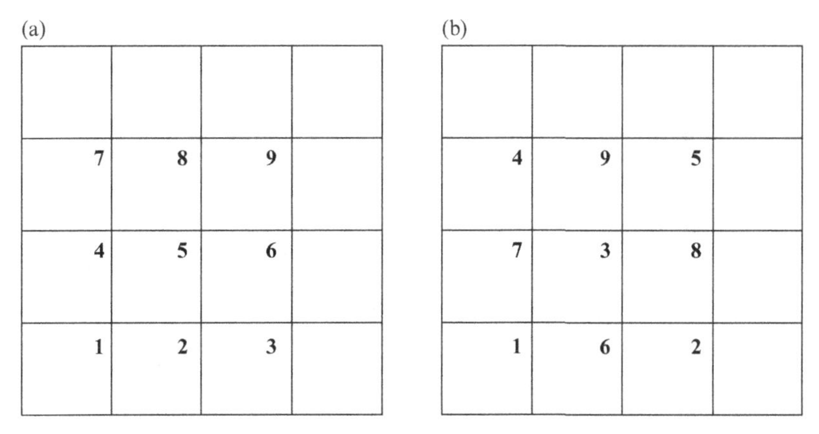

Figure 3.2. (a) The natural ordering and (b) the red–black ordering.

relation:

$$k = i + (m-1)(j-1), \quad i = 1, 2, \ldots, m-1, \quad j = 1, 2, \ldots, n-1 \tag{3.21}$$

(see Figure 3.2(a)).

Referring to Figure 3.2(a) that $h_x = h_y = h$, $m = n = 4$. Then there are nine equations and nine unknowns, so the coefficient matrix is 9 by 9. To write down the matrix-vector form, use a 1D array $\mathbf{x}$ to express the unknown U_{ij} according to the ordering, we should have

$$\begin{aligned} x_1 = U_{11}, \quad x_2 = U_{21}, \quad x_3 = U_{31}, \quad x_4 = U_{12}, \quad x_5 = U_{22}, \\ x_6 = U_{32}, \quad x_7 = U_{13}, \quad x_8 = U_{23}, \quad x_9 = U_{33}\,. \end{aligned} \tag{3.22}$$

Now if the algebraic equations are ordered in the same way as the unknowns, the nine equations from the standard central finite difference scheme using the five-point stencil are

$$\begin{aligned}
&Eqn.1: \quad && \frac{1}{h^2}(-4x_1 + x_2 + x_4) = f_{11} - \frac{u_{01} + u_{10}}{h^2} \\
&Eqn.2: \quad && \frac{1}{h^2}(x_1 - 4x_2 + x_3 + x_5) = f_{21} - \frac{u_{20}}{h^2} \\
&Eqn.3: \quad && \frac{1}{h^2}(x_2 - 4x_3 + x_6) = f_{31} - \frac{u_{30} + u_{41}}{h^2} \\
&Eqn.4: \quad && \frac{1}{h^2}(x_1 - 4x_4 + x_5 + x_7) = f_{12} - \frac{u_{02}}{h^2}
\end{aligned}$$

$$\textit{Eqn.5}: \qquad \frac{1}{h^2}(x_2 + x_4 - 4x_5 + x_6 + x_8) = f_{22}$$

$$\textit{Eqn.6}: \qquad \frac{1}{h^2}(x_3 + x_5 - 4x_6 + x_9) = f_{32} - \frac{u_{42}}{h^2}$$

$$\textit{Eqn.7}: \qquad \frac{1}{h^2}(x_4 - 4x_7 + x_8) = f_{13} - \frac{u_{03} + u_{14}}{h^2}$$

$$\textit{Eqn.8}: \qquad \frac{1}{h^2}(x_5 + x_7 - 4x_8 + x_9) = f_{23} - \frac{u_{24}}{h^2}$$

$$\textit{Eqn.9}: \qquad \frac{1}{h^2}(x_6 + x_8 - 4x_9) = f_{33} - \frac{u_{34} + u_{43}}{h^2}.$$

The corresponding coefficient matrix is *block tridiagonal*,

$$A = \frac{1}{h^2}\begin{bmatrix} B & I & 0 \\ I & B & I \\ 0 & I & B \end{bmatrix}, \tag{3.23}$$

where I is the 3×3 identity matrix and

$$B = \begin{bmatrix} -4 & 1 & 0 \\ 1 & -4 & 1 \\ 0 & 1 & -4 \end{bmatrix}.$$

In general, for an $n+1$ by $n+1$ grid we obtain

$$A = \frac{1}{h^2}\begin{bmatrix} B & I & & \\ I & B & I & \\ & \ddots & \ddots & \ddots \\ & & I & B \end{bmatrix}_{n^2 \times n^2}, \qquad B = \begin{bmatrix} -4 & 1 & & \\ 1 & -4 & 1 & \\ & \ddots & \ddots & \ddots \\ & & 1 & -4 \end{bmatrix}_{n^2 \times n^2}.$$

Since $-A$ is symmetric positive definite and weakly diagonally dominant, the coefficient matrix A is a nonsingular, and hence the solution of the system of the finite difference equations is unique.

The matrix-vector form is useful to understand the structure of the linear system of algebraic equations, and as mentioned it is required when a direct method (such as Gaussian elimination or a sparse matrix technique) is used to solve the system. However, it can sometimes be more convenient to use a two-index system, especially when an iterative method is preferred but also as more intuitive and to visualize the data. The eigenvalues and eigenvectors of A

can also be indexed by two parameters p and k, corresponding to wave numbers in the x and y directions. Assume $m=n$ for simplicity, then the (p,k)-th eigenvector $u^{p,k}$ has $n-1$ components,

$$u_{ij}^{p,k} = \sin(p\pi ih)\,\sin(k\pi jh), \quad i,j=1,2,\ldots,n-1 \tag{3.24}$$

for $p,k=1,2,\ldots,n-1$; and the corresponding (p,k)-th eigenvalue is

$$\lambda^{p,k} = \frac{2}{h^2}\Big(\cos(p\pi h)-1)+\cos(k\pi h)-1)\Big)\,. \tag{3.25}$$

The least dominant (smallest magnitude) eigenvalue is

$$\lambda^{1,1} = -2\pi^2 + O(h^2), \tag{3.26}$$

obtained from the Taylor expansion of (3.25) in terms of $h\sim 1/n$; and the dominant (largest magnitude) eigenvalue is

$$\lambda^{int(n/2),int(n/2)} \sim -\frac{8}{h^2}. \tag{3.27}$$

It is noted that the dominant and least dominant eigenvalues are twice the magnitude of those in 1D representation, so we have the following estimates:

$$\begin{aligned} &\|A\|_2 \sim \max|\lambda^{p,k}| = \frac{8}{h^2}, \qquad \|A^{-1}\|_2 = \frac{1}{\min|\lambda^{p,k}|} \sim \frac{1}{2\pi^2}, \\ &cond_2(A) = \|A\|_2\|A^{-1}\|_2 \sim \frac{4}{\pi^2 h^2} = O(n^2)\,. \end{aligned} \tag{3.28}$$

Note that the condition number is about the same order magnitude as that in the 1D case; and since it is large, double precision is recommended to reduce the effect of round-off errors.

3.3 The Maximum Principle and Error Analysis

Consider an elliptic differential operator

$$L = a\frac{\partial^2}{\partial x^2} + 2b\frac{\partial^2}{\partial x\partial y} + c\frac{\partial^2}{\partial y^2}, \quad b^2-ac<0, \quad \text{for} \quad (x,y)\in\Omega\,,$$

and without loss of generality assume that $a>0,\ c>0$. The maximum principle is given in the following theorem.

Theorem 3.1. *If $u(x,y)\in C^3(\Omega)$ satisfies $Lu(x,y)\ge 0$ in a bounded domain Ω, then $u(x,y)$ has its maximum on the boundary of the domain.*

Proof If the theorem is not true, then there is an interior point $(x_0, y_0) \in \Omega$ such that $u(x_0, y_0) \geq u(x, y)$ for all $(x, y) \in \Omega$. The necessary condition for a local extremum (x_0, y_0) is

$$\frac{\partial u}{\partial x}(x_0, y_0) = 0, \quad \frac{\partial u}{\partial y}(x_0, y_0) = 0 .$$

Now since (x_0, y_0) is not on the boundary of the domain and $u(x, y)$ is continuous, there is a neighborhood of (x_0, y_0) within the domain Ω where we have the Taylor expansion,

$$u(x_0 + \Delta x, y_0 + \Delta y) = u(x_0, y_0) + \frac{1}{2}\left((\Delta x)^2 u_{xx}^0 + 2\Delta x \Delta y u_{xy}^0 + (\Delta y)^2 u_{yy}^0 \right) + O((\Delta x)^3, (\Delta y)^3) ,$$

with superscript of 0 indicating that the functions are evaluated at (x_0, y_0), *i.e.*, $u_{xx}^0 = \frac{\partial^2 u}{\partial x^2}(x_0, y_0)$ evaluated at (x_0, y_0), and so on.

Since $u(x_0 + \Delta x, y_0 + \Delta y) \leq u(x_0, y_0)$ for all sufficiently small Δx and Δy,

$$\frac{1}{2}\left((\Delta x)^2 u_{xx}^0 + 2\Delta x \Delta y u_{xy}^0 + (\Delta y)^2 u_{yy}^0 \right) \leq 0 . \tag{3.29}$$

On the other hand, from the given condition

$$Lu^0 = a^0 u_{xx}^0 + 2b^0 u_{xy}^0 + c^0 u_{yy}^0 \geq 0 , \tag{3.30}$$

where $a^0 = a(x_0, y_0)$ and so forth. In order to match the Taylor expansion to get a contradiction, we rewrite the inequality above as

$$\left(\sqrt{\frac{a^0}{M}}\right)^2 u_{xx}^0 + 2\sqrt{\frac{a^0}{M}} \frac{b^0}{\sqrt{a^0 M}} u_{xy}^0 + \left(\frac{b^0}{\sqrt{a^0 M}}\right)^2 u_{yy}^0 + \frac{u_{yy}^0}{M}\left(c^0 - \frac{(b^0)^2}{a_0}\right) \geq 0 , \tag{3.31}$$

where $M > 0$ is a constant. The role of M is to make some choices of Δx and Δy that are small enough.

Let us now set

$$\Delta x = \sqrt{\frac{a^0}{M}}, \quad \Delta y = \frac{b^0}{\sqrt{a^0 M}} .$$

From (3.29), we know that

$$\frac{a^0}{M} u_{xx}^0 + \frac{2b^0}{M} u_{xy}^0 + \frac{b^0}{a^0 M} u_{yy}^0 \leq 0. \tag{3.32}$$

Now we take

$$\Delta x = 0, \qquad \Delta y = \sqrt{\left(c^0 - \frac{(b^0)^2}{a^0}\right) / M}\,;$$

and from (3.29) again,

$$(\Delta y)^2 u_{yy}^0 = \frac{1}{M}\left(c^0 - \frac{(b^0)^2}{a^0}\right) u_{yy}^0 \le 0\,. \tag{3.33}$$

Thus from (3.32) and (3.33), the left-hand side of (3.31) should not be positive, which contradicts the condition

$$Lu^0 = a^0 u_{xx}^0 + 2b^0 u_{xy}^0 + c^0 u_{yy}^0 \ge 0,$$

and with this the proof is completed. □

On the other hand, if $Lu \le 0$ then the minimum value of u is on the boundary of Ω. For general elliptic equations the maximum principle is as follows. Let

$$Lu = au_{xx} + 2bu_{xy} + cu_{yy} + d_1 u_x + d_2 u_y + eu = 0, \quad (x, y) \in \Omega,$$

$$b^2 - ac < 0, \quad a > 0,\ c > 0, \quad e \le 0,$$

where Ω is a bounded domain. Then from Theorem 3.1, $u(x, y)$ cannot have a positive local maximum or a negative local minimum in the interior of Ω.

3.3.1 The Discrete Maximum Principle

Theorem 3.2. *Consider a grid function* U_{ij}, $i = 0, 1, \ldots, m$, $j = 0, 1, 2, \ldots, n$. *If the discrete Laplacian operator (using the central five-point stencil) satisfies*

$$\Delta_h U_{ij} = \frac{U_{i-1,j} + U_{i+1,j} + U_{i,j-1} + U_{i,j+1} - 4U_{ij}}{h^2} \ge 0, \tag{3.34}$$
$$i = 1, 2, \ldots, m-1, \qquad j = 1, 2, \ldots, n-1\,,$$

then U_{ij} *attains its maximum on the boundary. On the other hand, if* $\Delta_h U_{ij} \le 0$ *then* U_{ij} *attains its minimum on the boundary.*

Proof Assume that the theorem is not true, so U_{ij} has its maximum at an interior grid point (i_0, j_0). Then $U_{i_0,j_0} \ge U_{i,j}$ for all i and j, and therefore

$$U_{i_0,j_0} \ge \frac{1}{4}\left(U_{i_0-1,j_0} + U_{i_0+1,j_0} + U_{i_0,j_0-1} + U_{i_0,j_0+1}\right).$$

On the other hand, from the condition $\Delta_h U_{ij} \ge 0$

$$U_{i_0,j_0} \le \frac{1}{4}\left(U_{i_0-1,j_0} + U_{i_0+1,j_0} + U_{i_0,j_0-1} + U_{i_0,j_0+1}\right),$$

in contradiction to the inequality above unless all U_{ij} at the four neighbors of (i_0, j_0) have the same value $U(i_0, j_0)$. This implies that neighboring U_{i_0-1,j_0} is also a maximum, and the same argument can be applied enough times until the boundary is reached. Then we would also know that U_{0,j_0} is a maximum. Indeed, if U_{ij} has its maximum in interior it follows that U_{ij} is a constant. Finally, if $\Delta_h U_{ij} \leq 0$ then we consider $-U_{ij}$ to complete the proof. □

3.3.2 Error Estimates of the Finite Difference Method for Poisson Equations

With the discrete maximum principle, we can easily get the following lemma.

Lemma 3.3. *Let U_{ij} be a grid function that satisfies*

$$\Delta_h U_{ij} = \frac{U_{i-1,j} + U_{i+1,j} + U_{i,j-1} + U_{i,j+1} - 4U_{ij}}{h^2} = f_{ij}\,, \tag{3.35}$$

$i,j = 0, 1, \ldots, n$ with an homogeneous boundary condition. Then we have

$$\|\mathbf{U}\|_\infty = \max_{0\leq i,j\leq n} |U_{ij}| \leq \frac{1}{8} \max_{1\leq i,j\leq n} |\Delta_h U_{ij}| = \frac{1}{8} \max_{0\leq i,j\leq n} |f_{ij}|\,. \tag{3.36}$$

Proof Define a grid function

$$w_{ij} = \frac{1}{4}\left(\left(x_i - \frac{1}{2}\right)^2 + \left(y_j - \frac{1}{2}\right)^2\right), \tag{3.37}$$

where

$$x_i = ih, \quad y_j = jh, \quad i,j = 0, 1, \ldots, n,\ h = \frac{1}{n},$$

corresponding to the continuous function $w(x) = \frac{1}{4}\left((x-1/2)^2 + (y-1/2)^2\right)$. Then

$$\Delta_h w_{ij} = \left. (w_{xx} + w_{yy})\right|_{(x_i,y_j)} + \frac{h^2}{12}\left(\frac{\partial^4 w}{\partial x^4} + \frac{\partial^4 w}{\partial y^4}\right)_{(x_i^*,y_j^*)} = 1\,, \tag{3.38}$$

where (x_i^*, y_j^*) is some point near (x_i, y_j), and consequently

$$\begin{aligned} \Delta_h\left(U_{ij} - \|f\|_\infty w_{ij}\right) &= \Delta_h U_{ij} - \|f\|_\infty = f_{ij} - \|f\|_\infty \leq 0, \\ \Delta_h\left(U_{ij} + \|f\|_\infty w_{ij}\right) &= \Delta_h U_{ij} + \|f\|_\infty = f_{ij} + \|f\|_\infty \geq 0\,. \end{aligned} \tag{3.39}$$

From the discrete maximum principle, $U_{ij} + \|f\|_\infty w_{ij}$ has its maximum on the boundary, while $U_{ij} - \|f\|_\infty w_{ij}$ has its minimum on the boundary, *i.e.*,

$$\min_{\partial\Omega}\left(U_{ij} - \|f\|_\infty w_{ij}\right) \le U_{ij} - \|f\|_\infty w_{ij}$$

$$\text{and} \qquad U_{ij} + \|f\|_\infty w_{ij} \le \max_{\partial\Omega}\left(U_{ij} + \|f\|_\infty w_{ij}\right),$$

for all i and j. Since U_{ij} is zero on the boundary and $\|f\|_\infty w_{ij} \ge 0$, we immediately have the following,

$$-\|f\|_\infty \min \|w_{ij}\|_{\partial\Omega} \le U_{ij} - \|f\|_\infty w_{ij} \le U_{ij},$$

$$\text{and} \qquad U_{ij} \le U_{ij} + \|f\|_\infty w_{ij} \le \|f\|_\infty \max \|w_{ij}\|_{\partial\Omega}.$$

It is easy to check that

$$\|w_{ij}\|_{\partial\Omega} = \frac{1}{8},$$

and therefore

$$-\frac{1}{8}\|f\|_\infty \le U_{ij} \le \frac{1}{8}\|f\|_\infty, \tag{3.40}$$

which completes the proof. □

Theorem 3.4. *Let U_{ij} be the solution of the finite difference equations using the standard central five-point stencil, obtained for a Poisson equation with a Dirichlet boundary condition. Assume that $u(x,y) \in C^4(\Omega)$, then the global error $\|\mathbf{E}\|_\infty$ satisfies:*

$$\begin{aligned} \|\mathbf{E}\|_\infty &= \|\mathbf{U} - \mathbf{u}\|_\infty = \max_{ij} |U_{ij} - u(x_i, y_j)| \\ &\le \frac{h^2}{96}\left(\max|u_{xxxx}| + \max|u_{yyyy}|\right), \end{aligned} \tag{3.41}$$

where $\max|u_{xxxx}| = \max_{(x,y)\in D}\left|\frac{\partial^4 u}{\partial x^4}(x,y)\right|$, *and so on.*

Proof We know that

$$\Delta_h U_{ij} = f_{ij} + T_{ij}, \qquad \Delta_h E_{ij} = T_{ij},$$

where T_{ij} is the local truncation error at (x_i, y_j) and satisfies

$$|T_{ij}| \le \frac{h^2}{12}\left(\max|u_{xxxx}| + \max|u_{yyyy}|\right),$$

so from lemma 3.3

$$\|\mathbf{E}\|_\infty \le \frac{1}{8}\|\mathbf{T}\|_\infty \le \frac{h^2}{96}\left(\max|u_{xxxx}| + \max|u_{yyyy}|\right).$$

3.4 Finite Difference Methods for General Second-order Elliptic PDEs

If the domain of the interest is a rectangle $[a, b] \times [c, d]$ and there is no mixed derivative term u_{xy} in the PDE, then the PDE can be discretized dimension by dimension. Consider the following example:

$$\nabla \cdot (p(x, y)\nabla u) - q(x, y)\, u = f(x, y), \quad \text{or} \quad (pu_x)_x + (pu_y)_y - qu = f,$$

with a Dirichlet boundary condition at $x = b$, $y = c$, and $y = d$ but a Neumann boundary condition $u_x = g(y)$ at $x = a$.

For simplicity, let us adopt a uniform Cartesian grid again

$$x_i = a + ih_x, \quad i = 0, 1, \ldots, m, \quad h_x = \frac{b - a}{m},$$

$$y_j = c + jh_y, \quad j = 0, 1, \ldots, n, \quad h_y = \frac{d - c}{n}.$$

If we discretize the PDE dimension by dimension, at a typical grid point (x_i, y_j) the finite difference equation is

$$\frac{p_{i+\frac{1}{2},j}U_{i+1,j} - (p_{i+\frac{1}{2},j} + p_{i-\frac{1}{2},j})U_{ij} + p_{i-\frac{1}{2},j}U_{i-1,j}}{(h_x)^2}$$
$$+ \frac{p_{i,j+\frac{1}{2}}U_{i,j+1} - (p_{i,j+\frac{1}{2}} + p_{i,j-\frac{1}{2}})U_{ij} + p_{i,j-\frac{1}{2}}U_{i,j-1}}{(h_y)^2} - q_{ij}U_{ij} = f_{ij} \quad (3.42)$$

for $i = 1, 2, \ldots, m - 1$ and $j = 1, 2, \ldots, n - 1$, where $p_{i\pm\frac{1}{2},j} = p(x_i \pm h_x/2, y_j)$ and so on.

For the indices $i = 0, j = 1, 2, \ldots, n - 1$, we can use the ghost point method to deal with the Neumann boundary condition. Using the central finite difference scheme for the flux boundary condition

$$\frac{U_{1,j} - U_{-1,j}}{2h_x} = g(y_j), \quad \text{or} \quad U_{-1,j} = U_{1,j} - 2h_x\, g(y_j), \quad j = 1, 2, \ldots, n - 1,$$

on substituting into the finite difference equation at $(0, j)$, we obtain

$$\frac{(p_{-\frac{1}{2},j} + p_{\frac{1}{2},j})U_{1,j} - (p_{\frac{1}{2},j} + p_{-\frac{1}{2},j})U_{0j}}{(h_x)^2}$$
$$+ \frac{p_{0,j+\frac{1}{2}}U_{0,j+1} - (p_{0,j+\frac{1}{2}} + p_{0,j-\frac{1}{2}})U_{0j} + p_{0,j-\frac{1}{2}}U_{0,j-1}}{(h_y)^2}$$
$$- q_{0j}U_{0j} = f_{0j} + \frac{2p_{-\frac{1}{2},j}\,g(y_j)}{h_x}. \quad (3.43)$$

For a general second-order elliptic PDE with no mixed derivative term u_{xy}, *i.e.*,

$$\nabla \cdot (p(x,y)\nabla u) + \mathbf{w} \cdot \nabla u - q(x,y)u = f(x,y),$$

the central finite difference scheme when $|w| \ll 1/h$ can be used, but an upwinding scheme may be preferred to deal with the advection term $\mathbf{w} \cdot \nabla u$.

3.4.1 A Finite Difference Formula for Approximating the Mixed Derivative u_{xy}

If there is a mixed derivative term u_{xy}, we cannot proceed dimension by dimension but a centered finite difference scheme (2.20) for u_{xy} can be used, *i.e.*,

$$u_{xy}(x_i, y_j) \approx \frac{u(x_{i-1}, y_{j-1}) + u(x_{i+1}, y_{j+1}) - u(x_{i+1}, y_{j-1}) - u(x_{i-1}, y_{j+1})}{4h_x h_y}. \tag{3.44}$$

From the Taylor expansion at (x_i, y_j), this finite difference formula can be shown to be consistent and the discretization is second-order accurate, and the consequent central finite difference formula for a second-order linear PDE involves nine grid points. The resulting linear system of algebraic equations for PDE is more difficult to solve, because it is no longer symmetric nor diagonally dominant. Furthermore, there is no known upwinding scheme to deal with the PDE with mixed derivatives.

3.5 Solving the Resulting Linear System of Algebraic Equations

The linear systems of algebraic equations resulting from finite difference discretizations for 2D or higher-dimensional problems are often very large, *e.g.*, the linear system from an $n \times n$ grid for an elliptic PDE has $O(n^2)$ equations, so the coefficient matrix is $O(n^2 \times n^2)$. Even for $n = 100$, a modest number, the $O(10^4 \times 10^4)$ matrix cannot be stored in most modern computers if the desirable double precision is used. However, the matrix from a self-adjoint elliptic PDE is sparse since the nonzero entries are about $O(5n^2)$, so an iterative method or sparse matrix technique may be used. For an elliptic PDE defined on a rectangle domain or a disk, frequently used methods are listed below.

- Fast Poisson solvers such as the fast Fourier transform (FFT) or cyclic reduction (Adams et al.). Usually the implementation is not so easy, and the use of existing software packages is recommended, *e.g.*, Fishpack, written in Fortran and free on the Netlib.

- Multigrid solvers, either structured multigrid, *e.g.*, MGD9V (De Zeeuw, 1990) that uses a nine-point stencil, or AMGs (algebraic multigrid solvers).
- Sparse matrix techniques.
- Simple iterative methods such as Jacobi, Gauss–Seidel, SOR(ω). They are easy to implement, but often slow to converge.
- Other iterative methods such as the conjugate gradient (CG) or preconditioned conjugate gradient (PCG), generalized minimized residual (GMRES), biconjugate gradient (BICG) method for nonsymmetric system of equations. We refer the reader to Saad (1986) for more information and references.

An important advantage of an iterative method is that zero entries play no role in the matrix-vector multiplications involved and there is no need to manipulate the matrix and vector forms, as the algebraic equations in the system are used directly. Assume that we are given a linear system of equation $A\mathbf{x}=b$ where A is nonsingular ($det(A)\neq 0$), if A can be written as $A=M-N$ where M is an invertible matrix, then $(M-N)\mathbf{x}=b$ or $M\mathbf{x}=N\mathbf{x}+b$ or $\mathbf{x}=M^{-1}N\mathbf{x}+M^{-1}b$. We may iterate starting from an initial guess $\mathbf{x}^0$, via

$$\mathbf{x}^{k+1}=M^{-1}N\mathbf{x}^k+M^{-1}b, \quad k=0,1,2,\ldots, \tag{3.45}$$

and the iteration converges or diverges depending the spectral radius of $\rho(M^{-1}N)=\max|\lambda_i(M^{-1}N)|$. Incidentally, if $T=M^{-1}N$ is a constant matrix, the iterative method is called stationary.

3.5.1 The Jacobi Iterative Method

The idea of the Jacobi iteration is to solve for the variables on the diagonals and then form the iteration. Solving for x_1 from the first equation in the algebraic system, x_2 from the second, and so forth, we have

$$\begin{aligned}
x_1 &= \frac{1}{a_{11}}\left(b_1-a_{12}x_2-a_{13}x_3\cdots-a_{1n}x_n\right)\\
x_2 &= \frac{1}{a_{22}}\left(b_2-a_{21}x_1-a_{23}x_3\cdots-a_{2n}x_n\right)\\
\vdots\ &\ \ \vdots\quad\ \vdots\quad\ \vdots\\
x_i &= \frac{1}{a_{ii}}\left(b_i-a_{i1}x_1-a_{i2}x_2\cdots-a_{i,i-1}x_{i-1}-a_{i,i+1}x_{i+1}-\cdots-a_{in}x_n\right)\\
\vdots\ &\ \ \vdots\quad\ \vdots\quad\ \vdots\\
x_n &= \frac{1}{a_{nn}}\left(b_n-a_{i1}x_1-a_{n2}x_2\cdots-a_{n,n-1}x_{n-1}\right).
\end{aligned}$$

Given some initial guess $\mathbf{x}^0$, the corresponding Jacobi iterative method is

$$
\begin{aligned}
x_1^{k+1} &= \frac{1}{a_{11}}\left(b_1 - a_{12}x_2^k - a_{13}x_3^k \cdots - a_{1n}x_n^k\right) \\
x_2^{k+1} &= \frac{1}{a_{22}}\left(b_2 - a_{21}x_1^k - a_{23}x_3^k \cdots - a_{2n}x_n^k\right) \\
&\vdots \quad \vdots \quad \vdots \quad \vdots \\
x_i^{k+1} &= \frac{1}{a_{ii}}\left(b_i - a_{i1}x_1^k - a_{i2}x_2^k \cdots - a_{in}x_n^k\right) \\
&\vdots \quad \vdots \quad \vdots \quad \vdots \\
x_n^{k+1} &= \frac{1}{a_{nn}}\left(b_n - a_{i1}x_1^k - a_{n2}x_2^k \cdots - a_{n,n-1}x_{n-1}^k\right).
\end{aligned}
$$

It can be written compactly as

$$
x_i^{k+1} = \frac{1}{a_{ii}}\left(b_i - \sum_{j=1, j\neq i}^{n} a_{ij}x_j^k\right), \quad i = 1, 2, \ldots, n, \tag{3.46}
$$

which is the basis for easy programming. Thus for the finite difference equations

$$
\frac{U_{i+1} - 2U_i + U_{i+1}}{h^2} = f_i
$$

with Dirichlet boundary conditions $U_0 = ua$ and $U_n = ub$, we have

$$
\begin{aligned}
U_1^{k+1} &= \frac{ua + U_2^k}{2} - \frac{h^2 f_1}{2} \\
U_i^{k+1} &= \frac{U_{i-1}^k + U_{i+1}^k}{2} - \frac{h^2 f_i}{2}, \quad i = 2, 3, \ldots, n-1 \\
U_{n-1}^{k+1} &= \frac{U_{n-2}^k + ub}{2} - \frac{h^2 f_{n-1}}{2};
\end{aligned}
$$

and for a 2D Poisson equation,

$$
U_{ij}^{k+1} = \frac{U_{i-1,j}^k + U_{i+1,j}^k + U_{i,j-1}^k + U_{i,j+1}^k}{4} - \frac{h^2 f_{ij}}{4},
$$

$i, j = 1, 2, \ldots, n-1$ assuming $m = n$.

3.5.2 The Gauss–Seidel Iterative Method

The idea of the Gauss–Seidel iteration is to solve for the variables on the diagonals, then form the iteration, and use the most updated information. In the Jacobi iterative method, all components $\mathbf{x}^{k+1}$ are updated based on $\mathbf{x}^k$, whereas in the Gauss–Seidel iterative method the *most updated* information is used as follows:

$$x_1^{k+1} = \frac{1}{a_{11}}\left(b_1 - a_{12}x_2^k - a_{13}x_3^k \cdots - a_{1n}x_n^k\right)$$

$$x_2^{k+1} = \frac{1}{a_{22}}\left(b_2 - a_{21}x_1^{k+1} - a_{23}x_3^k \cdots - a_{2n}x_n^k\right)$$

$$\vdots \quad \vdots \quad \vdots \quad \vdots$$

$$x_i^{k+1} = \frac{1}{a_{ii}}\left(b_i - a_{i1}x_1^{k+1} - a_{i2}x_2^{k+1} \cdots - a_{i,i-1}x_{i-1}^{k+1} - a_{i,i+1}x_{i+1}^k - \cdots - a_{in}x_n^k\right)$$

$$\vdots \quad \vdots \quad \vdots \quad \vdots$$

$$x_n^{k+1} = \frac{1}{a_{nn}}\left(b_n - a_{i1}x_1^{k+1} - a_{n2}x_2^{k+1} \cdots - a_{n,n-1}x_{n-1}^{k+1}\right),$$

or in a compact form

$$x_i^{k+1} = \frac{1}{a_{ii}}\left(b_i - \sum_{j=1}^{i-1} a_{ij}x_j^{k+1} - \sum_{j=i+1}^{n} a_{ij}x_j^k\right), \quad i = 1, 2, \ldots, n. \tag{3.47}$$

Below is a pseudo-code, where the Gauss–Seidel iterative method is used to solve the finite difference equations for the Poisson equation, assuming u_0 is an initial guess:

```
% Give u0(i,j) and a tolerance tol, say 1e-6.

err = 1000; k = 0;  u = u0;
while err > tol
  for i=1:n
     for j=1:n
        u(i,j) = ( (u(i-1,j)+u(i+1,j)+u(i,j-1)+u(i,j+1))
                   -h^2*f(i,j) )/4;
     end
  end
  err = max(max(abs(u-u0)));
  u0 = u;  k = k + 1;     % Next iteration if err > tol
end
```

Note that this pseudo-code has a framework generally suitable for iterative methods.

3.5.3 The Successive Overrelaxation Method SOR(ω)

The idea of the successive overrelaxation (SOR(ω)) iteration is based on an extrapolation technique. Suppose $\mathbf{x}_{GS}^{k+1}$ denotes the update from $\mathbf{x}^k$ in the

Gauss–Seidel method. Intuitively, one may anticipate the update

$$\mathbf{x}^{k+1} = (1-\omega)\mathbf{x}^k + \omega\mathbf{x}_{GS}^{k+1}, \tag{3.48}$$

a linear combination of $\mathbf{x}^k$ and $\mathbf{x}_{GS}^{k+1}$, may give a better approximation for a suitable choice of the relaxation parameter ω. If the parameter $0<\omega<1$, the combination above is an interpolation, and if $\omega>1$ it is an extrapolation or *overrelaxation*. For $\omega=1$, we recover the Gauss–Seidel method. For elliptic problems, we usually choose $1\leq\omega<2$. In component form, the SOR(ω) method can be represented as

$$x_i^{k+1} = (1-\omega)x_i^k + \frac{\omega}{a_{ii}}\left(b_i - \sum_{j=1}^{i-1} a_{ij}x_j^{k+1} - \sum_{j=i+1}^{n} a_{ij}x_j^k\right), \tag{3.49}$$

for $i=1,2,\ldots,n$. Therefore, only one line in the pseudo-code of the Gauss–Seidel method above need be changed, namely,

```
u(i,j) = (1-omega)*u0(i,j) + omega*( u(i-1,j) + u(i+1,j)
              + u(i,j-1) + u(i,j+1) -h^2*f(i,j))/4
```

The convergence of the SOR(ω) method depends on the choice of ω. For the linear system of algebraic equations obtained from the standard five-point stencil applied to a Poisson equation with $h=h_x=h_y=1/n$, it can be shown that the optimal ω is

$$\omega_{opt} = \frac{2}{1+\sin(\pi/n)} \sim \frac{2}{1+\pi/n}, \tag{3.50}$$

which approaches two as n approaches infinity. Although the optimal ω is unknown for general elliptic PDEs, we can use the optimal ω for the Poisson equation as a trial value, and in fact larger rather than smaller ω values are recommended. If ω is so large that the iterative method diverges, this is soon evident because the solution will "blow-up." Incidentally, the optimal choice of ω is independent of the right-hand side.

3.5.4 Convergence of Stationary Iterative Methods

For a stationary iterative method, the following theorem provides a necessary and sufficient condition for convergence.

Theorem 3.5. *Given a stationary iteration*

$$\mathbf{x}^{k+1} = T\mathbf{x}^k + c, \tag{3.51}$$

where T is a constant matrix and c is a constant vector, the vector sequence $\{\mathbf{x}^k\}$ *converges for arbitrary* $\mathbf{x}^0$ *if and only if* $\rho(T) < 1$ *where* $\rho(T)$ *is the spectral radius of T defined as*

$$\rho(T) = \max|\lambda_i(T)|, \tag{3.52}$$

i.e., the largest magnitude of all the eigenvalues of T.

Another commonly used sufficient condition to check the convergence of a stationary iterative method is given in the following theorem.

Theorem 3.6. *If there is a matrix norm* $\|\cdot\|$ *such that* $\|T\| < 1$*, then the stationary iterative method converges for arbitrary initial guess* $\mathbf{x}^0$.

We often check whether $\|T\|_p < 1$ for $p = 1, 2, \infty$, and if there is just one norm such that $\|T\| < 1$, then the iterative method is convergent. However, if $\|T\| \geq 1$ there is no conclusion about the convergence.

Let us now briefly discuss the convergence of the Jacobi, Gauss–Seidel, and SOR(ω) methods. Given a linear system $A\mathbf{x} = b$, let D denote the diagonal matrix formed from the diagonal elements of A, $-L$ the lower triangular part of A, and $-U$ the upper triangular part of A. The iteration matrices for the three basic iteration methods are thus

- Jacobi method: $T = D^{-1}(L + U)$, $c = D^{-1}b$.
- Gauss–Seidel method: $T = (D - L)^{-1}U$, $c = (D - L)^{-1}b$.
- SOR(ω) method: $T = (I - \omega D^{-1}L)^{-1}\left((1-\omega)I + \omega D^{-1}U\right)$, $c = \omega(I - \omega L)^{-1}D^{-1}b$.

Theorem 3.7. *If A is strictly row diagonally dominant, i.e.,*

$$|a_{ii}| > \sum_{j=1, j\neq n}^{n} |a_{ij}|, \tag{3.53}$$

then both the Jacobi and Gauss–Seidel iterative methods converge. The conclusion is also true when (1): A is weakly row diagonally dominant

$$|a_{ii}| \geq \sum_{j=1, j\neq n}^{n} |a_{ij}|; \tag{3.54}$$

(2): the inequality holds for at least one row; (3) A is irreducible.

We refer the reader to Golub and Van Loan (1989) for the definition of irreducibility. From this theorem, both the Jacobi and Gauss–Seidel iterative methods converge when they are applied to the linear system of algebraic equations obtained from the standard central finite difference method for Poisson equations. In general, Jacobi and Gauss–Seidel methods need $O(n^2)$ number of iterations for solving the resulting linear system of finite difference equations for a Poisson equations, while it is $O(n)$ for the SOR(ω) method with the optimal choice of ω.

3.6 A Fourth-Order Compact FD Scheme for Poisson Equations

A compact fourth-order accurate scheme ($\|\mathbf{u} - \mathbf{U}\| \le Ch^4$) can be applied to Poisson equations, using a nine-point discrete Laplacian. An advantage of a higher-order method is that fewer grid points are used for the same order accuracy as a lower-order method; therefore a smaller resulting system of algebraic equations needs to be solved. A disadvantage is that the resulting system of algebraic equations is denser.

Although other methods may be used, let us follow a symbolic derivation from the second-order central scheme for u_{xx}. Recalling that, (*cf.* (2.18) on page 20)

$$\begin{aligned} \delta_{xx}^2 u &= \frac{\partial^2 u}{\partial x^2} + \frac{h^2}{12}\frac{\partial^4 u}{\partial x^4} + O(h^4) \\ &= \left(1 + \frac{h^2}{12}\frac{\partial^2}{\partial x^2}\right)\frac{\partial^2}{\partial x^2} u + O(h^4), \end{aligned} \tag{3.55}$$

and substituting the operator relation

$$\frac{\partial^2}{\partial x^2} = \delta_{xx}^2 + O(h^2)$$

into (3.55), we obtain

$$\begin{aligned} \delta_{xx}^2 u &= \left(1 + \frac{h^2}{12}\left(\delta_{xx}^2 + O(h^2)\right)\right)\frac{\partial^2}{\partial x^2} u + O(h^4) \\ &= \left(1 + \frac{h^2}{12}\delta_{xx}^2\right)\frac{\partial^2}{\partial x^2} u + O(h^4), \end{aligned}$$

from which we further have

$$\frac{\partial^2}{\partial x^2} = \left(1 + \frac{h^2}{12}\delta_{xx}^2\right)^{-1}\delta_{xx}^2 u + \left(1 + \frac{h^2}{12}\delta_{xx}^2\right)^{-1} O(h^4).$$

It is noted that

$$\left(1 + \frac{h^2}{12}\delta_{xx}^2\right)^{-1} = 1 - \frac{h^2}{12}\delta_{xx}^2 + O(h^4),$$

if h is sufficiently small. Thus we have the symbolic relation

$$\frac{\partial^2}{\partial x^2} = \left(1 + \frac{h^2}{12}\delta_{xx}^2\right)^{-1} \delta_{xx}^2 + O(h^4), \quad \text{or}$$

$$\frac{\partial^2}{\partial x^2} = \left(1 - \frac{h^2}{12}\delta_{xx}^2\right) \delta_{xx}^2 + O(h^4).$$

On a Cartesian grid and invoking this fourth-order operator, the Poisson equation $\Delta u = f$ can be approximated by

$$\left(1 + \frac{h_x^2}{12}\delta_{xx}^2\right)^{-1} \delta_{xx}^2 u + \left(1 + \frac{h_y^2}{12}\delta_{yy}^2\right)^{-1} \delta_{yy}^2 u = f(x, y) + O(h^4),$$

where $h = \max(h_x, h_y)$. On multiplying this by

$$\left(1 + \frac{h_x^2}{12}\delta_{xx}^2\right)\left(1 + \frac{h_y^2}{12}\delta_{yy}^2\right)$$

and using the commutativity

$$\left(1 + \frac{(\Delta x)^2}{12}\delta_{xx}^2\right)\left(1 + \frac{(\Delta y)^2}{12}\delta_{yy}^2\right) = \left(1 + \frac{(\Delta y)^2}{12}\delta_{yy}^2\right)\left(1 + \frac{(\Delta x)^2}{12}\delta_{xx}^2\right),$$

we get:

$$\left(1 + \frac{h_y^2}{12}\delta_{yy}^2\right)\delta_{xx}^2 u + \left(1 + \frac{h_x^2}{12}\delta_{xx}^2\right)\delta_{yy}^2 u = \left(1 + \frac{h_x^2}{12}\delta_{xx}^2\right)\left(1 + \frac{h_y^2}{12}\delta_{yy}^2\right) f(x, y)$$

$$+ O(h^4) = \left(1 + \frac{h_x^2}{12}\delta_{xx}^2 + \frac{h_y^2}{12}\delta_{yy}^2\right) f(x, y) + O(h^4).$$

Expanding this expression above and ignoring the high-order terms, we obtain the nine-point scheme

$$\left(1 + \frac{h_y^2}{12}\delta_{yy}^2\right)\delta_{xx}^2 U_{ij} + \left(1 + \frac{h_y^2}{12}\delta_{yy}^2\right)\frac{U_{i-1,j} - 2U_{ij} + U_{i+1,j}}{(h_x)^2}$$

$$= \frac{U_{i-1,j} - 2U_{ij} + U_{i+1,j}}{(h_x)^2} + \frac{1}{12(h_y)^2}\Big(U_{i-1,j-1}, -2U_{i-1,j} + U_{i-1,j+1}$$

$$- 2U_{i,j-1} + 4U_{ij} - 2U_{i,j+1} + U_{i+1,j-1} - 2U_{i+1,j} + U_{i+1,j+1}\Big).$$

For the special case when $h_x = h_y = h$, the finite difference coefficients and linear combination of f are given in Figure 3.3. The above discretization is called

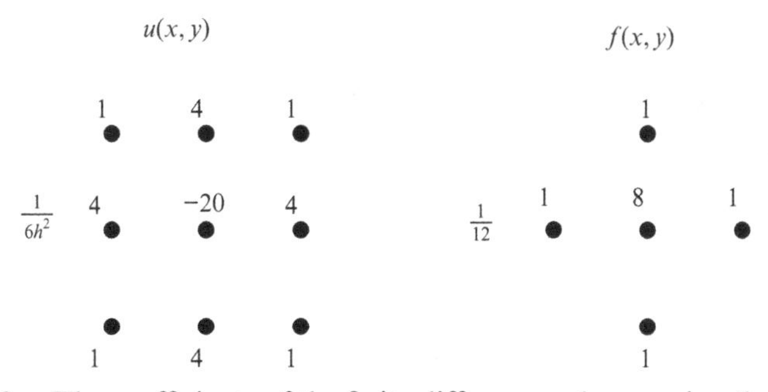

Figure 3.3. The coefficients of the finite difference scheme using the compact nine-point stencil (a) and the linear combination of f (b).

the nine-point discrete Laplacian, and it is a fourth-order compact scheme because the distance is the least between the grid points in the finite difference stencil and the master grid point (of all fourth-order finite difference schemes).

The advantages and disadvantages of nine-point finite difference schemes for Poisson equations include:

- It is fourth-order accurate and it is still compact. The coefficient matrix is still block tridiagonal.
- Less grid orientation effects compared with the standard five-point finite difference scheme.
- It seems that there are no FFT-based fast Poisson solvers for the fourth-order compact finite difference scheme.

Incidentally, if we apply

$$\frac{\partial^2}{\partial x^2} = \left(1 - \frac{h^2}{12}\delta_{xx}^2\right)\delta_{xx}^2 u + O(h^4)$$

to the Poisson equation directly, we obtain another nine-point finite difference scheme, which is not compact, and has stronger grid orientation effects.

3.7 A Finite Difference Method for Poisson Equations in Polar Coordinates

If the domain of interest is a circle, an annulus, or a fan, *etc.* (*cf.* Figure 3.4 for some illustrations), it is natural to use plane polar coordinates

$$x = r\cos\theta, \qquad y = r\sin\theta. \tag{3.56}$$

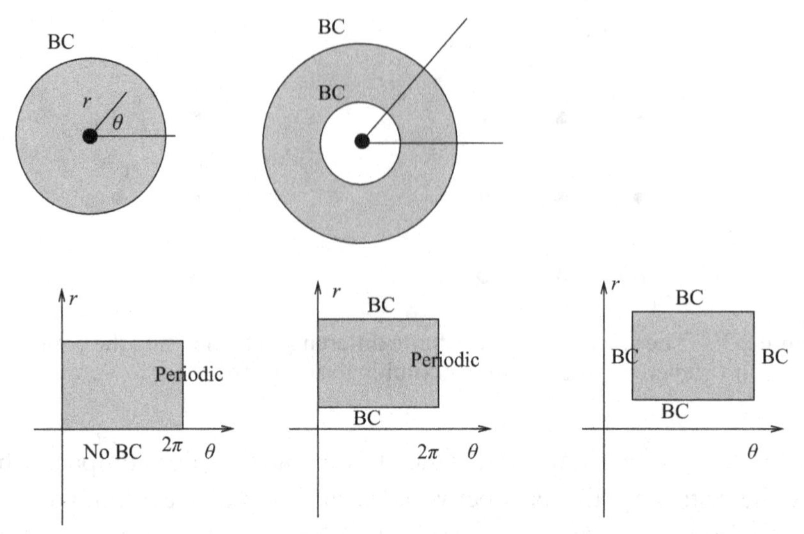

Figure 3.4. Diagrams of domains and boundary conditions that may be better solved in polar coordinates.

The Poisson equation in the polar coordinates is

$$
\begin{aligned}
&\frac{1}{r}\frac{\partial}{\partial r}\left(r\frac{\partial u}{\partial r}\right)+\frac{1}{r^2}\frac{\partial^2 u}{\partial \theta^2}=f(r,\theta), \quad \text{conservative form} \\
\text{or} \quad &\frac{\partial^2 u}{\partial r^2}+\frac{1}{r}\frac{\partial u}{\partial r}+\frac{1}{r^2}\frac{\partial^2 u}{\partial \theta^2}=f(r,\theta), \quad \text{nonconservative form.}
\end{aligned}
\tag{3.57}
$$

For $0<R_1\le r\le R_2$ and $\theta_l\le\theta\le\theta_r$, where the origin is not in the domain of interest, using a uniform grid in the polar coordinates

$$
r_i=R_1+i\Delta r, \quad i=0,1,\ldots,m, \quad \Delta r=\frac{R_2-R_1}{m},
$$

$$
\theta_j=\theta_l+j\Delta\theta, \quad j=0,1,\ldots,n, \quad \Delta\theta=\frac{\theta_r-\theta_l}{n},
$$

the discretized finite difference equation (in conservative form) is

$$
\begin{aligned}
&\frac{1}{r_i}\frac{r_{i-\frac{1}{2}}U_{i-1,j}-(r_{i-\frac{1}{2}}+r_{i+\frac{1}{2}})U_{ij}+r_{i+\frac{1}{2}}U_{i+1,j}}{(\Delta r)^2} \\
&\qquad+\frac{1}{r_i^2}\frac{U_{i,j-1}-2U_{ij}+U_{i,j+1}}{(\Delta\theta)^2}=f(r_i,\theta_j),
\end{aligned}
\tag{3.58}
$$

where again U_{ij} is an approximation to the solution $u(r_i,\theta_j)$.

3.7.1 Treating the Polar Singularity

If the origin is within the domain and $0 \le \theta < 2\pi$, we have a *periodic boundary condition* in the θ direction (*i.e.*, $u(r,\theta)=u(r,\theta+2\pi)$), but in the radial ($r$) direction the origin $R_1=0$ needs special attention. The PDE is singular at $r=0$, which is called a pole singularity. There are different ways of dealing with a singularity at the origin, but some methods lead to an undesirable structure in the coefficient matrix from the finite difference equations. One approach discussed is to use a staggered grid:

$$r_i = \left(i - \frac{1}{2}\right)\Delta r, \quad \Delta r = \frac{R_2}{m - \frac{1}{2}}, \quad i = 1, 2, \ldots, m, \tag{3.59}$$

where $r_1 = \Delta r/2$ and $r_m = R_2$. Except for $i=1$ (*i.e.*, at $i=2,\ldots,m-1$), the conservative form of the finite difference discretization can be used. At $i=1$, the following nonconservative form is used to deal with the pole singularity at $r=0$:

$$\frac{U_{0j} - 2U_{1j} + U_{2j}}{(\Delta r)^2} + \frac{1}{r_1}\frac{U_{2j} - U_{0j}}{2\Delta r} + \frac{1}{r_1^2}\frac{U_{1,j-1} - 2U_{1j} + U_{1,j+1}}{(\Delta\theta)^2} = f(r_1,\theta_j).$$

Note that $r_0 = -\Delta r/2$ and $r_1 = \Delta r/2$. The coefficient of U_{0j} in the above finite difference equation, the approximation at the ghost point r_0, is zero! The above finite difference equation simplifies to

$$\frac{-2U_{1j} + U_{2j}}{(\Delta r)^2} + \frac{1}{r_1}\frac{U_{2j}}{2\Delta r} + \frac{1}{r_1^2}\frac{U_{1,j-1} - 2U_{1j} + U_{1,j+1}}{(\Delta\theta)^2} = f(r_1,\theta_j),$$

and we still have a diagonally dominant system of linear algebraic equations.

3.7.2 Using the FFT to Solve Poisson Equations in Polar Coordinates

When the solution $u(r,\theta)$ is periodic in θ, we can approximate $u(r,\theta)$ by the truncated Fourier series

$$u(r,\theta) = \sum_{n=-N/2}^{N/2-1} u_n(r)e^{in\theta}, \tag{3.60}$$

where $i=\sqrt{-1}$ and $u_n(r)$ is the complex Fourier coefficient given by

$$u_n(r) = \frac{1}{N}\sum_{k=0}^{N-1} u(r,\theta)e^{-ink\theta}. \tag{3.61}$$

Substituting (3.60) into the Poisson equation (in polar coordinates) (3.57) gives

$$\frac{1}{r}\frac{\partial}{\partial r}\left(\frac{1}{r}\frac{\partial u_n}{\partial r}\right) - \frac{n^2}{r^2}u_n = f_n(r), \quad n = -N/2, \ldots, N/2-1, \tag{3.62}$$

where $f_n(r)$ is the n-th coefficient of the Fourier series of $f(r,\theta)$ defined in (3.61). For each n, we can discretize the above ODE in the r direction using the staggered grid, to get a tridiagonal system of equations that can be solved easily.

With a Dirichlet boundary condition $u(r_{max},\theta) = u^{BC}(\theta)$ at $r = r_{max}$, the Fourier transform

$$u_n^{BC}(r_{max}) = \frac{1}{N}\sum_{k=0}^{N-1} u^{BC}(\theta)e^{-ink\theta} \tag{3.63}$$

can be invoked to find $u_n^{BC}(r_{max})$, the corresponding boundary condition for the ODE. After the Fourier coefficient u_n is obtained, the inverse Fourier transform (3.60) can be applied to get an approximate solution to the problem.

3.8 Programming of 2D Finite Difference Methods

After discretization of an elliptic PDE, there are a variety of approaches that can be used to solve the system of the finite difference equations. Below we list some of them according to our knowledge.

- Sparse matrix techniques. In Matlab, one can form the coefficient matrix, A and the right-hand side $\mathbf{F}$, then get the finite difference solution using $\mathbf{U} = A\backslash\mathbf{F}$. Note that, the solution will be expressed as a 1D array. For the visualization purpose, it is desirable to convert between the 1D vector and the 2D array.
- Fast Poisson solvers. For Poisson equations defined on rectangular domains and linear boundary conditions (Dirichlet, Neumann, Robin) on four sides of the domain, one can apply a fast Poisson solver, for example, the Fishpack (Adams et al.). The Fishpack includes a variety of solvers for Cartesian, polar, or cylindrical coordinates.
- Iterative solvers. One can apply stationary iterative methods such as Jacobi, Gauss–Seidel, SOR(ω) for the linear system of equations obtained for general elliptic PDEs. The programming is easy and requires least storage. However, the convergence speed is often slow and the number of iterations is $O((mn)^2) = O(N^2)$ $(N = mn)$ assuming that m and n are the number of grid lines in each coordinate direction. One can also apply more advanced iterative methods such as the CG method, PCG method if the coefficient

matrix A is symmetric, or GMRES (Saad, 1986) iterative method is A is nonsymmetric.

- Multigrid solvers. There are two kinds of multigrid solvers that one can use. One is a structured multigrid. A very good package written in Fortran is DMG9V (De Zeeuw, 1990) that uses a nine-point stencil which is suitable for the finite difference methods applied to second-order linear elliptic PDEs described in this book, either centered five-point stencil or the compact nine-point stencil. Another type is algebraic multigrid solvers, for example, (Ruge and Stuben, 1987; Stuben, 1999). In general, the structured multigrid solvers work better than algebraic multigrid solvers for structured meshes.

3.8.1 A Matlab Code for Poisson Equations using $A\backslash F$

The accompanying Matlab code, poisson_matlab.m, solves the Poisson equation on a rectangular domain $[a, b] \times [c, d]$ with a Neumann boundary condition on $x = a$ and Dirichelt boundary conditions on other three sides. The mesh parameters are m and n; and the total number of unknowns is $M = (n - 1)m$. The conversion between the 1D solution $U(k)$ and 2D array using the natural row ordering is $k = i + (j - 1)m$. The files include poisson_matlab.m (the main code), f.m (function $f(x, y)$), ux.m (the Neumann boundary condition $\frac{\partial u}{\partial x}(a, y)$ at $x = a$), and ue.m (the true solution for testing purpose). In Figure 3.5, we show a mesh plot of the solution of the finite difference method and its error plot.

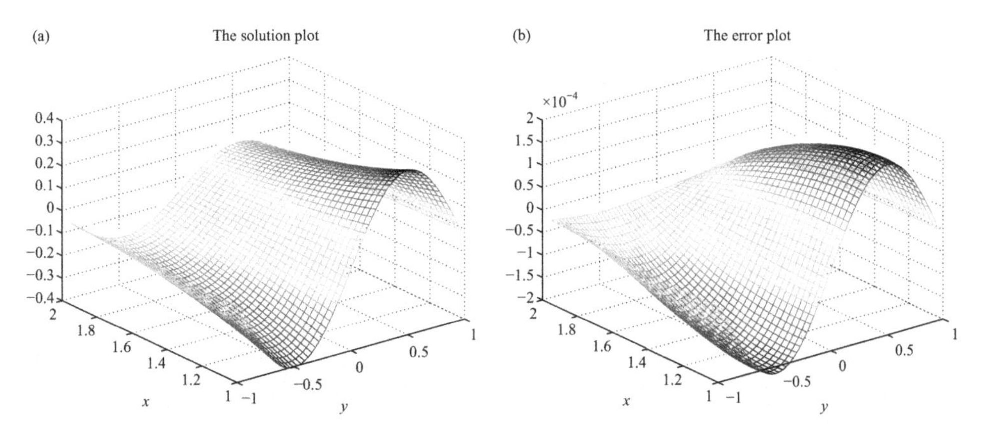

Figure 3.5. (a) The mesh plot of the computed finite difference solution $[1, 2] \times [-1, 1]$ and (b) the error plot. Note that we can see the errors are zeros for Dirichlet boundary conditions, and the errors are not zero for Neumann boundary condition at $x = 1$.

```
%%%%%%%%%%%%%%%%%%%%%%%%%%%%%%%%%%%%%%%%%%%%%%%%%%%%%%%%%%%%%%%%%%%%%%%%%%%%%%%%%%%%%%%%%%%%%%
clear;  close all
a = 1; b=2; c = -1; d=1;
m=32; n=64;

hx = (b-a)/m; hx1 = hx*hx; x=zeros(m+1,1);
for i=1:m+1,
  x(i) = a + (i-1)*hx;
end
hy = (d-c)/n; hy1 = hy*hy; y=zeros(n+1,1);
for i=1:n+1,
  y(i) = c + (i-1)*hy;
end

M = (n-1)*m; A = sparse(M,M); bf = zeros(M,1);

for j = 1:n-1,
  for i=1:m,
    k = i + (j-1)*m;
    bf(k) = f(x(i),y(j+1));
    A(k,k) = -2/hx1 -2/hy1;
    if i == 1
        A(k,k+1) = 2/hx1;
        bf(k) = bf(k) + 2*ux(y(j+1))/hx;
    else
       if i==m
         A(k,k-1) = 1/hx1;
         bf(k) = bf(k) - ue(x(i+1),y(j+1))/hx1;
    else
             A(k,k-1) = 1/hx1; A(k,k+1) = 1/hx1;
    end
  end

%-- y direction --------------

    if j == 1
        A(k,k+m) = 1/hy1;
        bf(k) = bf(k) - ue(x(i),c)/hy1;
    else
       if j==n-1
         A(k,k-m) = 1/hy1;
         bf(k) = bf(k) - ue(x(i),d)/hy1;
       else
          A(k,k-m) = 1/hy1; A(k,k+m) = 1/hy1;
       end
     end
  end
end

  U = A \bf;
```

```
%--- Transform back to (i,j) form to plot the solution ---

j = 1;
for k=1:M
  i = k - (j-1)*m ;
  u(i,j) = U(k);
  u2(i,j) = ue(x(i),y(j+1));
  j = fix(k/m) + 1;
end

% Analyze and Visualize the result.

e = max( max( abs(u-u2)))         % The maximum error
x1=x(1:m); y1=y(2:n);

mesh(y1,x1,u); title('The solution plot'); xlabel('y');
ylabel('x'); figure(2); mesh(y1,x1,u-u2); title('The error plot');
xlabel('y'); ylabel('x');
```

3.8.2 A Matlab Code for Poisson Equations using the SOR(ω) Iteration

The accompanying Matlab code, poisson_jacobi.m, and poisson_sor.m, provide interactive Jacobi and SOR(ω) iterative methods for the Poisson equation on a square domain $[a, b] \times [c, d]$ with $(b - a) = (d - c)$ and Dirichelt boundary conditions on all sides. When $\omega = 1$, the SOR(ω) method becomes the Gauss-Seidel iteration. The other Matlab functions involved are fcn.m, the source term $f(x, y)$, and uexact.m, the solution for the testing purpose.

We test an example with true solution $u(x, y) = e^x \sin(\pi y)$, the source term then is $f(x, y) = e^x \sin(\pi y)(1 - \pi^2)$. We take $m = n = 40$, and the domain is $[-1, 1] \times [-1, 1]$. Thus $h = 1/40$ and $h^2 = 6.25 \times 10^{-4}$. So we take the tolerance as $tol = 10^{-5}$. The Jacobi iteration takes 2105 iterations, the Gauss–Seidel takes 1169 iterations, the SOR(ω) with the optimal $\omega = 2/(1 + \sin \pi/n)) = 1.8545$ takes 95 iterations, SOR(1.8) takes 158 iterations, and SOR(1.9) takes 133 iterations. Usually we would rather take ω larger than smaller in the range $1 \leq \omega < 2$.

Exercises

1. State the maximum principle for 1D elliptic problems. Use the maximum principle to show that the three-point central finite difference scheme for $u''(x) = f(x)$ with a Dirichlet boundary condition is second-order accurate in the maximum norm.
2. Write down the coefficient matrix of the finite difference method using the standard central five-point stencil with both the Red–Black ***and*** the Natural row ordering for the Poisson equation defined on the rectangle $[a, b] \times [c, d]$. Take $m = n = 3$ and assume a Dirichlet

boundary condition at $x=a$, $y=c$ and $y=d$, and a Neumann boundary condition $\frac{\partial u}{\partial n}=g(y)$ at $x=b$. Use the ghost point method to deal with the Neumann boundary condition.

3. Implement and compare the Gauss–Seidel method and the SOR method (trying to find the best ω by testing), for the elliptic equation

$$u_{xx}+p(x,y)u_{yy}+r(x,y)u(x,y)=f(x,y)$$

$$a<x<b, \qquad c<y<d,$$

subject to the boundary conditions

$$u(a,y)=0, \quad u(x,c)=0, \quad u(x,d)=0, \quad \frac{\partial u}{\partial x}(b,y)=-\pi\sin(\pi y).$$

Test and debug your code for the case: $0<x,y<1$, and

$$p(x,y)=(1+x^2+y^2), \qquad r(x,y)=-xy.$$

The source term $f(x,y)$ is determined from the exact solution

$$u(x,y)=\sin(\pi x)\sin(\pi y).$$

Do the grid refinement analysis for $n=16$, $n=32$, and $n=64$ (if possible) in the infinity norm (**Hint:** In Matlab, use $max(max(abs(e)))$). Take the tolerance as 10^{-8}. Does the method behave like a second-order method? Compare the number of iterations, and test the optimal relaxation factor ω. Plot the solution and the error for $n=32$.
Having ensured your code is working correctly, introduce a point source $f(x,y)=\delta(x-0.5)\delta(y-0.5)$ and $u_x=-1$ at $x=1$, with $p(x,y)=1$ and $r(x,y)=0$. Along other three sides $x=0$, $y=0$, and $y=1$, $u=0$. The $u(x,y)$ can be interpreted as the steady state temperature distribution in a room with an insulated wall on three sides, a constant heat flow on one side, and a point source such as a heater in the room. Note that the heat source can be expressed as $f(n/2,n/2)=1/h^2$, and $f(i,j)=0$ for other grid points. Use the mesh and contour plots to visualize the solution for $n=36$ ($mesh(x,y,u)$, $contour(x,y,u,30)$).

4. (a) Show the eigenvalues and eigenvectors for the Laplace equation

$$\Delta u+\lambda u=0, \quad 0<x,y<1,$$
$$u=0, \quad \text{on the boundaries,}$$

are

$$\lambda_{k,l}=\pi^2\left(l^2+k^2\right), \quad l,k=1,2,\ldots,\infty,$$
$$u_{k,l}(x,y)=\sin(k\pi x)\sin(l\pi y).$$

(b) Show that the standard central finite difference scheme using the five-point stencil is stable for the Poisson equation.

Hint: The eigenvectors for $A_hU_h=F$ (grid functions) are

$$u_{k,l,i,j}=\sin(il\pi/N)\sin(jk\pi/N), \quad i,j,l,k=1,2,\ldots,N-1.$$

The 2-norm of A_h^{-1} is $1/\min\{|\lambda_i(A_h)|\}$.

5. Modify your SOR code for Problem 4, so that either the mixed boundary condition used at $x=b$ or the periodic boundary condition is used at $x=a$ and $x=b$.

6. Modify the Matlab code poisson_matlab.m for Poisson equations to solve the general self-adjoint elliptic PDEs

$$\nabla \cdot (p\nabla u) - qu = f(x, y), \qquad a < x < b, \quad c < y < d.$$

Validate your code with analytic solutions $u(x, y) = x^2 + y^2$ and $u(x, y) = \cos x \sin y$ and $p(x, y) = 1$ and $p(x, y) = e^{x+y}$; $q(x, y) = 1$ and $q(x, y) = x^2 + y^2$. Analyze your numerical results and plot the absolute and relate errors.

7. Search the Internet to find a fast Poisson solver or multigrid method, and test it.

4
FD Methods for Parabolic PDEs

A linear PDE of the form

$$u_t = Lu, \tag{4.1}$$

where t usually denotes the time and L is a linear elliptic differential operator in one or more spatial variables, is called parabolic. Furthermore, the second-order canonical form

$$a(x,t)u_{tt} + 2b(x,t)u_{xt} + c(x,t)u_{xx} + \text{lower-order terms} = f(x,t)$$

is parabolic if $b^2 - ac \equiv 0$ in the entire x and t domain. Note that, we can transform this second-order PDE into a system of two PDEs by setting $v = u_t$, where the t-derivative is first order. Some important parabolic PDEs are as follows.

- 1D heat equation with a source

  $$u_t = u_{xx} + f(x,t).$$

  The dimension refers to the space variable (x direction).
- General heat equation

  $$u_t = \nabla \cdot (\beta \nabla u) + f(\mathbf{x},t), \tag{4.2}$$

  where β is the diffusion coefficient and $f(\mathbf{x},t)$ is the source (or sink) term.
- Diffusion–advection equation

  $$u_t = \nabla \cdot (\beta \nabla u) + \mathbf{w} \cdot \nabla u + f(\mathbf{x},t),$$

  where $\nabla \cdot (\beta \nabla u)$ is the diffusion term and $\mathbf{w} \cdot \nabla u$ the advection term.
- Canonical form of diffusion–reaction equation

  $$u_t = \nabla \cdot (\beta \nabla u) + f(\mathbf{x},t,u).$$

  The nonlinear source term $f(\mathbf{x},t,u)$ is a reaction term.

The *steady-state solutions* (when $u_t = 0$) are the solutions of the corresponding elliptic PDEs, *i.e.*,

$$\nabla \cdot (\beta \nabla u) + \bar{f}(\mathbf{x}, u) = 0$$

for the last case, assuming $\lim_{t \to \infty} f(\mathbf{x}, t, u) = \bar{f}(\mathbf{x}, u)$ exists.

Initial and Boundary Conditions

In time-dependent problems, there is an initial condition that is usually specified at $t = 0$, *i.e.*, $u(\mathbf{x}, 0) = u_0(\mathbf{x})$ for the above PDEs, in addition to relevant boundary conditions. If the initial condition is given at $t = T \neq 0$, it can of course be rendered at $t = 0$ by a translation $t' = t - T$. Thus for the 1D heat equation $u_t = u_{xx}$ on $a < x < b$ for example, we expect to have an initial condition at $t = 0$ in addition to boundary conditions at $x = a$ and $x = b$. Note that the boundary conditions at $t = 0$ may or may not be consistent with the initial condition, *e.g.*, if a Dirichlet boundary condition is prescribed at $x = a$ and $x = b$ such that $u(a, t) = g_1(t)$ and $u(b, t) = g_2(t)$, then $u_0(a) = g_1(0)$ and $u_0(b) = g_2(0)$ for consistency.

Dynamical Stability

The fundamental solution $u(x.t) = e^{-x^2/4t}/\sqrt{4\pi t}$ for the 1D heat equation $u_t = u_{xx}$ is uniformly bounded. However, for the backward heat equation $u_t = -u_{xx}$, if $u(x, 0) \neq 0$ then $\lim_{t \to \infty} u(x, t) = \infty$. The solution is said to be dynamically *unstable* if it is not uniformly bounded, *i.e.*, if there is no constant $C > 0$ such that $|u(x, t)| \leq C$. Some applications are dynamically unstable and "blow up," but we do not discuss how to solve such dynamically unstable problems in this book, *i.e.*, we only consider the numerical solution of dynamically stable problems.

Some Commonly Used FD Methods

We discuss the following finite difference methods for parabolic PDE in this chapter:

- the forward and backward Euler methods;
- the Crank–Nicolson and θ methods;
- the method of lines (MOL), provided a good ODE solver can be applied; and
- the alternating directional implicit (ADI) method, for high-dimensional problems.

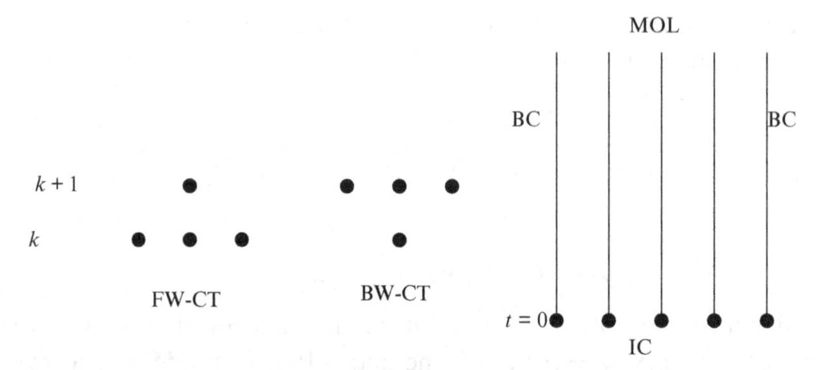

Figure 4.1. Diagram of the finite difference stencil for the forward and backward Euler methods, and the MOL.

Finite difference methods applicable to elliptic PDEs can be used to treat the spatial discretization and boundary conditions, so let us focus on the time discretization and initial condition(s). To consider the stability of the consequent numerical methods, we invoke a Fourier transformation and von Neumann stability analysis.

4.1 The Euler Methods

For the following problem involving the heat equation with a source term,

$$u_t = \beta u_{xx} + f(x,t), \quad a < x < b, \quad t > 0,$$

$$u(a,t) = g_1(t), \quad u(b,t) = g_2(t), \qquad u(x,0) = u_0(x),$$

let us seek a numerical solution for $u(x,t)$ at a particular time $T > 0$ or at certain times in the interval $0 < t < T$.

As the first step, we expect to generate a grid

$$x_i = a + ih, \quad i = 0, 1, \ldots, m, \quad h = \frac{b-a}{m},$$

$$t^k = k\Delta t, \quad k = 0, 1, \ldots, n, \quad \Delta t = \frac{T}{n}.$$

It turns out that we cannot use arbitrary Δt (even it may be small) for explicit methods because of numerical instability concerns. The second step is to approximate the derivatives with finite difference approximations. Since we already know how to discretize the spatial derivatives, let us focus on possible finite difference formulas for the time derivative. In Figure 4.1, we sketch the stencils of several finite difference methods.

4.1.1 Forward Euler Method (FW-CT)

At a grid point (x_i, t^k), $k > 0$, on using the forward finite difference approximation for u_t and central finite difference approximation for u_{xx} we have

$$\frac{u(x_i, t^k + \Delta t) - u(x_i, t^k)}{\Delta t} = \beta \frac{u(x_{i-1}, t^k) - 2u(x_i, t^k) + u(x_{i+1}, t^k)}{h^2} + f(x_i, t^k) + T(x_i, t^k) \, .$$

The local truncation error is

$$T(x_i, t^k) = -\frac{h^2 \beta}{12} u_{xxxx}(x_i, t^k) + \frac{\Delta t}{2} u_{tt}(x_i, t^k) + \cdots ,$$

where the dots denote higher-order terms, so the discretization is $O(h^2 + \Delta t)$. The discretization is first order in time and second order in space, when the finite difference equation is

$$\frac{U_i^{k+1} - U_i^k}{\Delta t} = \beta \, \frac{U_{i-1}^k - 2U_i^k + U_{i+1}^k}{h^2} + f_i^k, \tag{4.3}$$

where $f_i^k = f(x_i, t^k)$, with U_i^k again denoting the approximate values for the true solution $u(x_i, t^k)$. When $k = 0$, U_i^0 is the initial condition at the grid point $(x_i, 0)$; and from the values U_i^k at the time level k the solution of the finite difference equation at the next time level $k + 1$ is

$$U_i^{k+1} = U_i^k + \Delta t \left(\beta \frac{U_{i-1}^k - 2U_i^k + U_{i+1}^k}{h^2} + f_i^k \right), \quad i = 1, 2, \ldots, m - 1 \, . \tag{4.4}$$

The solution of the finite difference equations is thereby directly obtained from the approximate solution at previous time steps and we do not need to solve a system of algebraic equations, so the method is called *explicit*. Indeed, we successively compute the solution at t^1 from the initial condition at t^0, and then at t^2 using the approximate solution at t^1. Such an approach is often called a time marching method.

Remark 4.1. The local truncation error of the FW-CT finite difference scheme under our definition is

$$\begin{aligned} T(x, t) &= \frac{u(x, t + \Delta t) - u(x, t)}{\Delta t} - \beta \, \frac{u(x - h, t) - 2u(x, t) + u(x + h, t)}{h^2} - f(x, t) \\ &= O(h^2 + \Delta t) \, . \end{aligned}$$

In passing, we note an alternative definition of the truncation error in the literature.

$$T(x,t) = u(x,t+\Delta t) - u(x,t) - \Delta t\left(\beta\frac{u(x-h,t)-2u(x,t)+u(x+h,t)}{h^2} - f(x,t)\right)$$

$$= O\left(\Delta t(h^2+\Delta t)\right)$$

introduces an additional factor Δt, so it is one order higher in Δt.

Remark 4.2. If $f(x,t) \equiv 0$ and β is a constant, then from $u_t = \beta u_{xx}$ and $u_{tt} = \beta\partial u_{xx}/\partial t = \beta\partial^2 u_t/\partial x^2 = \beta^2 u_{xxxx}$, the local truncation error is

$$T(x,t) = \left(\frac{\beta^2 \Delta t}{2} - \frac{\beta h^2}{12}\right) u_{xxxx} + O\left((\Delta t)^2 + h^4\right). \tag{4.5}$$

Thus if β is constant we can choose $\Delta t = h^2/(6\beta)$ to get $O(h^4 + (\Delta t)^2) = O(h^4)$, *i.e.*, the local truncation error is fourth-order accurate without further computational complexity, which is significant for an *explicit* method.

It is easy to implement the forward Euler method compared with other methods. Below we list some scripts of the Matlab file called *FW_Euler_heat.m*:

```
a = 0;   b=1; m = 10; n=20;
h = (b-a)/m;
k = h^2/2;   % Try k = h^2/1.9 to see what happens;

t = 0;   tau = k/h^2;
for i=1:m+1,
   x(i) = a + (i-1)*h;   y1(i) = uexact(t,x(i));   y2(i) = 0;
end
plot(x,y1); hold

for j=1:n,
   y1(1)=0; y1(m+1)=0;
   for i=2:m
     y2(i) = y1(i) + tau*(y1(i-1)-2*y1(i)+y1(i+1)) + k*f(t,x(i));
   end
   plot(x,y2); pause(0.25)
   t = t + k;   y1 = y2;
end
```

In the code above, we also plot the history of the solution. On testing the forward Euler method with different Δt and checking the error in a problem with a known exact solution, we find the method works well when $0 < \Delta t \le \frac{h^2}{2\beta}$ but blows up when $\Delta t > \frac{h^2}{2\beta}$. Since the method is consistent, we anticipate that

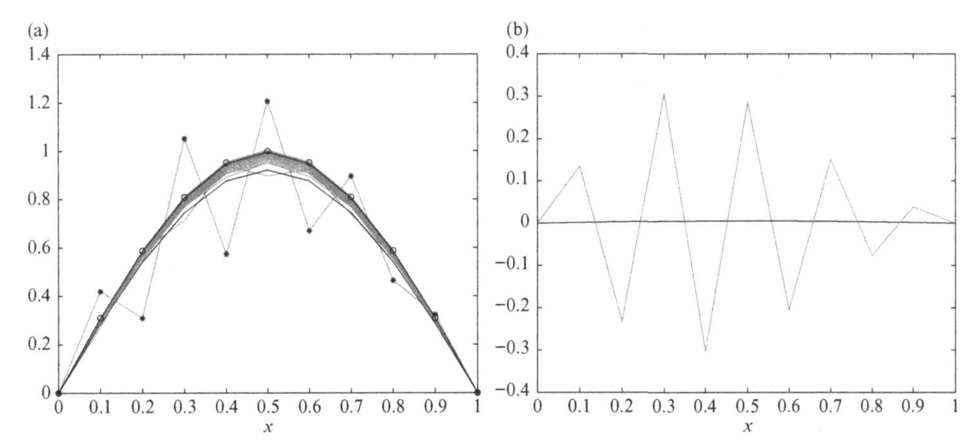

Figure 4.2. (a) Plot of the computed solutions using two different time step sizes and the exact solution at some time for the test problem. (b) Error plots of the computed solution using the two different step sizes: one is stable and the error is small; the other one is unstable and the error grows rapidly.

this is a question of numerical stability. Intuitively, to prevent the errors in u_i^k being amplified, one can set

$$0 < \frac{2\beta\Delta t}{h^2} \leq 1\,, \quad \text{or} \quad 0 < \Delta t \leq \frac{h^2}{2\beta}\,. \tag{4.6}$$

This is a time step constraint, often called the CFL (Courant–Friedrichs–Lewy) stability condition, which can be verified numerically. In Figure 4.2, we plot the computed solutions and errors for a testing problem with $\beta = 1$, $f(x) = -\sin t \sin(\pi x) + \cos t \sin(\pi x)\pi^2$. The true solution is $u(x,t) = \cos t \sin(\pi x)$. We take 20 time marching steps using two different time steps, one is $\Delta t_1 = h^2/2$ (stable), and the other one is $\Delta t_2 = 2h^2$ (unstable). The left plots are the true and computed solutions at $t_1 = 20\Delta t_1$ and $t_2 = 20\Delta t_2$. The gray lines are the history of the solution computed using $\Delta t_2 = 2h^2$, and the "*" indicates the computed solution at the grid points for the final step. We see that the solution begins to grow and oscillate. The plot of the black line is the true solution at $t_1 = 20\Delta t_1$ with the little "o" as the finite difference solution at the grid points, which is first-order accurate. The right figure is the error plots with the black one (the error is small) for the computed solution using $\Delta t_1 = h^2/2$; while the gray one for the computed solution using $\Delta t_2 = 2h^2$, whose error grows rapidly and begins to oscillate. If the final time T gets larger, so does the error, a phenomenon we call a blow-up due to the instability of the algorithm. The stability and the CFL condition of the time step constraint are very important for explicit or semi-explicit numerical algorithms.

4.1.2 The Backward Euler Method (BW-CT)

If the backward finite difference formula is used for u_t and the central finite difference approximation for u_{xx} at (x_i, t^k), we get

$$\frac{U_i^k - U_i^{k-1}}{\Delta t} = \beta\,\frac{U_{i-1}^k - 2U_i^k + U_{i+1}^k}{h^2} + f_i^k, \quad k = 1, 2, \ldots,$$

which is conventionally reexpressed as

$$\frac{U_i^{k+1} - U_i^k}{\Delta t} = \beta\,\frac{U_{i-1}^{k+1} - 2U_i^{k+1} + U_{i+1}^{k+1}}{h^2} + f_i^{k+1}, \quad k = 0, 1, \ldots. \tag{4.7}$$

The backward Euler method is also consistent, and the discretization error is again $O(\Delta t + h^2)$.

Using the backward Euler method, we cannot get U_i^{k+1} with a few simple algebraic operations because all of the U_i^{k+1}'s are coupled together. Thus we need to solve the following tridiagonal system of equations, in order to get the approximate solution at the time level $k+1$:

$$\begin{bmatrix} 1+2\mu & -\mu & & & & \\ -\mu & 1+2\mu & -\mu & & & \\ & -\mu & 1+2\mu & -\mu & & \\ & & \ddots & \ddots & \ddots & \\ & & & -\mu & 1+2\mu & -\mu \\ & & & & -\mu & 1+2\mu \end{bmatrix} \begin{bmatrix} U_1^{k+1} \\ U_2^{k+1} \\ U_3^{k+1} \\ \vdots \\ U_{m-2}^{k+1} \\ U_{m-1}^{k+1} \end{bmatrix}$$

$$= \begin{bmatrix} U_1^k + \Delta t f_1^{k+1} + \mu g_1^{k+1} \\ U_2^k + \Delta t f_2^{k+1} \\ U_3^k + \Delta t f_3^{k+1} \\ \vdots \\ U_{m-2}^k + \Delta t f_{m-2}^{k+1} \\ U_{m-1}^k + \Delta t f_{m-1}^{k+1} + \mu g_2^{k+1} \end{bmatrix}, \tag{4.8}$$

where $\mu = \frac{\beta\Delta t}{h^2}$ and $f_i^{k+1} = f(x_i, t^{k+1})$. Note that we can use $f(x_i, t^k)$ instead of $f(x_i, t^{k+1})$, since the method is first-order accurate in time. Such a numerical method is called *implicit*, because the solution at time level $k+1$ are coupled together. The advantage of the backward Euler method is that it is stable for any choice of Δt. For 1D problems, the computational cost is only slightly more than the explicit Euler method if we can use an efficient tridiagonal solver, such as the Grout factorization method at cost $O(5n)$ (*cf.* Burden and Faires, 2010, for example).

4.2 The Method of Lines

With a good solver for ODEs or systems of ODEs, we can use the MOL to solve parabolic PDEs. In Matlab, we can use the ODE Suite to solve a system of ODEs. The ODE Suite contains Matlab functions such as ode23, ode23s, ode15s, ode45, and others. The Matlab function ode23 uses a combination of Runge–Kutta methods of order 2 and 3 and uses an adaptive time step size. The Matlab function ode23s is designed for a stiff system of ODE.

Consider a general parabolic equation of the form

$$u_t(x,t) = Lu(x,t) + f(x,t)\,,$$

where L is an elliptic operator. Let L_h be a corresponding finite difference operator acting on a grid $x_i = a + ih$. We can form a semidiscrete system of ODEs of form

$$\frac{\partial U_i}{\partial t} = L_h U_i(t) + f_i(t)\,,$$

where $U_i(t) \simeq u(x_i, t)$ is the spatial discretization of $u(x,t)$ along the line $x = x_i$, *i.e.*, we only discretize the spatial variable. For example, the heat equation with a source $u_t = \beta u_{xx} + f$ where $L = \beta\partial^2/\partial x^2$ is represented by $L_h = \beta\delta_{xx}^2$ produces the discretized system of ODE

$$\begin{aligned}
\frac{\partial U_1(t)}{\partial t} &= \beta\frac{-2U_1(t) + U_2(t)}{h^2} + \beta\frac{g_1(t)}{h^2} + f(x_1,t)\,,\\
\frac{\partial U_i(t)}{\partial t} &= \beta\frac{U_{i-1}(t) - 2U_i(t) + U_{i+1}(t)}{h^2} + f(x_i,t), \quad i = 2,3,\ldots,m-2\,,\\
\frac{\partial U_{m-1}(t)}{\partial t} &= \beta\frac{U_{m-2}(t) - 2U_{m-1}(t)}{h^2} + \beta\frac{g_2(t)}{h^2} + f(x_{m-1},t)\,,
\end{aligned} \tag{4.9}$$

and the initial condition is

$$U_i(0) = u_0(x_i, 0), \quad i = 1, 2, \ldots, m-1\,. \tag{4.10}$$

The ODE system can be written in the vector form

$$\frac{d\mathbf{y}}{dt} = f(\mathbf{y}, t), \qquad \mathbf{y}(0) = \mathbf{y}_0\,. \tag{4.11}$$

The MOL is especially useful for nonlinear PDEs of the form $u_t = f(\partial/\partial x, u, t)$. For linear problems, we typically have

$$\frac{d\mathbf{y}}{dt} = A\mathbf{y} + \mathbf{c}\,,$$

where A is a matrix and $\mathbf{c}$ is a vector. Both A and $\mathbf{c}$ may depend on t.

There are many efficient solvers for a system of ODEs. Most are based on high-order Runge–Kutta methods with adaptive time steps, *e.g.*, ODE suite in Matlab, or *dsode.f* available through *Netlib*. However, it is important to recognise that the ODE system obtained from the MOL is typically *stiff*, *i.e.*, the eigenvalues of A have very different scales. For example, for the heat equation the magnitude of the eigenvalues range from $O(1)$ to $O(1/h^2)$. In Matlab, we can call an ODE solver using the format

```
[t,y] = ode23s('yfun-mol', [0, t_final], y0);
```

The solution is stored in the last row of $\mathbf{y}$, which can be extracted using

```
[mr,nc] = size(y);
ysol = y(mr,:);
```

Then *ysol* is the approximate solution at time $t = t\_final$. To define the ODE system of the MOL, we can create a Matlab file, say yfun-mol.m whose contents contain the following

```
function yp = yfun-mol(t,y)
global m h x
k = length(y); yp=size(k,1);
yp(1) = (-2*y(1) + y(2))/(h*h) + f(t,x(1)) + g1(t)/(h*h);
for i=2:m-2
   yp(i) = (y(i-1) -2*y(1) + y(2))/(h*h) + f(t,x(i));
end
yp(m-1) = (y(m-2) -2*y(m-1) )/(h*h) + f(t,x(i)) + g2(t)/(h*h);
```

where $g1(t)$ and $g2(t)$ are two Matlab functions for the boundary conditions at $x = a$ and $x = b$; and $f(t, x)$ is the source term.

The initial condition can be defined as

```
global m h x
for i=1:m-1
    y0(i) = u_0(x(i));
end
```

where $u_0(x)$ is a Matlab function of the initial condition.

4.3 The Crank–Nicolson Scheme

The time step constraint $\Delta t = h^2/(2\beta)$ for the explicit Euler method is generally considered to be a severe restriction, *e.g.*, if $h = 0.01$, the final time is $T = 10$ and $\beta = 100$, then we need 2×10^7 steps to get the solution at the final time. The backward Euler method does not have the time step constraint, but it is only first-order accurate. If we want second-order accuracy $O(h^2)$, we need to take $\Delta t = O(h^2)$. One finite difference scheme that is second-order accurate both in space and time, without compromising stability and computational complexity, is the Crank–Nicolson scheme.

The Crank–Nicolson scheme is based on the following lemma, which can be proved easily using the Taylor expansion.

Lemma 4.3. *Let $\phi(t)$ be a function that has continuous first- and second-order derivatives, i.e., $\phi(t) \in C^2$. Then*

$$\phi(t) = \frac{1}{2}\left(\phi\left(t - \frac{\Delta t}{2}\right) + \phi\left(t + \frac{\Delta t}{2}\right)\right) + \frac{(\Delta t)^2}{8} u''(t) + h.o.t. \quad (4.12)$$

Intuitively, the Crank–Nicolson scheme approximates the PDE

$$u_t = (\beta u_x)_x + f(x,t)$$

at $(x_i, t^k + \Delta t/2)$, by averaging the time level t^k and t^{k+1} of the spatial derivative $\nabla \cdot (\beta \nabla u))$ and $f(x,t)$. Thus it has the following form

$$\frac{U_i^{k+1} - U_i^k}{\Delta t} = \frac{\beta_{i-\frac{1}{2}}^k U_{i-1}^k - (\beta_{i-\frac{1}{2}}^k + \beta_{i+\frac{1}{2}}^k) U_i^k + \beta_{i+\frac{1}{2}}^k U_{i+1}^k}{2h^2}$$

$$+ \frac{\beta_{i-\frac{1}{2}}^{k+1} U_{i-1}^{k+1} - (\beta_{i-\frac{1}{2}}^{k+1} + \beta_{i+\frac{1}{2}}^{k+1}) U_i^{k+1} + \beta_{i+\frac{1}{2}}^{k+1} U_{i+1}^{k+1}}{2h^2} + \frac{1}{2}\left(f_i^k + f_i^{k+1}\right). \quad (4.13)$$

The discretization is second order in time (central at $t + \Delta t/2$ with step size $\Delta t/2$) and second order in space. This can easily be seen using the following relations, taking $\beta = 1$ for simplicity:

$$\frac{u(x,t+\Delta t)-u(x,t)}{\Delta t} = u_t(x,t+\Delta t/2) + \frac{1}{3}\left(\frac{\Delta t}{2}\right)^2 u_{ttt}(x,t+\Delta t/2) + O((\Delta t)^4),$$

$$\frac{u(x-h,t)-2u(x,t)+u(x+h,t)}{2h^2} = u_{xx}(x,t) + O(h^2),$$

$$\frac{u(x-h,t+\Delta t)-2u(x,t+\Delta t)+u(x+h,t+\Delta t)}{2h^2} = u_{xx}(x,t+\Delta t) + O(h^2),$$

$$\frac{1}{2}\Big(u_{xx}(x,t)+u_{xx}(x,t+\Delta t)\Big) = u_{xx}(x,t+\Delta t/2) + O((\Delta t)^2),$$

$$\frac{1}{2}\Big(f(x,t)+f(x,t+\Delta t)\Big) = f(x,t+\Delta t/2) + O((\Delta t)^2).$$

At each time step, we need to solve a tridiagonal system of equations to get U_i^{k+1}. The computational cost is only slightly more than that of the explicit Euler method in one space dimension, and we can take $\Delta t \simeq h$ and have second-order accuracy. Although the Crank–Nicolson scheme is an implicit method, it is much more efficient than the explicit Euler method since it is second-order accurate both in time and space with the same computational complexity. A sample Matlab code crank.m is accompanied with the book. If we use a fixed time step $\Delta t = h$, given a final time T, we can easily get the number of time marking steps as $N_T = int(T/h)$ as used in the crank.m. In the next section, we will prove it is unconditionally stable for the heat equation.

4.3.1 A Class of One-step FD Methods: The θ-Method

The θ-method for the heat equation $u_t = u_{xx} + f(x,t)$ has the following form:

$$\frac{U_i^{k+1}-U_i^k}{\Delta t} = \theta\delta_{xx}^2 U_i^k + (1-\theta)\delta_{xx}^2 U_i^{k+1} + \theta f_i^k + (1-\theta) f_i^{k+1}.$$

When $\theta = 1$, the method is the explicit Euler method; when $\theta = 0$, the method is the backward Euler method; and when $\theta = 1/2$, it is the Crank–Nicolson scheme. If $0 < \theta \le 1/2$, then the method is unconditionally stable, and otherwise it is conditionally stable, *i.e.*, there is a time step constraint. The θ-method is generally first order in time and second order in space, except for $\theta = 1/2$.

The accompanying Matlab code for this chapter included Euler, Crank–Nicolson, ADI, and MOL methods.

4.4 Stability Analysis for Time-dependent Problems

A standard approach to stability analysis of finite difference methods for time-dependent problems is named after John von Neumann and based on the discrete Fourier transform (FT).

4.4.1 Review of the Fourier Transform

Let us first consider the Fourier transform in continuous space. Consider $u(x) \in L^2(-\infty, \infty)$, *i.e.*, $\int_{-\infty}^{\infty} u^2 dx < \infty$ or $\|u\|_2 < \infty$. The Fourier transform is defined as

$$\hat{u}(\omega) = \frac{1}{\sqrt{2\pi}} \int_{-\infty}^{\infty} e^{-i\omega x} u(x) dx \tag{4.14}$$

where $i = \sqrt{-1}$, mapping $u(x)$ in the space domain into $\hat{u}(\omega)$ in the frequency domain. Note that if a function is defined in the domain $(0, \infty)$ instead of $(-\infty, \infty)$, we can use the Laplace transform. The inverse Fourier transform is

$$u(x) = \frac{1}{\sqrt{2\pi}} \int_{-\infty}^{\infty} e^{i\omega x} \hat{u}(\omega) d\omega \,. \tag{4.15}$$

Parseval's relation: Under the Fourier transform, we have $\|\hat{u}\|_2 = \|u\|_2$ or

$$\int_{-\infty}^{\infty} |\hat{u}|^2 d\omega = \int_{-\infty}^{\infty} |u|^2 dx \,. \tag{4.16}$$

From the definition of the Fourier transform we have

$$\widehat{\left(\frac{\partial \hat{u}}{\partial \omega}\right)} = -ixu \,, \qquad \widehat{\frac{\partial u}{\partial x}} = i\omega\hat{u} \,. \tag{4.17}$$

To show this we invoke the inverse Fourier transform

$$\frac{\partial u(x)}{\partial x} = \frac{1}{\sqrt{2\pi}} \int_{-\infty}^{\infty} e^{i\omega x} \widehat{\frac{\partial u}{\partial x}} \, d\omega$$

so that, since $u(x)$ and $\hat{u}(\omega)$ are both in $L^2(-\infty, \infty)$, on taking the partial derivative of the inverse Fourier transform with respect to x we have

$$\frac{\partial u(x)}{\partial x} = \frac{1}{\sqrt{2\pi}} \int_{-\infty}^{\infty} \frac{\partial}{\partial x} \left(e^{i\omega x} \,\widehat{u}\right) d\omega = \frac{1}{\sqrt{2\pi}} \int_{-\infty}^{\infty} i\omega\widehat{u} e^{i\omega x} \, d\omega \,.$$

Then as the Fourier transform and its inverse are unique, $\widehat{\partial u/\partial x} = i\omega\widehat{u}$. The proof of the first equality is left as an exercise. It is easy to generalize the equality, to get

$$\widehat{\frac{\partial^m u}{\partial x^m}} = (i\omega)^m \, \hat{u} \tag{4.18}$$

i.e., we remove the derivatives of one variable.

The Fourier transform is a powerful tool to solve PDEs, as illustrated below.

Example 4.4. Consider

$$u_t + a u_x = 0\,, \qquad -\infty < x < \infty, \quad t > 0\,, \quad u(x,0) = u_0(x)$$

which is called an advection equation, or a one-way wave equation. This is a Cauchy problem since the spatial variable is defined in the entire space and $t \geq 0$. On applying the FT to the equation and the initial condition,

$$\widehat{u_t} + \widehat{a u_x} = 0, \quad \text{or} \quad \widehat{u}_t + a i \omega \hat{u} = 0, \qquad \hat{u}(\omega, 0) = \hat{u}_0(\omega)$$

i.e., we get an ODE for $\hat{u}(\omega)$ whose solution is

$$\hat{u}(\omega, t) = \hat{u}(\omega, 0)\, e^{-ia\omega t} = \hat{u}_0(\omega)\, e^{-ia\omega t}.$$

The solution to the original advection equation is thus

$$\begin{aligned} u(x,t) &= \frac{1}{\sqrt{2\pi}} \int_{-\infty}^{\infty} e^{i\omega x}\, \hat{u}_0(\omega)\, e^{-ia\omega t}\, d\omega \\ &= \frac{1}{\sqrt{2\pi}} \int_{-\infty}^{\infty} e^{i\omega (x - at)}\, \hat{u}_0(\omega)\, d\omega \\ &= u(x - at, 0)\,, \end{aligned}$$

on taking the inverse Fourier transform. It is noted that the solution for the advection equation does not change shape, but simply propagates along the characteristic line $x - at = 0$, and that

$$\|u\|_2 = \|\hat{u}\|_2 = \|\hat{u}(\omega, 0) e^{-ia\omega t}\|_2 = \|\hat{u}(\omega, 0)\|_2 = \|u_0\|_2\,.$$

Example 4.5. Consider

$$u_t = \beta u_{xx}, \qquad -\infty < x < \infty, \quad t > 0, \quad u(x,0) = u_0(x), \quad \lim_{|x| \to \infty} u = 0\,,$$

involving the heat (or diffusion) equation. On again applying the Fourier transform to the PDE and the initial condition,

$$\widehat{u_t} = \widehat{\beta u_{xx}}, \quad \text{or} \quad \widehat{u}_t = \beta (i\omega)^2 \hat{u} = -\beta \omega^2 \hat{u}, \qquad \hat{u}(\omega, 0) = \hat{u}_0(\omega)\,,$$

and the solution of this ODE is

$$\hat{u}(\omega, t) = \hat{u}(\omega, 0)\, e^{-\beta \omega^2 t}\,.$$

Consequently, if $\beta > 0$, from the Parseval's relation, we have

$$\|u\|_2 = \|\hat{u}\|_2 = \|\hat{u}(\omega,0)e^{-\beta\omega^2 t}\|_2 \le \|u_0\|_2 \,.$$

Actually, it can be seen that $\lim_{t\to\infty} \|u\|_2 = 0$ and the second-order partial derivative term is called a diffusion or dissipative. If $\beta < 0$, then $\lim_{t\to\infty} \|u\|_2 = \infty$, the PDE is dynamically unstable.

Example 4.6. Dispersive waves.

Consider

$$u_t = \frac{\partial^{2m+1} u}{\partial x^{2m+1}} + \frac{\partial^{2m} u}{\partial x^{2m}} + l.o.t.,$$

where m is a nonnegative integer. For the simplest case $u_t = u_{xxx}$, we have

$$\widehat{u_t} = \widehat{\beta u_{xxx}}, \quad \text{or} \quad \widehat{u_t} = \beta(i\omega)^3 \hat{u} = -i\omega^3 \hat{u}\,,$$

and the solution of this ODE is

$$\hat{u}(\omega,t) = \hat{u}(\omega,0)\, e^{-i\omega^3 t}\,.$$

Therefore,

$$\|u\|_2 = \|\hat{u}\|_2 = \|\hat{u}(\omega,0)\|_2 = \|u(\omega,0)\|_2\,,$$

and the solution to the original PDE can be expressed as

$$\begin{aligned} u(x,t) &= \frac{1}{\sqrt{2\pi}} \int_{-\infty}^{\infty} e^{i\omega x}\, \hat{u}_0(\omega)\, e^{-i\omega^3 t}\, d\omega \\ &= \frac{1}{\sqrt{2\pi}} \int_{-\infty}^{\infty} e^{i\omega(x-\omega^2 t)}\, \hat{u}_0(\omega)\, d\omega\,. \end{aligned}$$

Evidently, the Fourier component with wave number ω propagates with velocity ω^2, so waves mutually interact but there is no diffusion.

Example 4.7. PDEs with even higher-order derivatives.

Consider

$$u_t = \alpha \frac{\partial^{2m} u}{\partial x^{2m}} + \frac{\partial^{2m-1} u}{\partial x^{2m-1}} + l.o.t.,$$

where m is a nonnegative integer. The Fourier transform yields

$$\widehat{u_t} = \alpha(i\omega)^{2m}\hat{u} + \cdots = \begin{cases} -\alpha\omega^{2m}\hat{u} + \cdots & \text{if } m = 2k+1, \\ \alpha\omega^{2m}\hat{u} + \cdots & \text{if } m = 2k, \end{cases}$$

hence

$$\widehat{u} = \begin{cases} \hat{u}(\omega,0)\, e^{-\alpha i \omega^{2m} t} + \cdots & \text{if } m = 2k+1, \\ \hat{u}(\omega,0)\, e^{\alpha i \omega^{2m} t} + \cdots & \text{if } m = 2k, \end{cases}$$

such that $u_t = u_{xx}$ and $u_t = -u_{xxxx}$ are dynamically stable, whereas $u_t = -u_{xx}$ and $u_t = u_{xxxx}$ are dynamically unstable.

4.4.2 The Discrete Fourier Transform

Motivations to study a discrete Fourier transform include the stability analysis of finite difference schemes, data analysis in the frequency domain, filtering techniques, *etc.*

Definition 4.8. If $\ldots, v_{-2}, v_{-1}, v_0, v_1, v_2, \ldots$ denote the values of a continuous function $v(x)$ at $x_i = i\,h$, the discrete Fourier transform is defined as

$$\hat{v}(\xi) = \frac{1}{\sqrt{2\pi}} \sum_{j=-\infty}^{\infty} h\, e^{-i\xi jh}\, v_j\,. \tag{4.19}$$

Remark 4.9.

- The definition is a quadrature approximation to the continuous case, *i.e.*, we approximate $\int$ by $\sum$, and replace dx by h.
- $\hat{v}(\xi)$ is a continuous and periodic function of ξ with period $2\pi/h$, since

$$e^{-ijh(\xi+2\pi/h)} = e^{-ijh\xi} e^{2ij\pi} = e^{-i\xi jh}\,, \tag{4.20}$$

so we can focus on $\hat{v}(\xi)$ in the interval $[-\pi/h, \pi/h]$, and consequently have the following definition.

Definition 4.10. The inverse discrete Fourier transform is

$$v_j = \frac{1}{\sqrt{2\pi}} \int_{-\pi/h}^{\pi/h} e^{i\xi jh}\, \hat{v}(\xi)\, d\xi\,. \tag{4.21}$$

Given any finite sequence not involving h,

$$v_1, v_2, \ldots, v_M\,,$$

we can extend the finite sequence according to the following

$$\ldots, 0, 0, v_1, v_2, \ldots, v_M, 0, 0, \ldots\,,$$

and alternatively define the discrete Fourier and inverse Fourier transform as

$$\hat{v}(\xi) = \frac{1}{\sqrt{2\pi}} \sum_{j=-\infty}^{\infty} e^{-i\xi j} v_j = \sum_{j=0}^{M} e^{-i\xi j}\, v_j \,, \tag{4.22}$$

$$v_j = \frac{1}{\sqrt{2\pi}} \int_{-\pi}^{\pi} e^{i\xi j}\, \hat{v}(\xi)\, d\xi \,. \tag{4.23}$$

We also define the *discrete norm* as

$$\|\mathbf{v}\|_h = \sqrt{\sum_{j=-\infty}^{\infty} v_j^2\, h} \,, \tag{4.24}$$

which is often denoted by $\|v\|_2$. Parseval's relation is also valid, *i.e.*,

$$\|\hat{v}\|_h^2 = \int_{-\pi/h}^{\pi/h} |\hat{v}(\xi)|^2 d\xi = \sum_{j=-\infty}^{\infty} h\, |v_j|^2 = \|v\|_h^2 \,. \tag{4.25}$$

4.4.3 Definition of the Stability of a FD Scheme

A finite difference scheme $P_{\Delta t,h} v_j^k = 0$ is stable in a stability region Λ if for any positive time T there is an integer J and a constant C_T independent of Δt and h such that

$$\|\mathbf{v}^n\|_h \le C_T \sum_{j=0}^{J} \|\mathbf{v}^j\|_h \,, \tag{4.26}$$

for any n that satisfies $0 \le n\Delta t \le T$ with $(\Delta t, h) \in \Lambda$.

Remark 4.11.

1. The stability is usually independent of source terms.
2. A stable finite difference scheme means that the growth of the solution is at most a constant multiple of the sum of the norms of the solution at the first $J+1$ steps.
3. The stability region corresponds to all possible Δt and h for which the finite difference scheme is stable.

The following theorem provides a simple way to check the stability of any finite difference scheme.

Theorem 4.12. *If $\|\mathbf{v}^{k+1}\|_h \le \|\mathbf{v}^k\|_h$ is true for any k, then the finite difference scheme is stable.*

Proof From the condition, we have

$$\|\mathbf{v}^n\|_h \le \|\mathbf{v}^{n-1}\|_h \le \cdots \le \|\mathbf{v}^1\|_h \le \|\mathbf{v}^0\|_h ,$$

and hence stability for $J=0$ and $C_T=1$.

4.4.4 The von Neumann Stability Analysis for FD Methods

The von Neumann stability analysis of a finite difference scheme can be sketched briefly as Discrete scheme $\Longrightarrow$ discrete Fourier transform $\Longrightarrow$ growth factor $g(\xi) \Longrightarrow$ stability ($|g(\xi)| \le 1$?). We will also explain a simplification of the von Neumann analysis.

Example 4.13. The forward Euler method (FW-CT) for the heat equation $u_t = \beta u_{xx}$ is

$$U_i^{k+1} = U_i^k + \mu\left(U_{i-1}^k - 2U_i^k + U_{i+1}^k\right), \quad \mu = \frac{\beta \Delta t}{h^2}. \tag{4.27}$$

From the discrete Fourier transform, we have the following

$$U_j^k = \frac{1}{\sqrt{2\pi}} \int_{-\pi/h}^{\pi/h} e^{i\xi jh} \hat{U}^k(\xi) d\xi , \tag{4.28}$$

$$U_{j+1}^k = \frac{1}{\sqrt{2\pi}} \int_{-\pi/h}^{\pi/h} e^{i\xi(j+1)h} \hat{U}^k(\xi) d\xi = \frac{1}{\sqrt{2\pi}} \int_{-\pi/h}^{\pi/h} e^{i\xi jh} e^{i\xi h} \hat{U}^k(\xi) d\xi , \tag{4.29}$$

and similarly

$$U_{j-1}^k = \frac{1}{\sqrt{2\pi}} \int_{-\pi/h}^{\pi/h} e^{i\xi jh} e^{-i\xi h} \hat{U}^k(\xi) d\xi . \tag{4.30}$$

Substituting these relations into the forward Euler finite difference scheme, we obtain

$$U_i^{k+1} = \frac{1}{\sqrt{2\pi}} \int_{-\pi/h}^{\pi/h} e^{i\xi jh} \left(1 + \mu(e^{-i\xi h} - 2 + e^{i\xi h})\right) \hat{U}^k(\xi) d\xi . \tag{4.31}$$

On the other hand, from the definition of the discrete Fourier transform, we also know that

$$U_i^{k+1} = \frac{1}{\sqrt{2\pi}} \int_{-\pi/h}^{\pi/h} e^{i\xi jh} \hat{U}^{k+1}(\xi) d\xi . \tag{4.32}$$

The discrete Fourier transform is unique, which implies

$$\hat{U}^{k+1}(\xi) = \left(1 + \mu(e^{-i\xi h} - 2 + e^{i\xi h})\right) \hat{U}^k(\xi) = g(\xi) \hat{U}^k(\xi) , \tag{4.33}$$

where

$$g(\xi) = 1 + \mu(e^{-i\xi h} - 2 + e^{i\xi h}) \tag{4.34}$$

is called the *growth factor*. If $|g(\xi)| \le 1$, then $|\hat{U}^{k+1}| \le |\hat{U}^k|$ and thus $\|\hat{\mathbf{U}}^{k+1}\|_h \le \|\hat{\mathbf{U}}^k\|_h$, so the finite difference scheme is stable.

Let us examine $|g(\xi)|$ now. We have

$$\begin{aligned} g(\xi) &= 1 + \mu\,(\cos(-\xi h) - i\sin(\xi h) - 2 + \cos(\xi h) + i\sin(\xi h)) \\ &= 1 + 2\mu\,(\cos(\xi h) - 1) = 1 - 4\mu\sin^2(\xi h)/2\,, \end{aligned} \tag{4.35}$$

but we need to know when $|g(\xi)| \le 1$, or $-1 \le g(\xi) \le 1$. Note that

$$-1 \le 1 - 4\mu \le 1 - 4\mu\sin^2(\xi h)/2 = g(\xi) \le 1\,, \tag{4.36}$$

so on taking $-1 \le 1 - 4\mu$ we can guarantee that $|g(\xi)| \le 1$, which implies the stability. Thus a sufficient condition for the stability of the forward Euler method is

$$-1 \le 1 - 4\mu \quad \text{or} \quad 4\mu \le 2, \quad \text{or} \quad \Delta t \le \frac{h^2}{2\beta}\,. \tag{4.37}$$

Although we cannot claim what will happen if this condition is violated, it provides an upper bound for the stability.

4.4.5 Simplification of the von Neumann Stability Analysis for One-step Time Marching Methods

Consider the one-step time marching method $\mathbf{U}^{k+1} = f(\mathbf{U}^k, \mathbf{U}^{k+1})$. The following theorem provides a simple way to determine the stability.

Theorem 4.14. *Let $\theta = h\xi$. A one-step finite difference scheme (with constant coefficients) is stable if and only if there is a constant K (independent of θ, Δt, and h) and some positive grid spacing Δt_0 and h_0 such that*

$$|g(\theta, \Delta t, h)| \le 1 + K\Delta t \tag{4.38}$$

for all θ and $0 < h \le h_0$. If $g(\theta, \Delta t, h)$ is independent of h and Δt, then the stability condition (4.38) can be replaced by

$$|g(\theta| \le 1. \tag{4.39}$$

Thus only the amplification factor $g(h\xi) = g(\theta)$ needs to be considered, as observed by von Neumann.

The von Neumann stability analysis usually involves the following steps:

1. set $U_j^k = e^{ijh\xi}$ and substitute it into the finite difference scheme;
2. express U_j^{k+1} as $U_j^{k+1} = g(\xi)e^{ijh\xi}$, *etc.*;
3. solve for $g(\xi)$ and determine whether or when $|g(\xi)| \le 1$ (for stability); but note that
4. if there are some ξ such that $|g(\xi)| > 1$, then the method is unstable.

Example 4.15. The stability of the backward Euler method for the heat equation $u_t = \beta u_{xx}$ is

$$U_i^{k+1} = U_i^k + \mu\left(U_{i-1}^{k+1} - 2U_i^{k+1} + U_{i+1}^{k+1}\right), \quad \mu = \frac{\beta\Delta t}{h^2}. \tag{4.40}$$

Following the procedure mentioned above, we have

$$\begin{aligned} g(\xi)e^{ijh\xi} &= e^{ijh\xi} + \mu\left(e^{i\xi(j-1)h} - 2e^{i\xi jh} + e^{i\xi(j+1)h}\right)g(\xi) \\ &= e^{i\xi jh}\left(1 + \mu\left(e^{-i\xi h} - 2 + e^{i\xi h}\right)g(\xi)\right), \end{aligned} \tag{4.41}$$

with solution

$$\begin{aligned} g(\xi) &= \frac{1}{1 - \mu(e^{-i\xi h} - 2 + e^{i\xi h})} \\ &= \frac{1}{1 - \mu(2\cos(h\xi) - 2)} = \frac{1}{1 + 4\mu\sin^2(h\xi)/2} \le 1, \end{aligned} \tag{4.42}$$

for any h and $\Delta t > 0$. Obviously, $-1 < 0 \le g(\xi)$ so $|g(\xi)| \le 1$ and the backward Euler method is unconditionally stable, *i.e.*, there is no constraint on Δt for stability.

Example 4.16. The Leapfrog scheme (two-stage method) for the heat equation $u_t = u_{xx}$ is

$$\frac{U_i^{k+1} - U_i^{k-1}}{2\Delta t} = \frac{U_{i-1}^k - 2U_i^k + U_{i+1}^k}{h^2}, \tag{4.43}$$

involving the central finite difference formula both in time and space. This method is unconditionally unstable! To show this, we use $U_j^{k-1} = e^{ijh\xi}/g(\xi)$ to get

$$\begin{aligned} g(\xi)e^{ijh\xi} &= \frac{1}{g(\xi)}e^{ijh\xi} + e^{i\xi jh}\left(\mu(e^{-i\xi h} - 2 + e^{i\xi h})\right) \\ &= \frac{1}{g(\xi)}e^{ijh\xi} - e^{ijh\xi}4\mu\sin^2(h\xi/2), \end{aligned}$$

yielding a quadratic equation for $g(\xi)$:

$$(g(\xi))^2 + 4\mu \sin^2(h\xi/2)\, g(\xi) - 1 = 0. \tag{4.44}$$

The two roots are

$$g(\xi) = -2\mu \sin^2(h\xi/2) \pm \sqrt{4\mu^2 \sin^4(h\xi/2) + 1}\,,$$

and one root

$$g(\xi) = -2\mu \sin^2(h\xi/2) - \sqrt{4\mu^2 \sin^4(h\xi/2) + 1}$$

has magnitude $|g(\xi)| \geq 1$. Thus there are ξ such that $|g(\xi)| > 1$, so the method is unstable.

4.5 FD Methods and Analysis for 2D Parabolic Equations

The general form of a parabolic PDE is

$$u_t + a_1 u_x + a_2 u_y = (\beta u_x)_x + (\beta u_y)_y + \kappa u + f(x, y, t)\,,$$

with boundary conditions and an initial condition. We need $\beta \geq \beta_0 > 0$ for the dynamic stability. The PDE can be written as

$$u_t = Lu + f,$$

where L is the spatial differential operator. The MOL can be used provided there is a good solver for the stiff ODE system. Note that the system is large ($O(mn)$), if the numbers of grid lines are $O(m)$ and $O(n)$ in the x- and y-directions, respectively.

For simplicity, let us consider the heat equation $u_t = \nabla \cdot (\beta \nabla u) + f(x, y, t)$ and assume β is a constant. The simplest method is the forward Euler method:

$$U_{lj}^{k+1} = U_{lj}^k + \mu \left(U_{l-1,j}^k + U_{l+1,j}^k + U_{l,j-1}^k + U_{l,j+1}^k - 4U_{l,j}^k \right) + \Delta t f_{lj}^k\,,$$

where $\mu = \beta\Delta t/h^2$. The method is first order in time and second order in space, and it is conditionally stable. The stability condition is

$$\Delta t \le \frac{h^2}{4\beta}. \tag{4.45}$$

Note that the factor is now 4, instead of 2 for 1D problems. To show stability using the von Neumann analysis with $f = 0$, set

$$u_{lj}^k = e^{i\,(lh_x\xi_1 + jh_y\xi_2)} = e^{i\,\xi\cdot\mathbf{x}} \tag{4.46}$$

where $\xi = [\xi_1,\ \xi_2]^T$ and $\mathbf{x} = [h_x l,\ h_y j]^T$,

$$U_{lj}^{k+1} = g(\xi_1, \xi_2)\, e^{i\,\xi\cdot\mathbf{x}}. \tag{4.47}$$

Note that the index is l instead of i in the x-direction, to avoid confusion with the imaginary unit $i = \sqrt{-1}$.

Substituting these expressions into the finite difference scheme, we obtain

$$g(\xi_1, \xi_2) = 1 - 4\mu\left(\sin^2(\xi_1 h/2) + \sin^2(\xi_2 h/2)\right),$$

where $h_x = h_y = h$ for simplicity. If we enforce

$$-1 \le 1 - 8\mu \le 1 - 4\mu\left(\sin^2(\xi_1 h/2) + \sin^2(\xi_2 h/2)\right) \le 1 - 8\mu,$$

and take $-1 \le 1 - 8\mu$, we can guarantee that $|g(\xi_1, \xi_2)| \le 1$, which implies the stability. Thus, a sufficient condition for the stability of the forward Euler method in 2D is

$$\frac{8\Delta t\beta}{h^2} \le 2, \quad \text{or} \quad \Delta t \le \frac{h^2}{4\beta},$$

in addition to the condition $\Delta t > 0$.

4.5.1 The Backward Euler Method (BW-CT) in 2D

The backward Euler scheme can be written as

$$\frac{U_{ij}^{k+1} - U_{ij}^k}{\Delta t} = \frac{U_{i-1,j}^{k+1} + U_{i+1,j}^{k+1} + U_{i,j-1}^{k+1} + U_{i,j+1}^{k+1} - 4U_{ij}^{k+1}}{h^2} + f_{ij}^{k+1}, \tag{4.48}$$

which is first order in time and second order in space, and it is unconditionally stable. The coefficient matrix for the unknown U_{ij}^{k+1} is block tridiagonal, and strictly row diagonally dominant if the natural row ordering is used to index the U_{ij}^{k+1} and the finite difference equations.

4.5.2 The Crank–Nicolson (C–N) Scheme in 2D

The Crank–Nicolson scheme can be written as

$$\frac{U_{ij}^{k+1} - U_{ij}^{k}}{\Delta t} = \frac{1}{2}\left(\frac{U_{i-1,j}^{k+1} + U_{i+1,j}^{k+1} + U_{i,j-1}^{k+1} + U_{i,j+1}^{k+1} - 4U_{ij}^{k+1}}{h^2} + f_{ij}^{k+1} \right.$$
$$\left. + \frac{U_{i-1,j}^{k} + U_{i+1,j}^{k} + U_{i,j-1}^{k} + U_{i,j+1}^{k} - 4U_{ij}^{k}}{h^2} + f_{ij}^{k}\right). \quad (4.49)$$

Both the local truncation error and global error are $O\left((\Delta t)^2 + h^2\right)$. The scheme is unconditionally stable for linear problems. However, we need to solve a system of equations with a strictly row diagonally dominant and block tridiagonal coefficient matrix, if we use the natural row ordering for both the equations and unknowns.

A structured multigrid method can be applied to solve the linear system of equations from the backward Euler method or the Crank–Nicolson scheme.

4.6 The ADI Method

The ADI is a *time splitting* or *fractional step* method. The idea is to use an implicit discretization in one direction and an explicit discretization in another direction. For the heat equation $u_t = u_{xx} + u_{yy} + f(x,y,t)$, the ADI method is

$$\frac{U_{ij}^{k+\frac{1}{2}} - U_{ij}^{k}}{(\Delta t)/2} = \frac{U_{i-1,j}^{k+\frac{1}{2}} - 2U_{ij}^{k+\frac{1}{2}} + U_{i+1,j}^{k+\frac{1}{2}}}{h_x^2} + \frac{U_{i,j-1}^{k} - 2U_{ij}^{k} + U_{i,j+1}^{k}}{h_y^2} + f_{ij}^{k+\frac{1}{2}},$$
$$\frac{U_{ij}^{k+1} - U_{ij}^{k+\frac{1}{2}}}{(\Delta t)/2} = \frac{U_{i-1,j}^{k+\frac{1}{2}} - 2U_{ij}^{k+\frac{1}{2}} + U_{i+1,j}^{k+\frac{1}{2}}}{h_x^2} + \frac{U_{i,j-1}^{k+1} - 2U_{ij}^{k+1} + U_{i,j+1}^{k+1}}{h_y^2} + f_{ij}^{k+\frac{1}{2}},$$
$$(4.50)$$

which is second order in time and in space if $u(x,y,t) \in C^4(\Omega)$, where Ω is the bounded domain where the PDE is defined. It is unconditionally stable for linear problems. We can use symbolic expressions to discuss the method by rewriting the ADI method as

$$
\begin{aligned}
U_{ij}^{k+\frac{1}{2}} &= U_{ij}^{k} + \frac{\Delta t}{2}\delta_{xx}^2 U_{ij}^{k+\frac{1}{2}} + \frac{\Delta t}{2}\delta_{yy}^2 U_{ij}^{k} + \frac{\Delta t}{2} f_{ij}^{k+\frac{1}{2}} , \\
U_{ij}^{k+1} &= U_{ij}^{k+\frac{1}{2}} + \frac{\Delta t}{2}\delta_{xx}^2 U_{ij}^{k+\frac{1}{2}} + \frac{\Delta t}{2}\delta_{yy}^2 U_{ij}^{k+1} + \frac{\Delta t}{2} f_{ij}^{k+\frac{1}{2}} .
\end{aligned}
\tag{4.51}
$$

Thus on moving unknowns to the left-hand side, in matrix-vector form we have

$$
\begin{aligned}
\left(I - \frac{\Delta t}{2} D_x^2\right) \mathbf{U}^{k+\frac{1}{2}} &= \left(I + \frac{\Delta t}{2} D_y^2\right) \mathbf{U}^{k} + \frac{\Delta t}{2} \mathbf{F}^{k+\frac{1}{2}} , \\
\left(I - \frac{\Delta t}{2} D_y^2\right) \mathbf{U}^{k+1} &= \left(I + \frac{\Delta t}{2} D_x^2\right) \mathbf{U}^{k+\frac{1}{2}} + \frac{\Delta t}{2} \mathbf{F}^{k+\frac{1}{2}} ,
\end{aligned}
\tag{4.52}
$$

leading to a simple analytically convenient result as follows. From the first equation we get

$$
\mathbf{U}^{k+\frac{1}{2}} = \left(I - \frac{\Delta t}{2} D_x^2\right)^{-1} \left(I + \frac{\Delta t}{2} D_y^2\right) \mathbf{U}^{k} + \left(I - \frac{\Delta t}{2} D_x^2\right)^{-1} \frac{\Delta t}{2} \mathbf{F}^{k+\frac{1}{2}} ,
$$

and substituting into the second equation to have

$$
\begin{aligned}
\left(I - \frac{\Delta t}{2} D_y^2\right) \mathbf{U}^{k+1} &= \left(I + \frac{\Delta t}{2} D_x^2\right) \left(I - \frac{\Delta t}{2} D_x^2\right)^{-1} \left(I + \frac{\Delta t}{2} D_y^2\right) \mathbf{U}^{k} \\
&\quad + \left(I + \frac{\Delta t}{2} D_x^2\right) \left(I - \frac{\Delta t}{2} D_x^2\right)^{-1} \frac{\Delta t}{2} \mathbf{F}^{k+\frac{1}{2}} + \frac{\Delta t}{2} \mathbf{F}^{k+\frac{1}{2}} .
\end{aligned}
$$

We can go further to get

$$
\begin{aligned}
\left(I - \frac{\Delta t}{2} D_x^2\right) \left(I - \frac{\Delta t}{2} D_y^2\right) \mathbf{U}^{k+1} &= \left(I + \frac{\Delta t}{2} D_x^2\right) \left(I + \frac{\Delta t}{2} D_y^2\right) \mathbf{U}^{k} \\
&\quad + \left(I + \frac{\Delta t}{2} D_x^2\right) \frac{\Delta t}{2} \mathbf{F}^{k+\frac{1}{2}} + \frac{\Delta t}{2} \mathbf{F}^{k+\frac{1}{2}} .
\end{aligned}
$$

This is the equivalent one step time marching form of the ADI method, which will be use to show the stability of the ADI method later. Note that in this derivation we have used

$$
\left(I + \frac{\Delta t}{2} D_x^2\right) \left(I + \frac{\Delta t}{2} D_y^2\right) = \left(I + \frac{\Delta t}{2} D_y^2\right) \left(I + \frac{\Delta t}{2} D_x^2\right)
$$

and other commutative operations.

4.6.1 Implementation of the ADI Algorithm

The key idea of the ADI method is to use the implicit discretization dimension by dimension by taking advantage of fast tridiagonal solvers. In the x-direction, the finite difference approximation is

$$U_{ij}^{k+\frac{1}{2}} = U_{ij}^{k} + \frac{\Delta t}{2}\delta_{xx}^{2} U_{ij}^{k+\frac{1}{2}} + \frac{\Delta t}{2}\delta_{yy}^{2} U_{ij}^{k} + \frac{\Delta t}{2} f_{ij}^{k+\frac{1}{2}} .$$

For a fixed j, we get a tridiagonal system of equations for $U_{1j}^{k+\frac{1}{2}}$, $U_{2j}^{k+\frac{1}{2}}$, ..., $U_{m-1,j}^{k+\frac{1}{2}}$, assuming a Dirichlet boundary condition at $x=a$ and $x=b$. The system of equations in matrix-vector form is

$$\begin{bmatrix} 1+2\mu & -\mu & & & & \\ -\mu & 1+2\mu & -\mu & & & \\ & -\mu & 1+2\mu & -\mu & & \\ & & \ddots & \ddots & \ddots & \\ & & & -\mu & 1+2\mu & -\mu \\ & & & & -\mu & 1+2\mu \end{bmatrix} \begin{bmatrix} U_{1j}^{k+\frac{1}{2}} \\ U_{2j}^{k+\frac{1}{2}} \\ U_{3j}^{k+\frac{1}{2}} \\ \vdots \\ U_{m-2,j}^{k+\frac{1}{2}} \\ U_{m-1,j}^{k+\frac{1}{2}} \end{bmatrix} = \widehat{\mathbf{F}},$$

where

$$\widehat{\mathbf{F}} = \begin{bmatrix} U_{1,j}^{k} + \frac{\Delta t}{2} f_{1j}^{k+\frac{1}{2}} + \mu\, u_{bc}(a, y_j)^{k+\frac{1}{2}} + \mu\left(U_{1,j-1}^{k} - 2U_{1,j}^{k} + U_{1,j+1}^{k}\right) \\ U_{2,j}^{k} + \frac{\Delta t}{2} f_{2j}^{k+\frac{1}{2}} + \mu\left(U_{2,j-1}^{k} - 2U_{2,j}^{k} + U_{2,j+1}^{k}\right) \\ U_{3j}^{k} + \frac{\Delta t}{2} f_{3j}^{k+\frac{1}{2}} + \mu\left(U_{3,j-1}^{k} - 2U_{3,j}^{k} + U_{3,j+1}^{k}\right) \\ \vdots \\ U_{m-2,j}^{k} + \frac{\Delta t}{2} f_{m-2,j}^{k+\frac{1}{2}} + \mu\left(U_{m-2,j-1}^{k} - 2U_{m-2,j}^{k} + U_{m-2,j+1}^{k}\right) \\ U_{m-1,j}^{k} + \frac{\Delta t}{2} f_{m-1,j}^{k+\frac{1}{2}} + \mu\left(U_{m-1,j-1}^{k} - 2U_{m-1,j}^{k} + U_{m-1,j+1}^{k}\right) + \mu\, u_{bc}(b, y_j)^{k+\frac{1}{2}} \end{bmatrix},$$

and $\mu = \frac{\beta \Delta t}{2h^2}$, and $f_i^{k+\frac{1}{2}} = f(x_i, t^{k+\frac{1}{2}})$. For each j, we need to solve a symmetric tridiagonal system of equations. The cost for the x-sweep is about $O(5mn)$.

4.6.1.1 A Pseudo-code of the ADI Method in Matlab

```
for j = 2:n,                                        % Loop for fixed j
   A = sparse(m-1,m-1); b=zeros(m-1,1);
   for i=2:m,
      b(i-1) = (u1(i,j-1) -2*u1(i,j) + u1(i,j+1))/h1 + ...
                 f(t2,x(i),y(j)) + 2*u1(i,j)/dt;
      if i == 2
        b(i-1) = b(i-1) + uexact(t2,x(i-1),y(j))/h1;
        A(i-1,i) = -1/h1;
      else
        if i==m
          b(i-1) = b(i-1) + uexact(t2,x(i+1),y(j))/h1;
          A(i-1,i-2) = -1/h1;
        else
           A(i-1,i) = -1/h1;
           A(i-1,i-2) = -1/h1;
        end
      end
      A(i-1,i-1) = 2/dt + 2/h1;
    end
    ut = A\b;                                      % Solve the diagonal matrix.

%------------- loop in the y direction --------------------------
for i = 2:m,
   A = sparse(m-1,m-1); b=zeros(m-1,1);
   for j=2:n,
      b(j-1) = (u2(i-1,j) -2*u2(i,j) + u2(i+1,j))/h1 + ...
                 f(t2,x(i),y(j)) + 2*u2(i,j)/dt;
      if j == 2
        b(j-1) = b(j-1) + uexact(t1,x(i),y(j-1))/h1;
        A(j-1,j) = -1/h1;
      else
        if j==n
          b(j-1) = b(j-1) + uexact(t1,x(i),y(j+1))/h1;
          A(j-1,j-2) =  -1/h1;
        else
           A(j-1,j) = -1/h1;
           A(j-1,j-2) = -1/h1;
        end
      end
      A(j-1,j-1) = 2/dt + 2/h1;                       % Solve the system
    end
    ut = A\b;
```

A Matlab test code adi.m can be found in the depository directory of this chapter.

4.6.2 Consistency of the ADI Method

Adding the two equations in (4.50) together, we get

$$\frac{U_{ij}^{k+1} - U_{ij}^{k}}{(\Delta t)/2} = 2\delta_{xx}^2 U_{ij}^{k+\frac{1}{2}} + \delta_{yy}^2 \left(U_{ij}^{k+1} + U_{ij}^{k} \right) + 2 f_{ij}^{k+\frac{1}{2}} ; \tag{4.53}$$

and if we subtract the first equation from the second equation, we get

$$4U_{ij}^{k+\frac{1}{2}} = 2\left(U_{ij}^{k+1} + U_{ij}^{k} \right) - \Delta t \delta_{yy}^2 \left(U_{ij}^{k+1} - U_{ij}^{k} \right) . \tag{4.54}$$

Substituting this into (4.53) we get

$$\left(1 + \frac{(\Delta t)^2}{4} \delta_{xx}^2 \delta_{yy}^2 \right) \frac{U_{ij}^{k+1} - U_{ij}^{k}}{\Delta t} = \left(\delta_{xx}^2 + \delta_{yy}^2 \right) \frac{U_{ij}^{k+1} + U_{ij}^{k}}{2} + f_{ij}^{k+\frac{1}{2}} , \tag{4.55}$$

and we can clearly see that the discretization is second-order accurate in both space and time, *i.e.*, $T_{ij}^{k} = O((\Delta t)^2 + h^2)$.

4.6.3 Stability Analysis of the ADI Method

Taking $f = 0$ and setting

$$U_{lj}^{k} = e^{i(\xi_1 h_1 l + \xi_2 h_2 j)}, \qquad U_{lj}^{k+1} = g(\xi_1, \xi_2)\, e^{i(\xi_1 h_1 l + \xi_2 h_2 j)}, \tag{4.56}$$

on using the operator form we have

$$\left(1 - \frac{\Delta t}{2} \delta_{xx}^2 \right) \left(1 - \frac{\Delta t}{2} \delta_{yy}^2 \right) \mathbf{U}_{jl}^{k+1} = \left(1 + \frac{\Delta t}{2} \delta_{xx}^2 \right) \left(1 + \frac{\Delta t}{2} \delta_{yy}^2 \right) \mathbf{U}_{jl}^{k},$$

which yields,

$$\begin{aligned} &\left(1 - \frac{\Delta t}{2} \delta_{xx}^2 \right) \left(1 - \frac{\Delta t}{2} \delta_{yy}^2 \right) g(\xi_1, \xi_2)\, e^{i(\xi_1 h_1 l + \xi_2 h_2 j)} \\ &\quad = \left(1 + \frac{\Delta t}{2} \delta_{xx}^2 \right) \left(1 + \frac{\Delta t}{2} \delta_{yy}^2 \right) e^{i(\xi_1 h_1 l + \xi_2 h_2 j)} . \end{aligned}$$

After some manipulations, we get

$$g(\xi_1, \xi_2) = \frac{\left(1 - 4\mu \sin^2(\xi_1 h/2) \right) \left(1 - 4\mu \sin^2(\xi_2 h/2) \right)}{\left(1 + 4\mu \sin^2(\xi_1 h/2) \right) \left(1 + 4\mu \sin^2(\xi_2 h/2) \right)} ,$$

where $\mu = \frac{\Delta t}{2h^2}$ and for simplicity we have set $h_x = h_y = h$. Thus $|g(\xi_1, \xi_2)| \le 1$, no matter what Δt and h are, so the ADI method is unconditionally stable for linear heat equations.

4.7 An Implicit–explicit Method for Diffusion and Advection Equations

Consider a diffusion and advection PDE in 2D

$$u_t + \mathbf{w} \cdot \nabla u = \nabla \cdot (\beta \nabla u) + f(x, y, t)$$

where $\mathbf{w}$ is a 2D vector, and ∇ is the gradient operator in 2D, see page 48. In this case, it is not so easy to get a second-order implicit scheme such that the coefficient matrix is diagonally dominant or symmetric positive or negative definite, due to the advection term $\mathbf{w} \cdot \nabla u$. One approach is to use an implicit scheme for the diffusion term and an explicit scheme for the advection term, of the following form from time level t^k to t^{k+1}:

$$\frac{u^{k+1} - u^k}{\Delta t} + (\mathbf{w} \cdot \nabla_h u)^{k+\frac{1}{2}} = \frac{1}{2}\left((\nabla_h \cdot \beta \nabla_h u)^k + (\nabla_h \cdot \beta \nabla_h u)^{k+1} \right) + f^{k+\frac{1}{2}}, \tag{4.57}$$

where

$$(\mathbf{w} \cdot \nabla_h u)^{k+\frac{1}{2}} = \frac{3}{2} (\mathbf{w} \cdot \nabla_h u)^k - \frac{1}{2} (\mathbf{w} \cdot \nabla_h u)^{k-1}, \tag{4.58}$$

where $\nabla_h u = [\delta_x u,\ \delta_y u]^T$, and at a grid point $(x_i,\ y_j)$, they are

$$\delta_x u = \frac{u_{i+1,j} - u_{i-1,j}}{2h_x}; \qquad \delta_x u = \frac{u_{i,j+1} - u_{i,j-1}}{2h_y}. \tag{4.59}$$

We treat the advection term explicitly, since the term only contains the first-order partial derivatives and the CFL constraint is not a main concern unless $\|\mathbf{w}\|$ is very large. The time step constraint is

$$\Delta t \le \frac{h}{2\|\mathbf{w}\|_2}. \tag{4.60}$$

At each time step, we need to solve a generalized Helmholtz equation

$$(\nabla \cdot \beta \nabla u)^{k+1} - \frac{2u^{k+1}}{\Delta t} = -\frac{2u^k}{\Delta t} + 2 (\mathbf{w} \cdot \nabla u)^{k+\frac{1}{2}} - (\nabla \cdot \beta \nabla u)^k - 2f^{k+\frac{1}{2}}. \tag{4.61}$$

We need $\mathbf{u}^1$ to get the scheme above started. We can use the explicit Euler method (FW-CT) to approximate $\mathbf{u}^1$, as this should not affect the stability and global error $O((\Delta t)^2 + h^2)$.

4.8 Solving Elliptic PDEs using Numerical Methods for Parabolic PDEs

We recall the steady-state solution of a parabolic PDE is the solution of the corresponding elliptic PDE, *e.g.*, the steady-state solution of the parabolic PDE

$$u_t = \nabla \cdot (\beta \nabla u) + \mathbf{w} \cdot \nabla u + f(\mathbf{x}, t)$$

is the solution to the elliptic PDE

$$\nabla \cdot (\beta \nabla u) + \mathbf{w} \cdot \nabla u + \bar{f}(\mathbf{x}) = 0,$$

if the limit

$$\bar{f}(\mathbf{x}) = \lim_{t \to \infty} f(\mathbf{x}, t)$$

exists. The initial condition is irrelevant to the steady-state solution, but the boundary condition is relevant. This approach has some advantages, especially for nonlinear problems where the solution is not unique. We can control the variation of the intermediate solutions, and the linear system of equations is more diagonally dominant. Since we only require the steady-state solution, we prefer to use implicit methods with large time steps since the accuracy in time is unimportant.

Exercises

1. Show that a scheme for

$$u_t = \beta\, u_{xx} \tag{4.62}$$

of the form

$$U_i^{k+1} = \alpha U_i^k + \frac{1-\alpha}{2}\left(U_{i+1}^k + U_{i-1}^k\right)$$

where $\alpha = 1 - 2\beta\mu$, $\mu = \Delta t/h^2$ is consistent with the heat equation (4.62). Find the order of the discretization.

2. Show that the implicit scheme

$$\left(1 - \frac{\Delta t \beta}{2}\,\delta_{xx}^2\right)\left(\frac{U_i^{k+1} - U_i^k}{\Delta t}\right) = \beta\left(1 - \frac{h^2}{12}\,\delta_{xx}^2\right)\delta_{xx}^2 U_i^k \tag{4.63}$$

for the heat equation (4.62) has order of accuracy $((\Delta t)^2, h^4)$, where

$$\delta_{xx}^2 U_i = \frac{U_{i-1} - 2U_i + U_{i+1}}{h^2},$$

and $\Delta t = O(h^2)$. Compare this method with FW-CT, BW-CT, and Crank–Nicolson schemes and explain the advantages and limitations. (**Note:** The stability condition of the scheme is $\beta\mu \le \frac{3}{2}$).

3. For the implicit Euler method applied to the heat equation $u_t = u_{xx}$, is it possible to choose Δt such that the discretization is $O((\Delta t)^2 + h^4)$?
4. Consider the diffusion and advection equation

$$u_t + u_x = \beta u_{xx}, \quad \beta > 0. \tag{4.64}$$

Use the von Neumann analysis to derive the time step restriction for the scheme

$$\frac{U_i^{k+1} - U_i^k}{\Delta t} + \frac{U_{i+1}^k - U_{i-1}^k}{2h} = \beta \frac{U_{i-1}^k - 2U_i^k + U_{i+1}^k}{h^2}.$$

5. Implement and compare the ***Crank–Nicolson*** and the ***MOL*** methods using Matlab for the heat equation:

$$u_t = \beta u_{xx} + f(x,t), \quad a < x < b, \quad t \ge 0,$$

$$u(x,0) = u_0(x); \quad u(a,t) = g_1(t); \quad u_x(b,t) = g_2(t),$$

where β is a constant. Use $u(x,t) = (\cos t)\, x^2 \sin(\pi x)$, $0 < x < 1$, *tfinal* $= 1.0$ to test and debug your code. Write a short report about these two methods. Your discussion should include the grid refinement analysis, error and solution plots for $m = 80$, comparison of cputime, and any conclusions you can draw from your results. You can use Matlab code *ode15s* or *ode23s* to solve the semidiscrete ODE system.
Assume that u is the temperature of a thin rod with one end ($x = b$) just heated. The other end of the rod has a room temperature ($70°\,C$). Solve the problem and find the history of the solution. Roughly how long does it take for the temperature of the rod to reach the steady state? What is the exact solution of the steady state? **Hint:** Take the initial condition as $u(x,0) = T_0\, e^{-(x-b)^2/\gamma}$, where T_0 and γ are two constants, $f(x,t) = 0$, and the Neumann boundary condition $u_x(b,t) = 0$.

6. Carry out the von Neumann analysis to determine the stability of the θ method

$$\frac{U_j^{(n+1)} - U_j^n}{k} = b\left(\theta\, \delta_x^2 U_j^{(n)} + (1-\theta)\, \delta_x^2 U_j^{(n+1)}\right) \tag{4.65}$$

for the heat equation $u_t = b u_{xx}$, where

$$\delta_x^2 U_j = \frac{U_{j-1} - 2U_j + U_{j+1}}{h^2} \quad \text{and} \quad 0 \le \theta \le 1.$$

7. Modify the Crank–Nicolson Matlab code for the backward Euler method and for variable $\beta(x,t)$'s in one space dimensions. Validate your code.
8. Implement and compare the ADI and Crank–Nicolson methods with the SOR(ω) (try to test optimal ω) for the following problem involving the 2D heat equation:

$$u_t = u_{xx} + u_{yy} + f(t,x,y), \quad a < x < b, \quad c \le y \le d, \quad t \ge 0,$$

$$u(0,x,y) = u_0(x,y),$$

and Dirichlet boundary conditions. Choose two examples with known exact solutions to test and debug your code. Write a short report about the two methods. Your discussion should include the grid refinement analysis (with a fixed final time, say $T = 0.5$), error and solution plots, comparison of cpu time and flops, and any conclusions you can draw from your results.

9. **Extra credit:** Modify the ADI Matlab code for variable heat conductivity $\beta(x, y)$.

5

Finite Difference Methods for Hyperbolic PDEs

In this chapter, we discuss finite difference methods for hyperbolic PDEs (see page 6 for the definition of hyperbolic PDEs). Let us first list a few typical model problems involving hyperbolic PDEs.

- Advection equation (one-way wave equation):

$$\begin{aligned} u_t + a u_x &= 0\,, \quad 0 < x < 1\,, \\ u(x,0) &= \eta(x)\,, \qquad \text{IC}\,, \\ u(0,t) = g_l(t) \quad \text{if} \quad a \geq 0\,, &\quad \text{or} \quad u(1,t) = g_r(t) \quad \text{if} \quad a \leq 0\,. \end{aligned} \tag{5.1}$$

  Here g_l and g_r are prescribed boundary conditions from the left and right, respectively.

- Second-order linear wave equation:

$$\begin{aligned} &u_{tt} = a u_{xx}\,, \quad 0 < x < 1\,, \\ &u(x,0) = \eta(x), \qquad \frac{\partial u}{\partial t}(x,0) = v(x), \qquad \text{IC}\,, \\ &u(0,t) = g_l(t)\,, \qquad u(1,t) = g_r(t), \qquad \text{BC}. \end{aligned} \tag{5.2}$$

- Linear first-order hyperbolic system:

$$\mathbf{u}_t = A\mathbf{u}_x + \mathbf{f}(x,t)\,, \tag{5.3}$$

  where $\mathbf{u}$ and $\mathbf{f}$ are two vectors and A is a matrix. The system is called *hyperbolic* if A is diagonalizable, *i.e.*, if there is a nonsingular matrix T such that $A = TDT^{-1}$, and all eigenvalues of A are real numbers.

- Nonlinear hyperbolic equation or system, notably conservation laws:

$$u_t + f(u)_x = 0\,, \quad e.g., \text{ Burgers' equation } u_x + \left(\frac{u^2}{2}\right)_x = 0\,; \tag{5.4}$$

$$\mathbf{u}_t + \mathbf{f}_x + \mathbf{g}_y = 0\,. \tag{5.5}$$

For nonlinear hyperbolic PDE, shocks (a discontinuous solution) can develop even if the initial data is smooth.

5.1 Characteristics and Boundary Conditions

We know the exact solution for the one-way wave equation

$$u_t + au_x = 0\,, \quad -\infty < x < \infty\,,$$
$$u(x,0) = \eta(x)\,, \quad t > 0$$

is $u(x,t) = \eta(x - at)$. If the domain is finite, we can also find the exact solution. We can solve the model problem

$$u_t + au_x = 0\,, \quad 0 < x < 1\,,$$
$$u(x,0) = \eta(x)\,, \quad t > 0\,, \qquad u(0,t) = g_l(t) \quad \text{if } a > 0$$

by the *method of characteristics* since the solution is constant along the characteristics. For any point (x,t) we can readily trace the solution back to the initial data. In fact, for the characteristic

$$z(s) = u(x + ks, t + s) \tag{5.6}$$

along which the solution is a constant ($z'(s) \equiv 0$), on substituting into the PDE we get

$$z'(s) = u_t + ku_x = 0\,,$$

which is always true if $k = a$. The solution at $(x + ks, t + s)$ is the same as at (x,t), so we can solve the problem by tracing back until the line hits the boundary, *i.e.*, $u(\bar{x},\bar{t}) = u(x + as, t + s) = u(x - at,\, 0)$ if $x - at \geq 0$, on tracing back to the initial condition. If $x - at < 0$, we can only trace back to $x = 0$ or $s = -\bar{x}/a$ and $t = \bar{x}/a$, and the solution is $u(\bar{x},\bar{t}) = u(0, t - \bar{x}/a) = g_l(t - \bar{x}/a)$. The solution for the case $a \geq 0$ can therefore be written as

$$u(x,t) = \begin{cases} \eta(x - at) & \text{if } x \geq at, \\ g_l\left(t - \dfrac{x}{a}\right) & \text{if } x < at\,. \end{cases} \tag{5.7}$$

Now we can see why we have to prescribe a boundary condition at $x=0$, but we cannot have any boundary condition at $x=1$. It is important to have correct boundary conditions for hyperbolic problems!

The one-way wave equation is often used as a benchmark problem for different numerical methods for hyperbolic problems.

5.2 Finite Difference Schemes

Simple numerical methods for hyperbolic problems include:

- Lax–Friedrichs method;
- Upwind scheme;
- Leap-frog method (note it does not work for the heat equation but works for linear hyperbolic equations);
- Box scheme;
- Lax–Wendroff method;
- Crank–Nicolson scheme (not recommended for hyperbolic problems, since there are no severe time step size constraints); and
- Beam–Warming method (one-sided second-order upwind scheme if the solution is smooth).

There are also some high-order methods in the literature. For linear hyperbolic problems, if the initial data is smooth (no discontinuities), it is recommended to use second-order accurate methods such as the Lax–Wendroff method. However, care has to be taken if the initial data has finite discontinuities, called shocks, as second- or high-order methods often lead to oscillations near the discontinuities (Gibbs phenomena). Some of the methods are the bases for numerical methods for a conservation law, a special conservative nonlinear hyperbolic system, for which shocks may develop in finite time even if the initial data is smooth. Also for hyperbolic differential equations, usually there is no strict time step constraint as for parabolic problems. Often explicit methods are preferred.

5.2.1 Lax–Friedrichs Method

Consider the one-way wave equation $u_t + au_x = 0$, and the simple finite difference scheme

$$\frac{U_j^{k+1} - U_j^k}{\Delta t} + \frac{a}{2h}\left(U_{j+1}^k - U_{j-1}^k\right) = 0,$$

$$\text{or} \quad U_j^{k+1} = U_j^k - \mu\left(U_{j+1}^k - U_{j-1}^k\right),$$

where $\mu = a\Delta t/(2h)$. The scheme has $O(\Delta t + h^2)$ local truncation error, but the method is unconditionally unstable from the von Neumann stability analysis. The growth factor for the FW-CT finite difference scheme is

$$\begin{aligned} g(\theta) &= 1 - \mu\left(e^{ih\xi} - e^{-ih\xi}\right) \\ &= 1 - \mu\, 2\, i\, \sin(h\xi), \end{aligned}$$

where $\theta = h\xi$, so

$$|g(\theta)|^2 = 1 + 4\mu^2 \sin^2(h\xi) \geq 1\,.$$

In the Lax–Friedrichs scheme, we average U_j^k using U_{j-1}^k and U_{j+1}^k to get

$$U_j^{k+1} = \frac{1}{2}\left(U_{j-1}^k + U_{j+1}^k\right) - \mu\left(U_{j+1}^k - U_{j-1}^k\right).$$

The local truncation error is $O(\Delta t + h)$ if $\Delta t \simeq h$. The growth factor is

$$\begin{aligned} g(\theta) &= \frac{1}{2}\left(e^{ih\xi} + e^{-ih\xi}\right) + \mu\left(e^{ih\xi} - e^{-ih\xi}\right) \\ &= \cos(h\xi) - 2\mu \sin(h\xi) i \end{aligned}$$

so

$$\begin{aligned} |g(\theta)|^2 &= \cos^2(h\xi) + 4\mu^2 \sin^2(h\xi) \\ &= 1 - \sin^2(h\xi) + 4\mu^2 \sin^2(h\xi) \\ &= 1 - (1 - 4\mu^2)\sin^2(h\xi)\,, \end{aligned}$$

and we conclude that $|g(\theta)| \leq 1$ if $1 - 4\mu^2 \geq 0$ or $1 - (a\Delta t/h)^2 \geq 0$, which implies that $\Delta t \leq h/|a|$. This is the CFL (Courant–Friedrichs–Lewy) condition.

For the Lax–Friedrichs scheme, we need a numerical boundary condition (NBC) at $x = 1$, as explained later. The Lax–Friedrichs scheme is the basis for several other popular schemes. A Matlab code called *lax_fred.m* can be found in the Matlab programming collections that accompany the book.

5.2.2 The Upwind Scheme

The upwind scheme for $u_t + au_x = 0$ is

$$\frac{U_j^{k+1} - U_j^k}{\Delta t} = \begin{cases} -\dfrac{a}{h}\left(U_j^k - U_{j-1}^k\right) & \text{if } a \geq 0\,, \\[2ex] -\dfrac{a}{h}\left(U_{j+1}^k - U_j^k\right) & \text{if } a < 0\,, \end{cases} \tag{5.8}$$

which is first-order accurate in time and in space. To find the CFL constraint, we conduct the von Neumann stability analysis. The growth factor for the case when $a \geq 0$ is

$$\begin{aligned} g(\theta) &= 1 - \mu\left(1 - e^{-ih\xi}\right) \\ &= 1 - \mu(1 - \cos(h\xi)) - i\mu\sin(h\xi) \end{aligned}$$

with magnitude

$$\begin{aligned} |g(\theta)|^2 &= (1 - \mu + \mu\cos(h\xi))^2 + \mu^2\sin^2(h\xi) \\ &= (1-\mu)^2 + 2(1-\mu)\mu\cos(h\xi) + \mu^2 \\ &= 1 - 2(1-\mu)\mu(1 - \cos(h\xi)), \end{aligned}$$

so if $1 - \mu \geq 0$ (*i.e.*, $\mu \leq 1$) or $\Delta t \leq h/a$ we have $|g(\theta)| \leq 1$.

Note that no NBC is needed for the upwind scheme, and there is no severe time step restriction, since $\Delta t \leq h/a$. If $a = a(x,t)$ is a variable function that does not change the sign, then the CFL condition is

$$0 < \Delta t \leq \frac{h}{\max|a(x,t)|}.$$

However, the upwind scheme is first-order accurate in time and in space, and there are some high-order schemes.

A Matlab code called upwind.m can be found in the Matlab programming collections accompanying this book. The main structure of the code is listed below:

```
a = 0;  b=1; tfinal = 0.5  % Input the domain and final time.
m = 20; h = (b-a)/m; k = h; mu = k/h;  % Set mesh and time step

n = fix(tfinal/k);  % Find the number of time steps
y1 = zeros(m+1,1); y2=y1; x=y1;     % Initialization

figure(1);  hold          % Open a plot window for solutions at
                          % different time.
axis([-0.1 1.1 -0.1 1.1]);

for i=1:m+1,              % Initialization.
  x(i) = a + (i-1)*h;
  y1(i) = uexact(t,x(i));  y2(i) = 0;
end

t = 0;      % Begin time marching.
for j=1:n,
  y1(1)=bc(t); y2(1)=bc(t+k);
  for i=2:m+1
    y2(i) = y1(i) - mu*(y1(i)-y1(i-1) );
  end
```

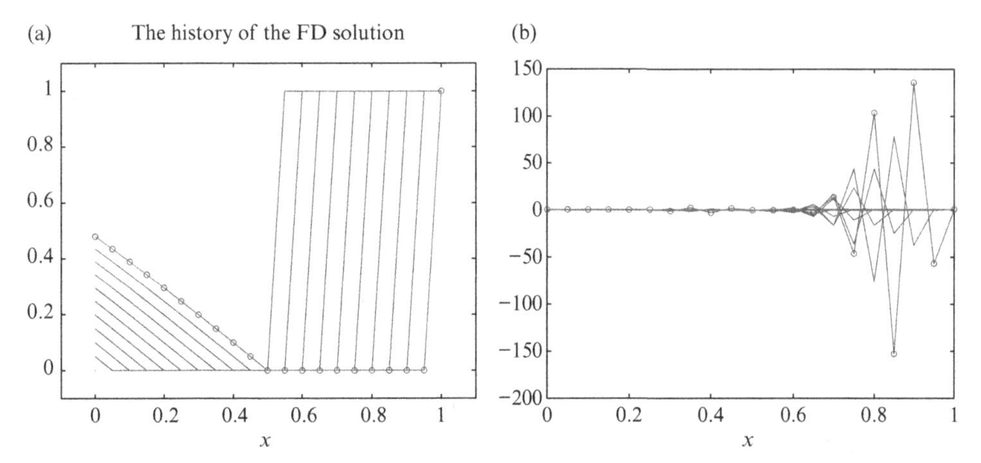

Figure 5.1. Plot of the initial and consecutive approximation of the upwinding method for an advection equation. (a) The time step is $\Delta t = h$ and the scheme is stable. (b) The time step is $\Delta t = 1.5h$ and the scheme is unstable which leads to a blowing-up quantity.

```
  t = t + k;  y1 = y2;        % Overwrite old solutions
  plot(x,y2); pause(0.5)      % Plot the current solution.
end
```

In Figure 5.1(a), we show the initial data and several consecutive finite difference approximations of the upwinding scheme applied to the advection equation $u_t + u_x = 0$ in the domain $0 < x < 1$. The initial condition is

$$u(x,0) = u_0(x) = \begin{cases} 0 & \text{if } 0 < x < 1/2, \\ 1 & \text{if } 1/2 \le x < 1 . \end{cases}$$

The boundary condition is $u(0,t) = \sin t$. The analytic solution is

$$u(x,t) = \begin{cases} u_0(x-t) & \text{if } 0 < t < x < 1, \\ \sin(t-x) & \text{if } 0 < x < t < 1 . \end{cases}$$

If we take $\Delta t \le h$, the scheme works well and we obtain the exact solution for this example (see Figure 5.1(a)). However, if we take $\Delta t > h$, say $\Delta t = 1.5h$ as in the plot of Figure 5.1(b), the solution blows up quickly since the scheme is unstable. Once again, it shows the importance of the time step constraint.

5.2.3 The Leap-frog Scheme

The leap-frog scheme for $u_t + au_x = 0$ is

$$\begin{aligned} \frac{U_j^{k+1} - U_j^{k-1}}{2\Delta t} + \frac{a}{2h}\left(U_{j+1}^k - U_{j-1}^k\right) = 0\,, \\ \text{or} \quad U_j^{k+1} = U_j^{k-1} - \mu\left(U_{j+1}^k - U_{j-1}^k\right), \end{aligned} \tag{5.9}$$

where $\mu = a\Delta t/(2h)$. The discretization is second-order in time and in space. It requires an NBC at one end and needs U_j^1 to get started. We know that the leap-frog scheme is unconditionally unstable for the heat equation. Let us consider the stability for the advection equation through the von Neumann analysis. Substituting

$$U_j^k = e^{ij\xi}, \quad U_j^{k+1} = g(\xi)e^{ij\xi}, \quad U_j^{k-1} = \frac{1}{g(\xi)}\,e^{ij\xi}$$

into the leap-frog scheme, we get

$$\begin{aligned} g^2 + \mu(e^{ih\xi} - e^{-ih\xi})g - 1 = 0\,, \\ \text{or} \quad g^2 + 2\mu i \sin(h\xi)\, g - 1 = 0\,, \end{aligned}$$

with solution

$$g_{\pm} = -i\mu \sin(h\xi) \pm \sqrt{1 - \mu^2 \sin^2(h\xi)}\,. \tag{5.10}$$

We distinguish three different cases.

1. If $|\mu| > 1$, then there are ξ such that at least one of $|g_-| > 1$ or $|g_+| > 1$ holds, so the scheme is unstable!
2. If $|\mu| < 1$, then $1 - \mu^2 \sin^2(h\xi) \geq 0$ such that

   $$|g_{\pm}|^2 = \mu^2 \sin^2(h\xi) + 1 - \mu^2 \sin^2(h\xi) = 1\,.$$

   However, since it is a two-stage method, we have to be careful about the stability. From linear finite difference theory, we know the general solution is

   $$\begin{aligned} U^k &= C_1 g_-^k + C_2 g_+^k \\ |U^k| &\leq \max\{C_1, C_2\}\left(|g_-^k| + |g_+^k|\right) \\ &\leq 2\max\{C_1, C_2\}\,, \end{aligned}$$

   so the scheme is neutral stable according to the definition $\|\mathbf{U}^k\| \leq C_T \sum_{j=0}^{J} \|\mathbf{U}^j\|$.

3. If $|\mu| = 1$, we still have $|g_{\pm}| = 1$, but we can find ξ such that $\mu \sin(h\xi) = 1$ and $g_+ = g_- = -i$, *i.e.*, $-i$ is a double root of the characteristic polynomial. The solution of the finite difference equation therefore has the form

$$U_j^k = C_1(-i)^k + C_2 k(-i)^k ,$$

where the possibly complex numbers C_1 and C_2 are determined from the initial conditions. Thus there are solutions such that $\|\mathbf{U}^k\| \simeq k$ which are unstable (slow growing).

In conclusion, the leap-frog scheme is stable if $\Delta t < \frac{h}{|a|}$. Note that we can use the upwind or other scheme (even unstable ones) to initialize the leap-frog scheme to get U_j^1. We call a numerical scheme (such as the Lax–Friedrichs and upwind schemes) dissipative if $|g(\xi)| < 1$, and otherwise (such as the leap-frog scheme) it is nondissipative.

5.3 The Modified PDE and Numerical Diffusion/Dispersion

A modified PDE is the PDE that a finite difference equation satisfies exactly at grid points. Consider the upwind method for the advection equation $u_t + au_x = 0$ in the case $a > 0$,

$$\frac{U_j^{k+1} - U_j^k}{\Delta t} + \frac{a}{h}\left(U_j^k - U_{j-1}^k\right) = 0 .$$

The derivation of a modified PDE is similar to computing the local truncation error, only now we insert $v(x,t)$ into the finite difference equation to derive a PDE that $v(x,t)$ satisfies exactly, thus

$$\frac{v(x.t + \Delta t) - v(x,t)}{\Delta t} + \frac{a}{h}\left(v(x,t) - v(x-h,t)\right) = 0 .$$

Expanding the terms in Taylor series about (x,t) and simplifying yields

$$v_t + \frac{1}{2}\Delta t\, v_{tt} + \cdots + a\left(v_x - \frac{1}{2}hv_{xx} + \frac{1}{6}h^2 v_{xxx} + \cdots\right) = 0 ,$$

which can be rewritten as

$$v_t + av_x = \frac{1}{2}(ahv_{xx} - \Delta t v_{tt}) - \frac{1}{6}\left(ah^2 v_{xxx} + (\Delta t)^2 v_{ttt}\right) + \cdots ,$$

which is the PDE that v satisfies. Consequently,

$$\begin{aligned} v_{tt} &= -av_{xt} + \frac{1}{2}(ahv_{xxt} - \Delta t v_{ttt}) \\ &= -av_{xt} + O(\Delta t, h) \\ &= -a\frac{\partial}{\partial x}\left(-av_x + O(\Delta t, h)\right), \end{aligned}$$

so the leading modified PDE is

$$v_t + av_x = \frac{1}{2}ah\left(1 - \frac{a\Delta t}{h}\right)v_{xx}. \tag{5.11}$$

This is an advection–diffusion equation. The grid values U_j^n can be viewed as giving a second-order accurate approximation to the true solution of this equation, whereas they only give a first-order accurate approximation to the true solution of the original problem. From the modified equation, we can conclude that:

- the computed solution smooths out discontinuities because of the diffusion term (the second-order derivative term is called numerical dissipation, or numerical viscosity);
- if a is a constant and $\Delta t = h/a$, then $1 - a\Delta t/h = 0$ (we have second-order accuracy);
- we can add the correction term to offset the leading error term to render a higher-order accurate method, but the stability needs to be checked. For instance, we can modify the upwind scheme to get

$$\frac{U_j^{k+1} - U_j^k}{\Delta t} + a\frac{U_j^k - U_{j-1}^k}{h} = \frac{1}{2}ah\left(1 - \frac{a\Delta t}{h}\right)\frac{U_{j-1}^k - 2U_j^k + U_{j+1}^k}{h^2},$$

  which is second-order accurate if $\Delta t \simeq h$;
- from the modified equation, we can see why some schemes are unstable, *e.g.*, the leading term of the modified PDE for the unstable scheme

$$\frac{U_j^{k+1} - U_j^k}{\Delta t} + a\frac{U_{j+1}^k - U_{j-1}^k}{2h} = 0 \tag{5.12}$$

  is

$$v_t + av_x = -\frac{a^2\Delta t}{2}v_{xx}, \tag{5.13}$$

  where the highest derivative is similar to the backward heat equation that is dynamically unstable!

5.4 The Lax–Wendroff Scheme and Other FD methods

To derive the Lax–Wendroff scheme, we notice that

$$\begin{aligned}\frac{u(x,t+\Delta t)-u(x,t)}{\Delta t} &= u_t + \frac{\Delta t}{2}u_{tt} + O((\Delta t)^2)\\ &= u_t + \frac{1}{2}a^2(\Delta t)u_{xx} + O((\Delta t)^2)\,.\end{aligned}$$

We can add the numerical viscosity $\frac{1}{2}a^2\Delta t\, u_{xx}$ term to improve the accuracy in time, to get the Lax–Wendroff scheme

$$\frac{U_j^{k+1}-U_j^k}{\Delta t} + a\frac{U_{j+1}^k - U_{j-1}^k}{2h} = \frac{1}{2}\frac{a^2\Delta t}{h^2}\left(U_{j-1}^k - 2U_j^k + U_{j+1}^k\right), \quad (5.14)$$

which is second-order accurate both in time and space. To show this, we investigate the corresponding local truncation error

$$\begin{aligned}T(x,t) &= \frac{u(x,t+\Delta t)-u(x,t)}{\Delta t} - \frac{a\,(u(x+h,t)-u(x-h,t))}{2h}\\ &\quad - \frac{a^2\Delta t\,(u(x-h,t)-2u(x,t)+u(x+h,t))}{2h^2}\\ &= u_t + \frac{\Delta t}{2}u_{tt} - au_x - \frac{a^2\Delta t}{2}u_{xx} + O((\Delta t)^2+h^2)\\ &= O((\Delta t)^2+h^2)\,,\end{aligned}$$

since $u_t = -au_x$ and $u_{tt} = -au_{xt} = -a\frac{\partial}{\partial x}u_t = a^2u_{xx}$.

To get the CFL condition for the Lax–Wendroff scheme, we carry out the von Neumann stability analysis. The growth factor of the Lax–Wendroff scheme is

$$\begin{aligned}g(\theta) &= 1 - \frac{\mu}{2}\left(e^{ih\xi} - e^{-ih\xi}\right) + \frac{\mu^2}{2}\left(e^{-ih\xi} - 2 + e^{ih\xi}\right)\\ &= 1 - \mu i\sin\theta - 2\mu^2\sin^2(\theta/2),\end{aligned}$$

where again $\theta = h\xi$, so

$$\begin{aligned}
|g(\theta)|^2 &= \left(1 - 2\mu^2 \sin^2 \frac{\theta}{2}\right)^2 + \mu^2 \sin^2 \theta \\
&= 1 - 4\mu^2 \sin^2 \frac{\theta}{2} + 4\mu^4 \sin^4 \frac{\theta}{2} + 4\mu^2 \sin^2 \frac{\theta}{2}\left(1 - \sin^2 \frac{\theta}{2}\right) \\
&= 1 - 4\mu^2\left(1 - \mu^2\right) \sin^4 \frac{\theta}{2} \\
&\le 1 - 4\mu^2\left(1 - \mu^2\right).
\end{aligned}$$

We conclude $|g(\theta)| \le 1$ if $\mu \le 1$, *i.e.*, $\Delta t \le h/|a|$. If $\Delta t > h/|a|$, there are ξ such that $|g(\theta)| > 1$ so the scheme is unstable.

The leading modified PDE for the Lax–Wendroff method is

$$v_t + av_x = -\frac{1}{6}ah^2\left(1 - \left(\frac{a\Delta t}{h}\right)^2\right) v_{xxx} \tag{5.15}$$

which is a dispersive equation. The group velocity for the wave number ξ is

$$c_g = a - \frac{1}{2}ah^2\left(1 - \left(\frac{a\Delta t}{h}\right)^2\right)\xi^2, \tag{5.16}$$

which is less than a for all wave numbers. Consequently, the numerical result can be expected to develop a train of oscillations behind the peak, with high wave numbers lagging farther behind the correct location (*cf.* Strikwerda, 1989 for more details). If we retain one more term in the modified equation for the Lax–Wendroff scheme, we get

$$v_t + av_x = \frac{1}{6}ah^2\left(\left(\frac{a\Delta t}{h}\right)^2 - 1\right) v_{xxx} - \epsilon v_{xxxx}, \tag{5.17}$$

where the ϵ in the fourth-order dissipative term is $O(h^3)$ and positive when the stability bound holds. This high-order dissipation causes the highest wave number to be damped, so that the oscillations are limited.

5.4.1 The Beam–Warming Method

The Beam–Warming method is a one-sided finite difference scheme for the modified equation

$$v_t + av_x = \frac{a^2\Delta t}{2} v_{xx}.$$

Recall the one-sided finite difference formulas, *cf.* page 21

$$u'(x) = \frac{3u(x) - 4u(x-h) + u(x-2h)}{2h} + O(h^2),$$

$$u''(x) = \frac{u(x) - 2u(x-h) + u(x-2h)}{h^2} + O(h).$$

The Beam–Warming method for $u_t + au_x = 0$ for $a > 0$ is

$$U_j^{k+1} = U_j^k - \frac{a\Delta t}{2h}\left(3U_j^k - 4U_{j-1}^k + U_{j-2}^k\right) + \frac{(a\Delta t)^2}{2h^2}\left(U_j^k - 2U_{j-1}^k + U_{j-2}^k\right), \tag{5.18}$$

which is second-order accurate in time and space if $\Delta t \simeq h$. The CFL constraint is

$$0 < \Delta t \leq \frac{2h}{|a|}. \tag{5.19}$$

For this method, we do not require an NBC at $x = 1$, but we need a scheme to compute the solution U_1^j. The leading terms of the modified PDE for the Beam–Warming method are

$$v_t + av_x = \frac{1}{6}ah^2\left(\left(\frac{a\Delta t}{h}\right)^2 - 1\right)v_{xxx}. \tag{5.20}$$

In this case, the group velocity is greater than a for all wave numbers when $0 \leq a\Delta t/h \leq 1$, so initial oscillations would move ahead of the main hump. On the other hand, if $1 \leq a\Delta t/h \leq 2$ the group velocity is less than a, so the oscillations fall behind.

5.4.2 The Crank–Nicolson Scheme

The Crank–Nicolson scheme for the advection equation $u_t + au_x = f$ is

$$\frac{U_j^{k+1} - U_j^k}{\Delta t} + a\frac{U_{j+1}^k - U_{j-1}^k + U_{j+1}^{k+1} - U_{j-1}^{k+1}}{4h} = f_j^{k+\frac{1}{2}}, \tag{5.21}$$

which is second-order accurate in time and in space, and unconditionally stable. An NBC is needed at $x = 1$. This method is effective for the 1D problem, since it is easy to solve the resulting tridiagonal system of equations. For higher-dimensional problems, the method is not recommended in general as for hyperbolic equations the time step constraint $\Delta t \simeq h$ is not a major concern.

5.4.3 The Method of Lines

Different method of lines (MOL) methods can be used, depending on how the spatial derivative term is discretized. For the advection equation $u_t + au_x = 0$, if we use

$$\frac{\partial U_i}{\partial t} + a\frac{U_{i+1} - U_{i-1}}{2h} = 0 \tag{5.22}$$

the ODE solver is likely to be implicit, since the leap-frog method is unstable!

5.5 Numerical Boundary Conditions

We need a numerical boundary condition (NBC) at one end for the one-way wave equation when we use any of the Lax–Friedrichs, Lax–Wendroff, or leap-frog schemes. There are several possible approaches.

- Extrapolation. One simplest first-order approximation is

  $$U_M^{k+1} = U_{M-1}^{k+1}.$$

  To get a second-order approximation, recall the Lagrange interpolation formula

  $$f(x) \simeq f(x_1)\frac{x - x_2}{x_1 - x_2} + f(x_2)\frac{x - x_1}{x_2 - x_1}.$$

  We can use the same time level for the interpolation to get

  $$U_M^{k+1} = U_{M-2}^{k+1}\frac{x_M - x_{M-1}}{x_{M-1} - x_M} + U_{M-1}^{k+1}\frac{x_M - x_{M-2}}{x_{M-2} - x_{M-1}}.$$

  If a uniform grid is used with spatial step size h, this formula becomes

  $$U_M^{k+1} = -U_{M-2}^{k+1} + 2U_{M-1}^{k+1}.$$

- Quasi-characteristics. If we use previous time levels for the interpolation, we get

  $$U_M^{k+1} = U_{M-1}^k, \quad \text{first order,}$$

  $$U_M^{k+1} = U_{M-2}^k\frac{x_M - x_{M-1}}{x_{M-1} - x_M} + U_{M-1}^k\frac{x_M - x_{M-2}}{x_{M-2} - x_{M-1}}, \qquad \text{second order.}$$

- We may use schemes that do not need an NBC at or near the boundary, *e.g.*, the upwind scheme or the Beam–Warming method to provide the boundary conditions.

The accuracy and stability of numerical schemes usually depend upon the NBC used. Usually, the main scheme and the scheme for an NBC should both be stable.

5.6 Finite Difference Methods for Second-order Linear Hyperbolic PDEs

In reality, a 1D sound wave propagates in two directions and can be modeled by the wave equation

$$u_{tt} = a^2 u_{xx}, \tag{5.23}$$

where $a > 0$ is the wave speed. We can find the general solution by changing variables as follows,

$$\begin{cases} \xi = x - at \\ \eta = x + at \end{cases} \quad \text{or} \quad \begin{cases} x = \dfrac{\xi + \eta}{2} \\ t = \dfrac{\eta - \xi}{2a} \end{cases} \tag{5.24}$$

and using the chain-rule, we have

$$\begin{aligned} u_t &= -a u_\xi + a u_\eta \,, \\ u_{tt} &= a^2 u_{\xi\xi} - 2a^2 u_{\xi\eta} + a^2 u_{\eta\eta} \,, \\ u_x &= u_\xi + u_\eta \,, \\ u_{xx} &= u_{\xi\xi} + 2u_{\xi\eta} + u_{\eta\eta} \,. \end{aligned}$$

Substituting these relations into the wave equation, we get

$$u_{\xi\xi} a^2 - 2a^2 u_{\xi\eta} + a^2 u_{\eta\eta} = a^2 \left(u_{\xi\xi} + 2u_{\xi\eta} + u_{\eta\eta} \right),$$

which simplifies to

$$4a^2 u_{\xi\eta} = 0 \,,$$

yielding the solution

$$\begin{aligned} &u_\xi = \tilde{F}(\xi), \quad \Longrightarrow \quad u(x,t) = F(\xi) + G(\eta) \,, \\ &u(x,t) = F(x - at) + G(x + at) \,, \end{aligned}$$

where $F(\xi)$ and $G(\eta)$ are two differential functions of one variable. The two functions are determined by initial and boundary conditions.

With the general solution above, we can get the analytic solution to the Cauchy problem below:

$$\begin{aligned} &u_{tt} = a^2 u_{xx} \,, \qquad -\infty < x < \infty \,, \\ &u(x,0) = u_0(x) \,, \quad u_t(x,0) = g(x) \,, \end{aligned}$$

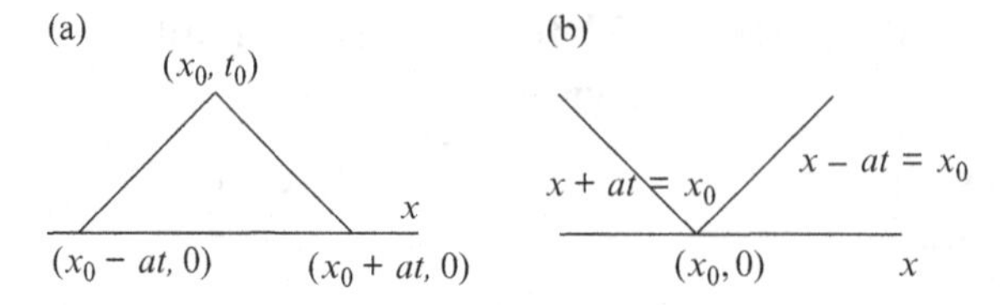

Figure 5.2. A diagram of the domain of dependence (a) and influence (b).

as

$$u(x,t) = \frac{1}{2}\left(u_0(x-at) + u_0(x+at)\right) + \frac{1}{2a}\int_{x-at}^{x+at} g(s)ds. \tag{5.25}$$

The solution is called the D'Alembert's formula. In particular, if $u_t(x,0)=0$, then the solution is

$$u(x,t) = \frac{1}{2}\Big(u(x-at,0) + u(x+at,0)\Big),$$

demonstrating that a signal (wave) propagates along each characteristic $x - at$ and $x + at$ with speed a at half of its original strength. The solution $u(x,t)$ at a point (x_0, t_0) depends on the initial conditions only in the interval of $(x_0 - at_0, x_0 + at_0)$. The initial values between $(x_0 - at_0, x_0 + at_0)$ not only determine the solution value of $u(x,t)$ at (x_0, t_0) but also all the values of $u(x,t)$ in the triangle formed by the three vertices (x_0, t_0), $(x_0 - at_0, 0)$, and $(x_0 + at_0, 0)$. This domain is called the domain of dependence (see Figure 5.2(a)).

Also we see that given any point $(x_0, 0)$, any solution value $u(x,t)$, $t > 0$, in the cone formed by the characteristic lines $x + at = x_0$ and $x - at = x_0$ depends on the initial values at $(x_0, 0)$. The domain formed by the cone is called the domain of influence (see Figure 5.2(b)).

5.6.1 An FD Method (CT–CT) for Second-order Wave Equations

Now we discuss how to solve the boundary value problems for which the analytic solution is difficult to obtain.

$$u_{tt} = a^2 u_{xx}, \qquad 0 < x < 1,$$

$$\text{IC:} \quad u(x,0) = u_0(x), \quad u_t(x,0) = u_1(x),$$

$$\text{BC:} \quad u(0,t) = g_1(t), \quad u(1,t) = g_2(t).$$

We can use the central finite difference discretization both in time and space to get

$$\frac{U_j^{k+1} - 2U_j^k + U_j^{k-1}}{(\Delta t)^2} = a^2 \frac{U_{j-1}^k - 2U_j^k + U_{j+1}^k}{h^2}, \tag{5.26}$$

which is second-order accurate both in time and space $((\Delta t)^2 + h^2)$. The CFL constraint for this method is $\Delta t \leq \frac{h}{|a|}$, as verified through the following discussion.

The scheme above cannot be used to obtain the values of U_j^1 since $U_j^{-1} \sim u(x_j, -\Delta t)$ is not explicitly defined. There are several ways to jump-start the process. We list two commonly used ones below.

- Apply the forward Euler method to the boundary condition $u_t(x,0) = u_1(x)$ to get $U_j^1 = U_j^0 + \Delta t\, u_1(x_j)$. The finite difference solution in a finite time $t = T$ will still be second-order accurate.
- Apply the ghost point method using $U_j^{-1} = U_j^1 - 2\Delta t\, u_1(x_j)$ to get

$$U_j^1 = U_j^0 + \Delta t\, u_1(x_j) + \left(\frac{a^2}{h}\right)^2 \left(U_{j-1}^0 - 2U_j^0 + U_{j+1}^0\right). \tag{5.27}$$

5.6.1.1 The Stability Analysis

The von Neumann analysis gives

$$\frac{g - 2 + 1/g}{(\Delta t)^2} = a^2 \frac{e^{-ih\xi} - 2 + e^{ih\xi}}{h^2}.$$

When $\mu = |a|\Delta t/h$, using $1 - \cos(h\xi) = 2\sin^2(h\xi/2)$, this equation becomes

$$g^2 - 2g + 1 = \left(-4\mu^2 \sin^2\theta\right) g,$$

or

$$g^2 - \left(2 - 4\mu^2 \sin^2\theta\right) g + 1 = 0,$$

where $\theta = h\xi/2$, with solution

$$g = 1 - 2\mu^2 \sin^2\theta \pm \sqrt{(1 - 2\mu^2 \sin^2\theta)^2 - 1}.$$

Note that $1 - 2\mu^2 \sin^2\theta \leq 1$. If we also have $1 - 2\mu^2 \sin^2\theta < -1$, then one of the roots is

$$g_1 = 1 - 2\mu^2 \sin^2\theta - \sqrt{(1 - 2\mu^2 \sin^2\theta)^2 - 1} < -1$$

so $|g_1| > 1$ for some θ, such that the scheme is unstable.

To have a stable scheme, we require $1-2\mu^2\sin^2\theta \geq -1$, or $\mu^2\sin^2\theta \leq 1$, which can be guaranteed if $\mu^2 \leq 1$ or $\Delta t \leq h/|a|$. This is the CFL condition expected. Under this CFL constraint,

$$|g|^2 = \left(1 - 2\mu^2 \sin^2\theta\right)^2 + \left(1 - \left(1 - 2\mu^2 \sin^2\theta\right)^2\right) = 1$$

since the second part in the expression of g is imaginary, so the scheme is neutrally stable.

Recall a finite difference scheme for a second-order PDE (in time) $P_{\Delta t,h} v_j^k = 0$ is stable in a stability region Λ if there is an integer J such that for any positive time T there is a constant C_T independent of Δt and h, such that

$$\|\mathbf{v}^n\|_h \leq \sqrt{1+n^2}\, C_T \sum_{j=0}^{J} \|\mathbf{v}^j\|_h \tag{5.28}$$

for any n that satisfies $0 \leq n\Delta t \leq T$ with $(\Delta t, h) \in \Lambda$. The definition allows linear growth in time. Once again, a finite difference scheme converges if it is consistent and stable.

5.6.2 *Transforming the Second-order Wave Equation to a First-Order System*

Although we can solve the second-order wave equation directly, in this section, let us discuss how to change this equation into a first-order system. The first-order linear hyperbolic system of interest has the form

$$\mathbf{u}_t = (A\mathbf{u})_x = A\mathbf{u}_x\,,$$

which is a special case of 1D conservation laws

$$\mathbf{u}_t + (\mathbf{f}(\mathbf{u}))_x = 0.$$

To transfer the second-order wave equation to a first-order system, let us consider

$$\begin{cases} p = u_t \\ q = u_x\,, \end{cases} \qquad u_{tt} = p_t, \quad q_x = u_{xx}\,,$$

then we have

$$\begin{cases} p_t = u_{tt} = u_{xx} = q_x \\ q_t = u_{xt} = (u_t)_x = p_x \end{cases}$$

or in matrix-vector form

$$\begin{bmatrix} p \\ q \end{bmatrix}_t = \begin{bmatrix} 0 & 1 \\ 1 & 0 \end{bmatrix} \begin{bmatrix} p \\ q \end{bmatrix}_x , \tag{5.29}$$

and the eigenvalues of A are -1 and 1.

5.6.2.1 Initial and Boundary Conditions for the System

From the given boundary conditions for $u(x,t)$, we get

$$u(0,t) = g_1(t), \qquad u_t(0,t) = g_1'(t) = p(0,t),$$

$$u(1,t) = g_2(t), \qquad u_t(0,t) = g_1'(t) = p(1,t),$$

and there is no boundary condition for $q(x,t)$. The initial conditions are

$$p(x,0) = u_t(x,0) = u_1(x), \quad \text{known},$$

$$q(x,0) = u_x(x,0) = \frac{\partial}{\partial x} u(x,0) = u_0'(x), \quad \text{known}.$$

To solve the first-order system $\mathbf{u}_t = A\mathbf{u}_x$ numerically, we usually diagonalize the system (corresponding to characteristic directions) and then determine the boundary conditions, and apply an appropriate numerical method (*e.g.*, the upwind method). Thus we write $A = T^{-1}DT$, where $D = diag(\lambda_1, \lambda_2, \ldots, \lambda_n)$ is the matrix containing the eigenvalues of A on the diagonal and T is a nonsingular matrix. From the following

$$\mathbf{u}_t = A\mathbf{u}_x, \qquad T\mathbf{u}_t = TAT^{-1}T\mathbf{u}_x, \qquad (T\mathbf{u})_t = D\,(T\mathbf{u})_x,$$

and writing $\tilde{\mathbf{u}} = T\mathbf{u}$, we get the new first-order system

$$\tilde{\mathbf{u}}_t = D\tilde{\mathbf{u}}_x$$

or $(\tilde{u}_i)_t = \lambda_i(\tilde{u}_i)_x$, $i = 1, 2, \ldots, n$, a simple system of equations that we can solve one by one. We also know at which end we should have a boundary condition, depending on the sign of λ_i.

For the second-order wave equation, let us recall that the eigenvalues are 1 and -1. The unit eigenvector (such that $\|x\|_2 = 1$) corresponding to the eigenvalue 1, found by solving $Ax = x$, is $x = [1,\ 1]^T/\sqrt{2}$. Similarly, the unit eigenvector corresponding to the eigenvalue -1 is $x = [-1,\ 1]^T/\sqrt{2}$, so

$$T = \begin{bmatrix} \frac{1}{\sqrt{2}} & -\frac{1}{\sqrt{2}} \\ \frac{1}{\sqrt{2}} & \frac{1}{\sqrt{2}} \end{bmatrix}, \qquad T^{-1} = \begin{bmatrix} \frac{1}{\sqrt{2}} & \frac{1}{\sqrt{2}} \\ -\frac{1}{\sqrt{2}} & \frac{1}{\sqrt{2}} \end{bmatrix}.$$

The transformed result is thus

$$\begin{bmatrix} \frac{1}{\sqrt{2}} & -\frac{1}{\sqrt{2}} \\ \frac{1}{\sqrt{2}} & \frac{1}{\sqrt{2}} \end{bmatrix} \begin{bmatrix} p \\ q \end{bmatrix}_t = \begin{bmatrix} \frac{1}{\sqrt{2}} & -\frac{1}{\sqrt{2}} \\ \frac{1}{\sqrt{2}} & \frac{1}{\sqrt{2}} \end{bmatrix} \begin{bmatrix} 0 & 1 \\ 1 & 0 \end{bmatrix} \begin{bmatrix} \frac{1}{\sqrt{2}} & \frac{1}{\sqrt{2}} \\ -\frac{1}{\sqrt{2}} & \frac{1}{\sqrt{2}} \end{bmatrix} \begin{bmatrix} p \\ q \end{bmatrix}_x$$

$$= \begin{bmatrix} -1 & 0 \\ 0 & 1 \end{bmatrix} \begin{bmatrix} \frac{1}{\sqrt{2}} & -\frac{1}{\sqrt{2}} \\ \frac{1}{\sqrt{2}} & \frac{1}{\sqrt{2}} \end{bmatrix} \begin{bmatrix} p \\ q \end{bmatrix}_x,$$

or in equivalent component form

$$\left(\frac{1}{\sqrt{2}}p - \frac{1}{\sqrt{2}}q\right)_t = -\left(\frac{1}{\sqrt{2}}p - \frac{1}{\sqrt{2}}q\right)_x$$

$$\left(\frac{1}{\sqrt{2}}p + \frac{1}{\sqrt{2}}q\right)_t = \left(\frac{1}{\sqrt{2}}p + \frac{1}{\sqrt{2}}q\right)_x .$$

By setting

$$\begin{cases} y_1 = \frac{1}{\sqrt{2}}p - \frac{1}{\sqrt{2}}q, \\ y_2 = \frac{1}{\sqrt{2}}p + \frac{1}{\sqrt{2}}q, \end{cases}$$

we get

$$\begin{cases} \frac{\partial}{\partial t}y_1 = -\frac{\partial}{\partial x}y_1, \\ \frac{\partial}{\partial t}y_2 = \frac{\partial}{\partial x}y_2, \end{cases}$$

i.e., two separate one-way wave equations, for which we can use various numerical methods.

We already know the initial conditions, but need to determine a boundary condition for y_1 at $x = 0$ and a boundary condition for y_2 at $x = 1$. Note that

$$y_1(0,t) = \frac{1}{\sqrt{2}}p(0,t) - \frac{1}{\sqrt{2}}q(0,t),$$

$$y_2(0,t) = \frac{1}{\sqrt{2}}p(0,t) + \frac{1}{\sqrt{2}}q(0,t),$$

and $q(0,t)$ is unknown. We do know that, however,

$$y_1(0,t) + y_2(0,t) = \frac{2}{\sqrt{2}} p(0,t)\,,$$

and can use the following steps to determine the boundary condition at $x=0$:

1. update $(y_1)_0^{k+1}$ first, for which we do not need a boundary condition; and
2. use $(y_2)_0^{k+1} = \frac{2}{\sqrt{2}} p_0^{k+1} - (y_1)_0^{k+1}$ to get the boundary condition for y_2 at $x=0$.

Similar steps likewise can be applied to the boundary condition at $x=1$ as well.

5.7 Some Commonly Used FD Methods for Linear System of Hyperbolic PDEs

We now list some commonly used finite difference methods for solving a linear hyperbolic system of PDEs

$$\mathbf{u}_t + A\mathbf{u}_x = 0\,, \qquad \mathbf{u}(x,0) = \mathbf{u}_0(x)\,, \tag{5.30}$$

where A is a matrix with real eigenvalues.

- The Lax–Friedrichs scheme

$$\mathbf{U}_j^{k+1} = \frac{1}{2}\left(\mathbf{U}_{j+1}^k + \mathbf{U}_{j-1}^k\right) - \frac{\Delta t}{2h} A \left(\mathbf{U}_{j+1}^k - \mathbf{U}_{j-1}^k\right). \tag{5.31}$$

- The leap-frog scheme

$$\mathbf{U}_j^{k+1} = \mathbf{U}_j^{k-1} - \frac{\Delta t}{2h} A \left(\mathbf{U}_{j+1}^k - \mathbf{U}_{j-1}^k\right). \tag{5.32}$$

- The Lax–Wendroff scheme

$$\begin{aligned}\mathbf{U}_j^{k+1} &= \mathbf{U}_j^k - \frac{\Delta t}{2h} A \left(\mathbf{U}_{j+1}^k - \mathbf{U}_{j-1}^k\right) \\ &\quad + \frac{(\Delta t)^2}{2h^2} A^2 \left(\mathbf{U}_{j-1}^k - 2\mathbf{U}_j^k + \mathbf{U}_{j+1}^k\right).\end{aligned} \tag{5.33}$$

To determine correct boundary conditions, we usually need to find the diagonal form $A = T^{-1}DT$ and the new system $\tilde{\mathbf{u}}_t = D\tilde{\mathbf{u}}_x$ with $\tilde{\mathbf{u}} = T\mathbf{u}$.

5.8 Finite Difference Methods for Conservation Laws

The canonical form for the 1D conservation law is

$$\mathbf{u}_t + \mathbf{f}(u)_x = 0\,, \tag{5.34}$$

and one famous benchmark problem is Burgers' equation

$$u_t + \left(\frac{u^2}{2}\right)_x = 0\,, \tag{5.35}$$

in which $f(u) = u^2/2$. The term $\mathbf{f}(u)$ is often called the flux. This equation can be written in the nonconservative form

$$u_t + uu_x = 0\,, \tag{5.36}$$

and the solution likely develops shock(s) where the solution is discontinuous,[1] even if the initial condition is arbitrarily differentiable, *i.e.*, $u_0(x) = \sin x$.

We can use the upwind scheme to solve Burgers' equation. From the nonconservative form, we obtain

$$\frac{U_j^{k+1} - U_j^k}{\Delta t} + U_j^k \frac{U_j^k - U_{j-1}^k}{h} = 0\,, \quad \text{if } U_j^k \geq 0\,,$$

$$\frac{U_j^{k+1} - U_j^k}{\Delta t} + U_j^k \frac{U_{j+1}^k - U_j^k}{h} = 0\,, \quad \text{if } U_j^k < 0\,,$$

or from the conservative form

$$\frac{U_j^{k+1} - U_j^k}{\Delta t} + \frac{(U_j^k)^2 - (U_{j-1}^k)^2}{2h} = 0\,, \quad \text{if } U_j^k \geq 0\,,$$

$$\frac{U_j^{k+1} - U_j^k}{\Delta t} + \frac{(U_{j+1}^k)^2 - (U_j^k)^2}{2h} = 0\,, \quad \text{if } U_j^k < 0\,.$$

If the solution is smooth, both methods work well (first-order accurate). However, if shocks develop the conservative form gives much better results than that of the nonconservative form.

We can derive the Lax–Wendroff scheme using the modified equation of the nonconservative form. Since $u_t = -uu_x$,

$$\begin{aligned} u_{tt} &= -u_t u_x - uu_{tx} \\ &= uu_x^2 + u(uu_x)_x \\ &= uu_x^2 + u\left(u_x^2 + uu_{xx}\right) \\ &= 2uu_x^2 + u^2 u_{xx}\,, \end{aligned}$$

so the leading term of the modified equation for the first-order method is

$$u_t + uu_x = \frac{\Delta t}{2}\left(2uu_x^2 + u^2 u_{xx}\right)\,, \tag{5.37}$$

[1] There is no classical solution to the PDE when shocks develop because u_x is not well defined. We need to look for weak solutions.

and the nonconservative Lax–Wendroff scheme for Burgers' equation is

$$U_j^{k+1} = U_j^k - \Delta t U_j^k \frac{U_{j+1}^k - U_{j-1}^k}{2h}$$
$$= + \frac{(\Delta t)^2}{2} \left(2U_j^k \left(\frac{U_{j+1}^k - U_{j-1}^k}{2h} \right)^2 + (U_j^k)^2 \frac{U_{j-1}^k - 2U_j^k + U_{j+1}^k}{h^2} \right).$$

5.8.1 Conservative FD Methods for Conservation Laws

Consider the conservation law

$$\mathbf{u}_t + \mathbf{f}(u)_x = 0 ,$$

and let us seek a numerical scheme of the form

$$\mathbf{u}_j^{k+1} = \mathbf{u}_j^k - \frac{\Delta t}{h} \left(\mathbf{g}_{j+\frac{1}{2}}^k - \mathbf{g}_{j-\frac{1}{2}}^k \right), \tag{5.38}$$

where

$$\mathbf{g}_{j+\frac{1}{2}} = \mathbf{g} \left(\mathbf{u}_{j-p+1}^k, \mathbf{u}_{j-p+2}^k, \dots, \mathbf{u}_{j+q+1}^k \right)$$

is called the numerical flux, satisfying

$$g(u, u, \dots, u) = f(u) . \tag{5.39}$$

Such a scheme is called conservative. For example, we have $g(u) = u^2/2$ for Burgers' equation.

We can derive general criteria which g should satisfy, as follows.

1. Integrate the equation with respect to x from $x_{j-\frac{1}{2}}$ to $x_{j+\frac{1}{2}}$, to get

$$\int_{x_{j-\frac{1}{2}}}^{x_{j+\frac{1}{2}}} u_t dx = - \int_{x_{j-\frac{1}{2}}}^{x_{j+\frac{1}{2}}} f(u)_x dx$$
$$= - \left(f(u(x_{j+\frac{1}{2}}, t)) - f(u(x_{j-\frac{1}{2}}, t)) \right) .$$

2. Integrate the equation above with respect to t from t^k to t^{k+1}, to get

$$\int_{t^k}^{t^{k+1}} \int_{x_{j-\frac{1}{2}}}^{x_{j+\frac{1}{2}}} u_t \, dx \, dt = - \int_{t^k}^{t^{k+1}} \left(f(u(x_{j+\frac{1}{2}}, t)) - f(u(x_{j-\frac{1}{2}}, t)) \right) dt ,$$

$$\int_{x_{j-\frac{1}{2}}}^{x_{j+\frac{1}{2}}} \left(u(x, t^{k+1}) - u(x, t^k) \right) dx = - \int_{t^k}^{t^{k+1}} \left(f(u(x_{j+\frac{1}{2}}, t)) - f(u(x_{j-\frac{1}{2}}, t)) \right) dt .$$

Define the average of $u(x,t)$ as

$$\bar{u}_j^k = \frac{1}{h}\int_{x_{j-\frac{1}{2}}}^{x_{j+\frac{1}{2}}} u(x,t^k)dx\,, \tag{5.40}$$

which is the cell average of $u(x,t)$ over the cell $(x_{j-\frac{1}{2}}, x_{j+\frac{1}{2}})$ at the time level k. The expression that we derived earlier can therefore be rewritten as

$$\begin{aligned}
\bar{u}_j^{k+1} &= \bar{u}_j^k - \frac{1}{h}\left(\int_{t^k}^{t^{k+1}} f(u(x_{j+\frac{1}{2}},t))dt - \int_{t^k}^{t^{k+1}} f(u(x_{j-\frac{1}{2}},t))dt\right)\\
&= \bar{u}_j^k - \frac{\Delta t}{h}\left(\frac{1}{\Delta t}\int_{t^k}^{t^{k+1}} f(u(x_{j+\frac{1}{2}},t))dt - \frac{1}{\Delta t}\int_{t^k}^{t^{k+1}} f(u(x_{j-\frac{1}{2}},t))dt\right)\\
&= \bar{u}_j^k - \frac{\Delta t}{h}\left(g_{j+\frac{1}{2}} - g_{j+\frac{1}{2}}\right),
\end{aligned}$$

where

$$g_{j+\frac{1}{2}} = \frac{1}{\Delta t}\int_{t^k}^{t^{k+1}} f(u(x_{j+\frac{1}{2}},t))dt\,.$$

Different conservative schemes can be obtained, if different approximations are used to evaluate the integral.

5.8.2 Some Commonly Used Numerical Scheme for Conservation Laws

Some commonly used schemes are:

- Lax–Friedrichs scheme

$$U_j^{k+1} = \frac{1}{2}\left(U_{j+1}^k + U_{j-1}^k\right) - \frac{\Delta t}{2h}\left(f(U_{j+1}^k) - f(U_{j-1}^k)\right); \tag{5.41}$$

- Lax–Wendroff scheme

$$\begin{aligned}
U_j^{k+1} &= U_j^k - \frac{\Delta t}{2h}\left(f(U_{j+1}^k) - f(U_{j-1}^k)\right)\\
&+ \frac{(\Delta t)^2}{2h^2}\left\{A_{j+\frac{1}{2}}\left(f(U_{j+1}^k) - f(U_j^k)\right) - A_{j-\frac{1}{2}}\left(f(U_j^k) - f(U_{j-1}^k)\right)\right\},
\end{aligned} \tag{5.42}$$

where $A_{j+\frac{1}{2}} = Df(u(x_{j+\frac{1}{2}},t))$ is the Jacobian matrix of $f(u)$ at $u(x_{j+\frac{1}{2}},t)$.

A modified version

$$\begin{cases} U_{j+\frac{1}{2}}^{k+\frac{1}{2}} = \frac{1}{2}\left(U_j^k + U_{j+1}^k\right) - \frac{\Delta t}{2h}\left(f(U_{j+1}^k) - f(U_j^k)\right) \\ U_j^{k+1} = U_j^k - \frac{\Delta t}{h}\left(f(U_{j+\frac{1}{2}}^{k+\frac{1}{2}}) - f(U_{j-\frac{1}{2}}^{k+\frac{1}{2}})\right), \end{cases} \tag{5.43}$$

called the Lax–Wendroff–Richtmyer scheme, does not need the Jacobian matrix.

Exercises

1. Show that the following scheme is consistent with the PDE $u_t + cu_{tx} + au_x = f$:

$$\frac{U_i^{n+1} - U_i^n}{k} + c\frac{U_{i+1}^{n+1} - U_{i-1}^{n+1} - U_{i+1}^n + U_{i-1}^n}{2kh} + a\frac{U_{i+1}^n - U_{i-1}^n}{2h} = f_i^n.$$

   Discuss also the stability, as far as you can.

2. Implement and test the upwind and the Lax–Wendroff schemes for the one-way wave equation

$$u_t + u_x = 0.$$

   Assume the domain is $-1 \le x \le 1$, and $t_{final} = 1$. Test your code for the following parameters:

   (a) $u(t,-1) = 0$, and $u(0,x) = (x+1)e^{-x/2}$.

   (b) $u(t,-1) = 0$, and $u(0,x) = \begin{cases} 0 & \text{if } x < -1/2, \\ 1 & \text{if } -1 \le x \le 1/2, \\ 0 & \text{if } x > 1/2. \end{cases}$

   Do the grid refinement analysis at $t_{final} = 1$ for case (a) where the exact solution is available, take $m = 10, 20, 40$, and 80. For problem (b), use $m = 40$. Plot the solution at $t_{final} = 1$ for both cases.

3. Use the upwind and Lax–Wendroff schemes for Burgers' equation

$$u_t + \left(\frac{u^2}{2}\right)_x = 0$$

   with the same initial and boundary conditions as in problem 2.

4. Solve the following wave equation numerically using a finite difference method.

$$\frac{\partial^2 u}{\partial t^2} = c^2\frac{\partial^2 u}{\partial x^2} + f(x,t), \qquad 0 < x < 1,$$

$$u(0,t) = u(1,t) = 0, \qquad u(x,0) = \begin{cases} x/4 & \text{if } 0 \le x < 1/2, \\ (1-x)/4 & \text{if } 1/2 \le x \le 1. \end{cases}$$

   (a) Test your code and convergence order using a problem that has the exact solution.

   (b) Test your code again by setting $f(x,t) = 0$.

   (c) Modify and validate your code to the PDE with a damping term

$$\frac{\partial^2 u}{\partial t^2} - \beta\frac{\partial u}{\partial t} = c^2\frac{\partial^2 u}{\partial x^2} + f(x,t).$$

(d) Modify and validate your code to the PDE with an advection term

$$\frac{\partial^2 u}{\partial t^2} = c^2 \frac{\partial^2 u}{\partial x^2} + \beta \frac{\partial u}{\partial x} + f(x,t),$$

where c and β are positive constants.

5. Download the Clawpack for conservation laws from the Internet. Run a test problem in 2D. Write 2 to 3 pages to detail the package and the numerical results.

Part II

Finite Element Methods

6

Finite Element Methods for 1D Boundary Value Problems

The finite element (FE) method was developed to solve complicated problems in engineering, notably in elasticity and structural mechanics modeling involving elliptic PDEs and complicated geometries. But nowadays the range of applications is quite extensive. We will use the following 1D and 2D model problems to introduce the finite element method:

$$\text{1D:} \quad -u''(x) = f(x), \quad 0 < x < 1, \quad u(0) = 0, \quad u(1) = 0;$$

$$\text{2D:} \quad -\left(u_{xx} + u_{yy}\right) = f(x, y), \quad (x, y) \in \Omega, \quad u(x, y)\Big|_{\partial\Omega} = 0,$$

where Ω is a bounded domain in (x, y) plane with the boundary $\partial\Omega$.

6.1 The Galerkin FE Method for the 1D Model

We illustrate the finite element method for the 1D two-point BVP

$$-u''(x) = f(x), \quad 0 < x < 1, \quad u(0) = 0, \quad u(1) = 0,$$

using the Galerkin finite element method described in the following steps.

1. *Construct a variational or weak formulation*, by multiplying both sides of the differential equation by a test function $v(x)$ satisfying the boundary conditions (BC) $v(0) = 0$, $v(1) = 0$ to get

$$-u''v = fv,$$

and then integrating from 0 to 1 (using integration by parts) to have the following,

$$\begin{aligned} \int_0^1 (-u''v)\,dx &= -u'v\Big|_0^1 + \int_0^1 u'v'dx \\ &= \int_0^1 u'v'dx \\ \Longrightarrow \quad \int_0^1 u'v'dx &= \int_0^1 fv\,dx\,, \text{ the weak form.} \end{aligned}$$

2. *Generate a mesh, e.g.*, a uniform Cartesian mesh $x_i = ih$, $i = 0, 1, \ldots, n$, where $h = 1/n$, defining the intervals (x_{i-1}, x_i), $i = 1, 2, \ldots, n$.
3. *Construct a set of basis functions* based on the mesh, such as the piecewise linear functions ($i = 1, 2, \ldots, n-1$)

$$\phi_i(x) = \begin{cases} \dfrac{x - x_{i-1}}{h} & \text{if } x_{i-1} \le x < x_i, \\ \dfrac{x_{i+1} - x}{h} & \text{if } x_i \le x < x_{i+1}, \\ 0 & \text{otherwise}\,, \end{cases}$$

x_{i-1} x_i x_{i+1}

often called the hat functions, see the right diagram for a hat function.
4. *Represent the approximate (FE) solution by a linear combination of the basis functions*

$$u_h(x) = \sum_{j=1}^{n-1} c_j \phi_j(x)\,,$$

where the coefficients c_j are the unknowns to be determined. On assuming the hat basis functions, obviously $u_h(x)$ is also a piecewise linear function, although this is not usually the case for the true solution $u(x)$. Other basis functions are considered later. We then derive a linear system of equations for the coefficients by substituting the approximate solution $u_h(x)$ for the exact solution $u(x)$ in the weak form $\int_0^1 u'v'dx = \int_0^1 fv\,dx$, *i.e.*,

$$\int_0^1 u_h'v'dx = \int_0^1 fvdx\,, \text{ (noting that errors are introduced!)}$$

$$\begin{aligned} \Longrightarrow \quad \int_0^1 \sum_{j=1}^{n-1} c_j\phi_j'v'dx &= \sum_{j=1}^{n-1} c_j \int_0^1 \phi_j'v'dx \\ &= \int_0^1 fv\,dx\,. \end{aligned}$$

Next, choose the test function $v(x)$ as ϕ_1, ϕ_2, ..., ϕ_{n-1} successively, to get the system of linear equations (noting that further errors are introduced):

$$\left(\int_0^1 \phi_1'\phi_1' dx\right) c_1 + \cdots + \left(\int_0^1 \phi_1'\phi_{n-1}' dx\right) c_{n-1} = \int_0^1 f\phi_1 dx$$

$$\left(\int_0^1 \phi_2'\phi_1' dx\right) c_1 + \cdots + \left(\int_0^1 \phi_2'\phi_{n-1}' dx\right) c_{n-1} = \int_0^1 f\phi_2 dx$$

$$\cdots \quad \cdots \quad \cdots \quad \cdots \quad \cdots \quad \cdots \quad \cdots \quad \cdots$$

$$\left(\int_0^1 \phi_i'\phi_1' dx\right) c_1 + \cdots + \left(\int_0^1 \phi_i'\phi_{n-1}' dx\right) c_{n-1} = \int_0^1 f\phi_i dx$$

$$\cdots \quad \cdots \quad \cdots \quad \cdots \quad \cdots \quad \cdots \quad \cdots \quad \cdots$$

$$\left(\int_0^1 \phi_{n-1}'\phi_1' dx\right) c_1 + \cdots + \left(\int_0^1 \phi_{n-1}'\phi_{n-1}' dx\right) c_{n-1} = \int_0^1 f\phi_{n-1} dx,$$

or in the matrix-vector form:

$$\begin{bmatrix} a(\phi_1,\phi_1) & a(\phi_1,\phi_2) & \cdots & a(\phi_1,\phi_{n-1}) \\ a(\phi_2,\phi_1) & a(\phi_2,\phi_2) & \cdots & a(\phi_2,\phi_{n-1}) \\ \vdots & \vdots & \vdots & \vdots \\ a(\phi_{n-1},\phi_1) & a(\phi_{n-1},\phi_2) & \cdots & a(\phi_{n-1},\phi_{n-1}) \end{bmatrix} \begin{bmatrix} c_1 \\ c_2 \\ \vdots \\ c_{n-1} \end{bmatrix} = \begin{bmatrix} (f,\phi_1) \\ (f,\phi_2) \\ \vdots \\ (f,\phi_{n-1}) \end{bmatrix},$$

where

$$a(\phi_i,\phi_j) = \int_0^1 \phi_i'\phi_j' dx, \qquad (f,\phi_i) = \int_0^1 f\phi_i dx.$$

The term $a(u, v)$ is called a bilinear form since it is linear with each variable (function), and (f, v) is called a linear form. If ϕ_i are the hat functions, then in particular we get

$$\begin{bmatrix} \frac{2}{h} & -\frac{1}{h} & & & & \\ -\frac{1}{h} & \frac{2}{h} & -\frac{1}{h} & & & \\ & -\frac{1}{h} & \frac{2}{h} & -\frac{1}{h} & & \\ & & \ddots & \ddots & \ddots & \\ & & & -\frac{1}{h} & \frac{2}{h} & -\frac{1}{h} \\ & & & & -\frac{1}{h} & \frac{2}{h} \end{bmatrix} \begin{bmatrix} c_1 \\ c_2 \\ c_3 \\ \vdots \\ c_{n-2} \\ c_{n-1} \end{bmatrix} = \begin{bmatrix} \int_0^1 f\phi_1 dx \\ \int_0^1 f\phi_2 dx \\ \int_0^1 f\phi_3 dx \\ \vdots \\ \int_0^1 f\phi_{n-2} dx \\ \int_0^1 f\phi_{n-1} dx \end{bmatrix}.$$

5. *Solve the linear system of equations* for the coefficients and hence obtain the approximate solution $u_h(x) = \sum_i c_i\phi_i(x)$.
6. *Carry out the error analysis* (*a priori* or *a posteriori* error analysis).

Questions are often raised about how to appropriately

- represent ODE or PDE problems in a weak form;
- choose the basis functions ϕ, *e.g.*, in view of ODE/PDE, mesh, and the boundary conditions, *etc.*;
- implement the finite element method;
- solve the linear system of equations; and
- carry out the error analysis,

which will be addressed in subsequent chapters.

6.2 Different Mathematical Formulations for the 1D Model

Let us consider the 1D model again,

$$\begin{aligned} -u''(x) &= f(x), \quad 0 < x < 1, \\ u(0) &= 0, \quad u(1) = 0. \end{aligned} \tag{6.1}$$

There are at least three different formulations to consider for this problem:

1. the (D)-form, the original differential equation;
2. the (V)-form, the variational form or weak form

$$\int_0^1 u'v'dx = \int_0^1 fv\,dx \tag{6.2}$$

for any test function $v \in H_0^1(0,1)$, the Sobolev space for functions in integral forms like the C^1 space for functions (see later), and as indicated above, the

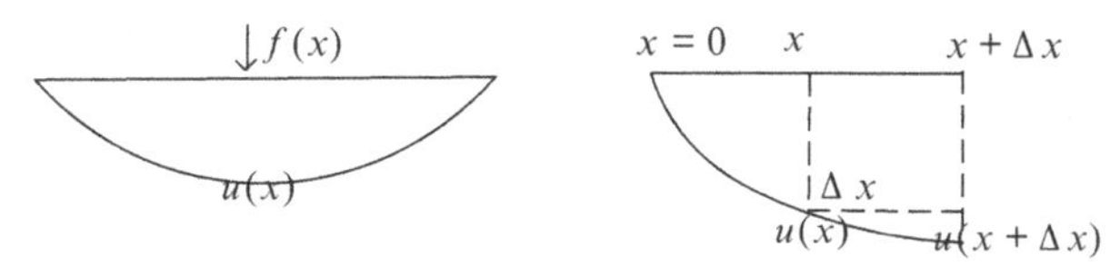

Figure 6.1. A diagram of elastic string with two ends fixed: the displacement and force.

corresponding finite element method is often called the Galerkin method; and

3. the (M)-form, the minimization form

$$\min_{v(x)\in H_0^1(0,1)} \left\{ \int_0^1 \left(\frac{1}{2}(v')^2 - fv \right) dx \right\}, \tag{6.3}$$

when the corresponding finite element method is often called the Ritz method.

As discussed in subsequent subsections, under certain assumptions these three different formulations are equivalent.

6.2.1 A Physical Example

From the viewpoint of mathematical modeling, both the variational (or weak) form and the minimization form are more natural than the differential formulation. For example, suppose we seek the equilibrium position of an elastic string of unit length, with two ends fixed and subject to an external force.

The equilibrium is the state that minimizes the total energy. Let $u(x)$ be the displacement of the string at a point x, and consider the deformation of an element of the string in the interval $(x, x + \Delta x)$ (see Figure 6.1 for an illustration). The potential energy of the deformed element is

$$\begin{aligned}
&\tau \cdot \text{increase in the element length} \\
&= \tau \left(\sqrt{(u(x+\Delta x) - u(x))^2 + (\Delta x)^2} - \Delta x \right) \\
&= \tau \left(\sqrt{\left(u(x) + u_x(x)\Delta x + \frac{1}{2}u_{xx}(x)(\Delta x)^2 + \cdots - u(x) \right)^2 + (\Delta x)^2} - \Delta x \right) \\
&\simeq \tau \left(\sqrt{[1 + u_x^2(x)]\,(\Delta x)^2} - \Delta x \right) \\
&\simeq \frac{1}{2}\tau u_x^2(x)\Delta x,
\end{aligned}$$

where τ is the coefficient of the elastic tension that we assume to be constant. If the external force is denoted by $f(x)$, the work done by this force is $-f(x)u(x)$ at every point x. Consequently, the total energy of the string (over $0 < x < 1$) is

$$F(u) = \int_0^1 \frac{1}{2}\tau u_x^2(x)\,dx - \int_0^1 f(x)u(x)\,dx\,,$$

from the work–energy principle: the change in the kinetic energy of an object is equal to the net work done on the object. Thus to minimize the total energy, we seek the extremal u^* such that

$$F(u^*) \le F(u)$$

for all admissible $u(x)$, *i.e.*, the "minimizer" u^* of the functional $F(u)$ (a function of functions).

Using the principle of virtual work, we also have

$$\int_0^1 u'v'dx = \int_0^1 fvdx$$

for any admissible function $v(x)$.

On the other hand, the force balance yields the relevant differential equation. The external force $f(x)$ is balanced by the tension of the elastic string given by Hooke's law, see Figure 6.1 for an illustration, such that

$$\tau\,(u_x(x+\Delta x) - u_x(x)) \simeq -f(x)\Delta x$$

$$\text{or} \qquad \tau\frac{u_x(x+\Delta x) - u_x(x)}{\Delta x} \simeq -f(x),$$

thus, for $\Delta x \to 0$ we get the PDE

$$-\tau u_{xx} = f(x),$$

along with the boundary condition $u(0) = 0$ and $u(1) = 0$ since the string is fixed at the two ends.

The three formulations are equivalent representations of the same problem. We show the mathematical equivalence in the next subsection.

6.2.2 Mathematical Equivalence

At the beginning of this chapter, we proved that (*D*) is equivalent to (*V*) using integration by parts. Let us now prove that under certain conditions (*V*) is equivalent to (*D*), and that (*V*) is equivalent to (*M*), and that (*M*) is equivalent (*V*).

Theorem 6.1. *(V) → (D). If u_{xx} exists and is continuous, then*

$$\int_0^1 u'v'dx = \int_0^1 fvdx, \quad \forall\, v(0) = v(1) = 0, \quad v \in H^1(0,1),$$

implies that $-u_{xx} = f(x)$.

Recall that $H^1(0,1)$ denotes a Sobolev space, which here we can regard as the space of all functions that have a first-order derivative.
Proof From integration by parts, we have

$$\int_0^1 u'v'dx = u'v\Big|_0^1 - \int_0^1 u''v\,dx,$$

$$\Longrightarrow \quad -\int_0^1 u''vdx = \int_0^1 fv\,dx,$$

$$\text{or} \quad \int_0^1 (u''+f)\,v\,dx = 0\,.$$

Since $v(x)$ is arbitrary and continuous, and u'' and f are continuous, we must have

$$u''+f=0, \qquad i.e., \quad -u''=f.$$

Theorem 6.2. *(V) → (M). Suppose $u^*(x)$ satisfies*

$$\int_0^1 u^{*\prime}v'dx = \int_0^1 vfdx$$

for any $v(x) \in H^1(0,1)$, and $v(0) = v(1) = 0$. Then

$$F(u^*) \le F(u) \quad or$$

$$\frac{1}{2}\int_0^1 (u^*)_x^2 dx - \int_0^1 fu^*dx \le \frac{1}{2}\int_0^1 u_x^2 dx - \int_0^1 fudx.$$

Proof

$$
\begin{aligned}
F(u) &= F(u^* + u - u^*) = F(u^* + w) \quad (\text{where } w = u - u^*, \quad w(0) = w(1) = 0), \\
&= \int_0^1 \left(\frac{1}{2}(u^* + w)_x^2 - (u^* + w)f\right) dx \\
&= \int_0^1 \left(\frac{(u^*)_x^2 + w_x^2 + 2(u^*)_x w_x}{2} - u^* f - wf\right) dx \\
&= \int_0^1 \left(\frac{1}{2}(u^*)_x^2 - u^* f\right) dx + \int_0^1 \frac{1}{2} w_x^2 dx + \int_0^1 \left((u^*)_x w_x - fw\right) dx \\
&= \int_0^1 \left(\frac{1}{2}(u^*)_x^2 - u^* f\right) dx + \int_0^1 \frac{1}{2} w_x^2 dx + 0 \\
&= F(u^*) + \int_0^1 \frac{1}{2} w_x^2 dx \\
&\geq F(u^*).
\end{aligned}
$$

The proof is completed.

Theorem 6.3. *(M) → (V). If $u^*(x)$ is the minimizer of $F(u^*)$, then*

$$\int_0^1 (u^*)_x v_x dx = \int_0^1 fv dx$$

for any $v(0) = v(1) = 0$ and $v \in H^1(0, 1)$.

Proof Consider the auxiliary function:

$$g(\epsilon) = F(u^* + \epsilon v)\,.$$

Since $F(u^*) \leq F(u^* + \epsilon v)$ for any ϵ, $g(0)$ is a global or local minimum such that $g'(0) = 0$. To obtain the derivative of $g(\epsilon)$, we have

$$
\begin{aligned}
g(\epsilon) &= \int_0^1 \left\{\frac{1}{2}(u^* + \epsilon v)_x^2 - (u^* + \epsilon v)f\right\} dx \\
&= \int_0^1 \left\{\frac{1}{2}\left((u^*)_x^2 + 2(u^*)_x v_x \epsilon + v_x^2 \epsilon^2\right) - u^* f - \epsilon v f\right\} dx \\
&= \int_0^1 \left(\frac{1}{2}(u^*)_x^2 - u^* f\right) dx + \epsilon \int_0^1 \left((u^*)_x v_x - fv\right) dx + \frac{\epsilon^2}{2} \int_0^1 v_x^2 dx\,.
\end{aligned}
$$

Thus we have

$$g'(\epsilon) = \int_0^1 \left((u^*)_x v_x - fv\right) dx + \epsilon \int_0^1 v_x^2 dx$$

and

$$g'(0) = \int_0^1 ((u^*)_x v_x - fv)\, dx = 0$$

since $v(x)$ is arbitrary, *i.e.*, the weak form is satisfied.

However, the three different formulations may not be equivalent for some problems, depending on the regularity of the solutions. Thus, although

$$(D) \Longrightarrow (M) \Longrightarrow (V),$$

in order for (V) to imply (M), the differential equation is usually required to be self-adjoint, and for (M) or (V) to imply (D); the solution of the differential equation must have continuous second-order derivatives.

6.3 Key Components of the FE Method for the 1D Model

In this section, we discuss the model problem (6.1) using the following methods:

- Galerkin method for the variational or weak formulation;
- Ritz method for the minimization formulation.

We also discuss another important aspect of finite element methods, namely, how to assemble the stiffness matrix using the element by element approach.

The first step is to choose an integral form, usually the weak form, say $\int_0^1 u'v'\,dx = \int_0^1 fv\,dx$ for any $v(x)$ in the Sobolev space $H^1(0,1)$ with $v(0) = v(1) = 0$.

6.3.1 Mesh and Basis Functions

For a 1D problem, a mesh is a set of points in the interval of interest, say, $x_0 = 0$, $x_1, x_2, \ldots, x_M = 1$ (see Figure 6.2 for an illustration). Let $h_i = x_{i+1} - x_i$, $i = 0, 1, \ldots, M-1$, then

- x_i is called a *node*, or nodal point.
- (x_i, x_{i+1}) is called an element.
- $h = \max\limits_{0 \le i \le M-1} \{h_i\}$ is the mesh size, a measure of how fine the partition is.

6.3.1.1 Define a Finite Dimensional Space on *the Mesh*

Let the solution be in the space V, which is $H_0^1(0,1)$ in the model problem. Based on the mesh, we wish to construct a subspace

$$V_h \,(\text{a finite dimensional space}) \subset V \,(\text{the solution space}),$$

such that the discrete problem is contained in the continuous problem.

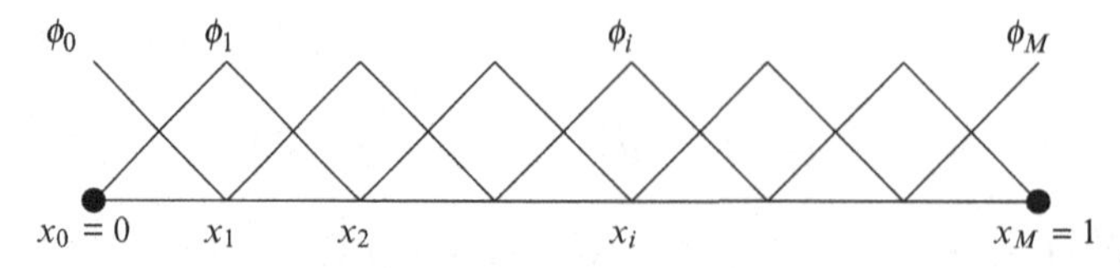

Figure 6.2. Diagram of a mesh and hat basis functions.

Any such finite element method is called a conforming one. Different *finite* dimensional spaces generate different finite element solutions. Since V_h has finite dimension, we can find a set of basis functions

$$\phi_1, \phi_2, \ldots, \phi_{M-1} \subset V_h$$

that are *linearly independent*, *i.e.*, if

$$\sum_{j=1}^{M-1} \alpha_j \phi_j = 0 ,$$

then $\alpha_1 = \alpha_0 = \cdots = \alpha_{M-1} = 0$. Thus V_h is the space *spanned* by the basis functions:

$$V_h = \left\{ v_h(x), \; v_h(x) = \sum_{j=1}^{M-1} \alpha_j \phi_j \right\}.$$

The simplest finite dimensional space is the *piecewise continuous linear* function space defined over the mesh:

$$V_h = \left\{ v_h(x), \; v_h(x) \text{ is continuous piecewise linear}, \; v_h(0) = v_h(1) = 0 \right\}.$$

It is easy to show that V_h has a finite dimension, even though there are an infinite number of elements in V_h.

6.3.1.2 *Find the Dimension of V_h*

A linear function $l(x)$ in an interval (x_i, x_{i+1}) is uniquely determined by its values at x_i and x_{i+1}:

$$l(x) = l(x_i) \frac{x - x_{i+1}}{x_i - x_{i+1}} + l(x_{i+1}) \frac{x - x_i}{x_{i+1} - x_i}.$$

There are $M - 1$ nodal values $l(x_i)$'s, $l(x_1)$, $l(x_2)$, ..., $l(x_{M-1})$ for a piecewise continuous linear function over the mesh, in addition to $l(x_0) = l(x_M) = 0$. Given a vector $[l(x_1), l(x_2), \ldots, l(x_{M-1})]^T \in R^{M-1}$, we can construct a $v_h(x) \in V_h$ by taking $v_h(x_i) = l(x_i)$, $i = 1, \ldots, M - 1$. On the other hand, given $v_h(x) \in$

V_h, we get a vector $[v(x_1), v(x_2), \ldots, v(x_{M-1})]^T \in R^{M-1}$. Thus there is a *one to one* relation between V_h and R^{M-1}, so V_h has the finite dimension $M-1$. Consequently, V_h is considered to be equivalent to R^{M-1}.

6.3.1.3 Find a Set of Basis Functions

The finite dimensional space can be spanned by a set of basis functions. There are infinitely many sets of basis functions, but we should choose one that:

- is simple;
- has compact (minimum) support, *i.e.*, zero almost everywhere except for a small region; and
- meets the regularity requirement, *i.e.*, continuous and differentiable, except at nodal points.

The simplest is the set of hat functions

$$
\begin{aligned}
&\phi_1(x_1) = 1\,, \quad \phi_1(x_j) = 0\,, \quad j = 0, 2, 3, \ldots, M,\\
&\phi_2(x_2) = 1\,, \quad \phi_2(x_j) = 0\,, \quad j = 0, 1, 3, \ldots, M,\\
&\cdots\\
&\phi_i(x_i) = 1\,, \quad \phi_i(x_j) = 0\,, \quad j = 0, 1, \ldots i-1,\ i+1, \ldots, M,\\
&\cdots\\
&\phi_{M-1}(x_{M-1}) = 1\,, \quad \phi_{M-1}(x_j) = 0\,, \quad j = 0, 1, \ldots, M-2,\ M.
\end{aligned}
$$

They can be represented simply as $\phi_i(x_j) = \delta_i^j$, *i.e.*,

$$
\phi_i(x_j) = \begin{cases} 1\,, & \text{if } i = j\,,\\ 0\,, & \text{otherwise}\,. \end{cases} \tag{6.4}
$$

The analytic form of the hat functions for $i = 1, 2, \ldots, m-1$ is

$$
\phi_i(x) = \begin{cases} 0\,, & \text{if } x < x_{i-1}\,,\\ \dfrac{x - x_{i-1}}{h_i}\,, & \text{if } x_{i-1} \le x < x_i\,,\\ \dfrac{x_{i+1} - x}{h_{i+1}}\,, & \text{if } x_i \le x < x_{i+1}\,,\\ 0\,, & \text{if } x_{i+1} \le x\,; \end{cases} \tag{6.5}
$$

and the finite element solution sought is

$$
u_h(x) = \sum_{j=1}^{M-1} \alpha_j \phi_j(x)\,, \tag{6.6}
$$

and either the minimization form (M) or the variational or weak form (V) can be used to derive a linear system of equations for the coefficients α_j. On using the hat functions, we have

$$u_h(x_i) = \sum_{j=1}^{M-1} \alpha_j \phi_j(x_i) = \alpha_i \phi_i(x_i) = \alpha_i \,, \tag{6.7}$$

so α_i is an approximate solution to the exact solution at $x = x_i$.

6.3.2 The Ritz Method

Although not every problem has a minimization form, the Ritz method was one of the earliest and has proven to be one of the most successful.

For the model problem (6.1), the minimization form is

$$\min_{v \in H_0^1(0,1)} F(v): \quad F(v) = \frac{1}{2} \int_0^1 (v_x)^2 dx - \int_0^1 fv \, dx \,. \tag{6.8}$$

As before, we look for an approximate solution of the form $u_h(x) = \sum_{j=1}^{M-1} \alpha_j \phi_j(x)$. Substituting this into the functional form gives

$$F(u_h) = \frac{1}{2} \int_0^1 \left(\sum_{j=1}^{M-1} \alpha_j \phi_j'(x) \right)^2 - \int_0^1 f \sum_{j=1}^{M-1} \alpha_j \phi_j(x) dx \,, \tag{6.9}$$

which is a multivariate function of $\alpha_1, \alpha_2, \ldots, \alpha_{M-1}$ and can be written as

$$F(u_h) = F(\alpha_1, \alpha_2, \ldots, \alpha_{M-1}) \,.$$

The necessary condition for a global minimum (also a local minimum) is

$$\frac{\partial F}{\partial \alpha_1} = 0 \,, \quad \frac{\partial F}{\partial \alpha_2} = 0 \,, \quad \cdots \,, \quad \frac{\partial F}{\partial \alpha_i} = 0 \,, \quad \cdots \,, \quad \frac{\partial F}{\partial \alpha_{M-1}} = 0 \,.$$

Thus taking the partial derivatives with respect to α_j we have

$$\frac{\partial F}{\partial \alpha_1} = \int_0^1 \left(\sum_{j=1}^{M-1} \alpha_j \phi_j' \right) \phi_1' dx - \int_0^1 f \phi_1 dx = 0$$

$$\cdots$$

$$\frac{\partial F}{\partial \alpha_i} = \int_0^1 \left(\sum_{j=1}^{M-1} \alpha_j \phi_j' \right) \phi_i' dx - \int_0^1 f \phi_i dx = 0, \quad i = 1, 2, \ldots, M-1 \,,$$

and on exchanging the order of integration and summation:

$$\sum_{j=1}^{M-1} \left(\int_0^1 \phi_j' \phi_i' dx \right) \alpha_j = \int_0^1 f \phi_i dx, \quad i = 1, 2, \ldots, M-1.$$

This is the same system of equations that follow from the Galerkin method with the weak form, *i.e.*,

$$\int_0^1 u'v'dx = \int_0^1 fv\,dx \quad \text{immediately gives}$$

$$\int_0^1 \left(\sum_{j=1}^{M-1} \alpha_j \phi_j' \right) \phi_i' dx = \int_0^1 f \phi_i \, dx, \quad i = 1, 2, \ldots, M-1 \,.$$

6.3.2.1 Comparison of the Ritz and the Galerkin FE Methods

For many problems, the Ritz and Galerkin methods are theoretically equivalent.

- The Ritz method is based on the minimization form, and optimization techniques can be used to solve the problem.
- The Galerkin method usually has weaker requirements than the Ritz method. Not every problem has a minimization form, whereas almost all problems have some kind of weak form. How to choose a suitable weak form and the convergence of different methods are all important issues for finite element methods.

6.3.3 Assembling the Stiffness Matrix Element by Element

Given a problem, say the model problem, after we have derived the minimization or weak form and constructed a mesh and a set of basis functions we need to form:

- the coefficient matrix $A = \{a_{ij}\} = \{\int_0^1 \phi_i' \phi_j' dx\}$, often called the *stiffness matrix for the first-order derivatives*, and
- the right-hand side vector $F = \{f_i\} = \{\int_0^1 f_i \phi_i dx\}$, often called the *load vector*.

The procedure to form A and F is a crucial part in the finite element method. For the model problem, one way is by *assembling element by element*:

$$\begin{array}{ccccccc} (x_0, x_1), & (x_1, x_2), & \cdots & (x_{i-1}, x_i) & \cdots & (x_{M-1}, x_M), \\ \Omega_1, & \Omega_2, & \cdots & \Omega_i, & \cdots & \Omega_M. \end{array}$$

The idea is to break up the integration element by element, so that for any integrable function $g(x)$ we have

$$\int_0^1 g(x)\,dx = \sum_{k=1}^{M} \int_{x_{k-1}}^{x_k} g(x)\,dx = \sum_{k=1}^{M} \int_{\Omega_k} g(x)\,dx\,.$$

The stiffness matrix can then be written

$$A = \begin{bmatrix} \int_0^1 (\phi_1')^2 dx & \int_0^1 \phi_1'\phi_2' dx & \cdots & \int_0^1 \phi_1'\phi_{M-1}' dx \\ \int_0^1 \phi_2'\phi_1' dx & \int_0^1 (\phi_2')^2 dx & \cdots & \int_0^1 \phi_2'\phi_{M-1}' dx \\ \vdots & \vdots & \vdots & \vdots \\ \int_0^1 \phi_{M-1}'\phi_1' dx & \int_0^1 \phi_{M-1}'\phi_2' dx & \cdots & \int_0^1 (\phi_{M-1}')^2 dx \end{bmatrix}$$

$$= \begin{bmatrix} \int_{x_0}^{x_1} (\phi_1')^2 dx & \int_{x_0}^{x_1} \phi_1'\phi_2' dx & \cdots & \int_{x_0}^{x_1} \phi_1'\phi_{M-1}' dx \\ \int_{x_0}^{x_1} \phi_2'\phi_1' dx & \int_{x_0}^{x_1} (\phi_2')^2 dx & \cdots & \int_{x_0}^{x_1} \phi_2'\phi_{M-1}' dx \\ \vdots & \vdots & \vdots & \vdots \\ \int_{x_0}^{x_1} \phi_{M-1}'\phi_1' dx & \int_{x_0}^{x_1} \phi_{M-1}'\phi_2' dx & \cdots & \int_{x_0}^{x_1} (\phi_{M-1}')^2 dx \end{bmatrix}$$

$$+ \begin{bmatrix} \int_{x_1}^{x_2} (\phi_1')^2 dx & \int_{x_1}^{x_2} \phi_1'\phi_2' dx & \cdots & \int_{x_1}^{x_2} \phi_1'\phi_{M-1}' dx \\ \int_{x_1}^{x_2} \phi_2'\phi_1' dx & \int_{x_1}^{x_2} (\phi_2')^2 dx & \cdots & \int_{x_1}^{x_2} \phi_2'\phi_{M-1}' dx \\ \vdots & \vdots & \vdots & \vdots \\ \int_{x_1}^{x_2} \phi_{M-1}'\phi_1' dx & \int_{x_1}^{x_2} \phi_{M-1}'\phi_2' dx & \cdots & \int_{x_1}^{x_2} (\phi_{M-1}')^2 dx \end{bmatrix}$$

$$+ \cdots$$

$$+ \begin{bmatrix} \int_{x_{M-1}}^{x_M} (\phi_1')^2 dx & \int_{x_{M-1}}^{x_M} \phi_1'\phi_2' dx & \cdots & \int_{x_{M-1}}^{x_M} \phi_1'\phi_{M-1}' dx \\ \int_{x_{M-1}}^{x_M} \phi_2'\phi_1' dx & \int_{x_{M-1}}^{x_M} (\phi_2')^2 dx & \cdots & \int_{x_{M-1}}^{x_M} \phi_2'\phi_{M-1}' dx \\ \vdots & \vdots & \vdots & \vdots \\ \int_{x_{M-1}}^{x_M} \phi_{M-1}'\phi_1' dx & \int_{x_{M-1}}^{x_M} \phi_{M-1}'\phi_2' dx & \cdots & \int_{x_{M-1}}^{x_M} (\phi_{M-1}')^2 dx \end{bmatrix}.$$

For the hat basis functions, it is noted that each interval has only *two* nonzero basis functions (*cf.* Figure 6.3). This leads to

$$A = \begin{bmatrix} \int_{x_0}^{x_1} (\phi_1')^2 dx & 0 & \cdots & 0 \\ 0 & 0 & \cdots & 0 \\ \vdots & \vdots & \vdots & \vdots \\ 0 & 0 & \cdots & 0 \end{bmatrix} + \begin{bmatrix} \int_{x_1}^{x_2} (\phi_1')^2 dx & \int_{x_1}^{x_2} \phi_1'\phi_2' dx & \cdots & 0 \\ \int_{x_1}^{x_2} \phi_2'\phi_1' dx & \int_{x_1}^{x_2} (\phi_2')^2 dx & \cdots & 0 \\ \vdots & \vdots & \vdots & \vdots \\ 0 & 0 & \cdots & 0 \end{bmatrix}$$

$$+ \begin{bmatrix} 0 & 0 & 0 & \cdots & 0 \\ 0 & \int_{x_2}^{x_3} (\phi_2')^2 dx & \int_{x_2}^{x_3} \phi_2'\phi_3' dx & \cdots & 0 \\ 0 & \int_{x_2}^{x_3} \phi_3'\phi_2' dx & \int_{x_2}^{x_3} (\phi_3')^2 dx & \cdots & 0 \\ \vdots & \vdots & \vdots & \vdots & \vdots \\ 0 & 0 & 0 & \cdots & 0 \end{bmatrix} + \begin{bmatrix} 0 & 0 & 0 & \cdots & 0 \\ 0 & 0 & 0 & \cdots & 0 \\ \vdots & \vdots & \vdots & \vdots & \vdots \\ 0 & 0 & 0 & \cdots & 0 \\ 0 & 0 & 0 & \cdots & \int_{x_{M-1}}^{x_M} (\phi_{M-1}')^2 dx \end{bmatrix}.$$

The nonzero contribution from a particular element is

$$K_i^e = \begin{bmatrix} \int_{x_i}^{x_{i+1}} (\phi_i')^2 dx & \int_{x_i}^{x_{i+1}} \phi_i'\phi_{i+1}' dx \\ \int_{x_i}^{x_{i+1}} \phi_{i+1}'\phi_i' dx & \int_{x_i}^{x_{i+1}} (\phi_{i+1}')^2 dx \end{bmatrix},$$

the two by two *local stiffness matrix*. Similarly, the *local load vector* is

$$F_i^e = \begin{bmatrix} \int_{x_i}^{x_{i+1}} f\phi_i dx \\ \int_{x_i}^{x_{i+1}} f\phi_{i+1} dx \end{bmatrix},$$

and the global load vector can also be assembled element by element:

$$F = \begin{bmatrix} \int_{x_0}^{x_1} f\phi_1 dx \\ 0 \\ 0 \\ \vdots \\ 0 \\ 0 \end{bmatrix} + \begin{bmatrix} \int_{x_1}^{x_2} f\phi_1 dx \\ \int_{x_1}^{x_2} f\phi_2 dx \\ 0 \\ \vdots \\ 0 \\ 0 \end{bmatrix} + \begin{bmatrix} 0 \\ \int_{x_2}^{x_3} f\phi_2 dx \\ \int_{x_2}^{x_3} f\phi_3 dx \\ \vdots \\ 0 \\ 0 \end{bmatrix} + \cdots + \begin{bmatrix} 0 \\ 0 \\ 0 \\ \vdots \\ 0 \\ \int_{x_{M-1}}^{x_M} f\phi_{M-1} dx \end{bmatrix}.$$

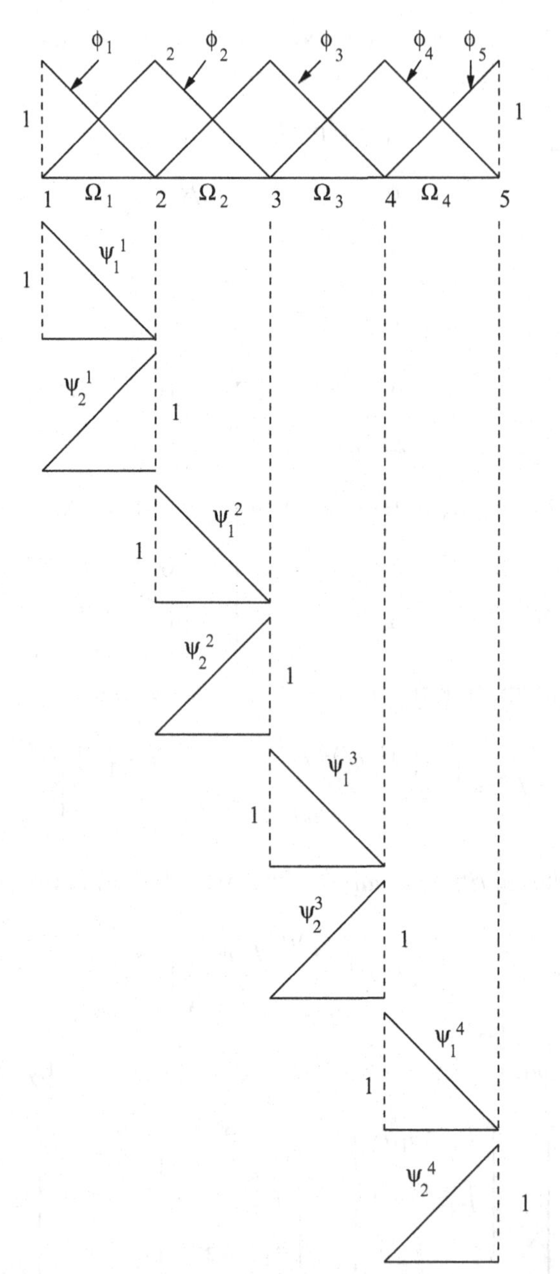

Figure 6.3. Continuous piecewise linear basis functions ϕ_i for a four-element mesh generated by linear shape functions ψ_1^e, ψ_2^e defined over each element. On each interior element, there are only two nonzero basis functions. The figure is adapted from Carey and Oden (1983).

6.3.4 Computing Local Stiffness Matrix K_i^e and Local Load Vector F_i^e

In the element (x_i, x_{i+1}), there are only two nonzero hat functions centered at x_i and x_{i+1} respectively:

$$\psi_i^e(x) = \frac{x_{i+1} - x}{x_{i+1} - x_i}, \qquad \psi_{i+1}^e(x) = \frac{x - x_i}{x_{i+1} - x_i},$$

$$(\psi_i^e)' = -\frac{1}{h_i}, \qquad (\psi_{i+1}^e)' = \frac{1}{h_i},$$

where ψ_i^e and ψ_{i+1}^e are defined only on one particular element. We can concentrate on the corresponding contribution to the stiffness matrix and load vector from the two nonzero hat functions. It is easy to verify that

$$\int_{x_i}^{x_{i+1}} (\psi_i')^2 \, dx = \int_{x_i}^{x_{i+1}} \frac{1}{h_i^2} \, dx = \frac{1}{h_i}, \quad \int_{x_i}^{x_{i+1}} \psi_i' \psi_{i+1}' \, dx = \int_{x_i}^{x_{i+1}} -\frac{1}{h_i^2} \, dx = -\frac{1}{h_i},$$

$$\int_{x_i}^{x_{i+1}} (\psi_{i+1}')^2 \, dx = \int_{x_i}^{x_{i+1}} \frac{1}{h_i^2} \, dx = \frac{1}{h_i}.$$

The local stiffness matrix K_i^e is therefore

$$K_i^e = \begin{bmatrix} \frac{1}{h_i} & -\frac{1}{h_i} \\ -\frac{1}{h_i} & \frac{1}{h_i} \end{bmatrix},$$

and the stiffness matrix A is assembled as follows:

$$A = 0^{(M-1)\times(M-1)}, \quad A = \begin{bmatrix} \frac{1}{h_0} & 0 & 0 & \cdots \\ 0 & 0 & 0 & \cdots \\ \vdots & \vdots & \vdots & \vdots \end{bmatrix}, \quad A = \begin{bmatrix} \frac{1}{h_0} + \frac{1}{h_1} & -\frac{1}{h_1} & 0 & \cdots \\ -\frac{1}{h_1} & \frac{1}{h_1} & 0 & \cdots \\ 0 & 0 & 0 & \cdots \\ \vdots & \vdots & \vdots & \vdots \end{bmatrix},$$

$$\cdots \quad A = \begin{bmatrix} \frac{1}{h_0} + \frac{1}{h_1} & -\frac{1}{h_1} & 0 & 0 & \cdots \\ -\frac{1}{h_1} & \frac{1}{h_1} + \frac{1}{h_2} & -\frac{1}{h_2} & 0 & \cdots \\ 0 & -\frac{1}{h_2} & \frac{1}{h_2} & 0 & \cdots \\ \vdots & \vdots & \vdots & \vdots \end{bmatrix}.$$

Thus we finally assemble the tridiagonal matrix

$$A = \begin{bmatrix} \frac{1}{h_0}+\frac{1}{h_1} & -\frac{1}{h_1} & & & & \\ -\frac{1}{h_1} & \frac{1}{h_1}+\frac{1}{h_2} & -\frac{1}{h_2} & & & \\ & -\frac{1}{h_2} & \frac{1}{h_2}+\frac{1}{h_3} & -\frac{1}{h_3} & & \\ & & \ddots & \ddots & \ddots & \\ & & & -\frac{1}{h_{M-3}} & \frac{1}{h_{M-3}}+\frac{1}{h_{M-2}} & -\frac{1}{h_{M-2}} \\ & & & & -\frac{1}{h_{M-2}} & \frac{1}{h_{M-2}}+\frac{1}{h_{M-1}} \end{bmatrix}.$$

Remark 6.4. For a uniform mesh $x_i = ih$, $h = 1/M$, $i = 0, 1, \ldots, M$ and the integral approximated by the mid-point rule

$$\begin{aligned} \int_0^1 f(x)\phi_i(x)dx &= \int_{x_{i-1}}^{x_{i+1}} f(x)\phi_i(x) \simeq \int_{x_{i-1}}^{x_{i+1}} f(x_i)\phi_i(x)dx \\ &= f(x_i)\int_{x_{i-1}}^{x_{i+1}} \phi_i(x)dx = f(x_i)\,, \end{aligned}$$

the resulting system of equations for the model problem from the finite element method is identical to that obtained from the FD method.

6.4 Matlab Programming of the FE Method for the 1D Model Problem

Matlab code to solve the 1D model problem

$$-u''(x) = f(x), \quad a < x < b; \qquad u(a) = u(b) = 0 \tag{6.10}$$

using the hat basis functions is available either through the link

www4.ncsu.edu/~zhilin/FD_FEM_book
or by e-mail request to the authors.

The Matlab code includes the following Matlab functions:

- $U = fem1d(x)$ is the main subroutine of the finite element method using the hat basis functions. The input x is the vector containing the nodal points. The output U, $U(0) = U(M) = 0$ is the finite element solution at the nodal points, where $M + 1$ is the total nodal points.
- $y = hat1(x, x1, x2)$ is the local hat function in the interval $[x1, x2]$ which takes one at $x = x2$ and zero at $x = x1$.

- $y = hat2(x, x1, x2)$ is the local hat function in the interval $[x1, x2]$ which takes one at $x = x1$ and zero at $x = x2$.
- $y = int\_hata1\_f(x1, x2)$ computes the integral $\int_{x1}^{x2} f(x) hat1 dx$ using the Simpson rule.
- $y = int\_hata2\_f(x1, x2)$ computes the integral $\int_{x1}^{x2} f(x) hat2 dx$ using the Simpson quadrature rule.
- The main function is *drive.m* which solves the problem, plots the solution and the error.
- $y = f(x)$ is the right-hand side of the differential equation.
- $y = soln(x)$ is the exact solution of differential equation.
- $y = fem\_soln(x, U, xp)$ evaluates the finite element solution at an arbitrary point xp in the solution domain.

We explain some of these Matlab functions in the following subsections.

6.4.1 Define the Basis Functions

In an element $[x_1, x_2]$ there are two nonzero basis functions: one is

$$\psi_1^e(x) = \frac{x - x_1}{x_2 - x_1} \tag{6.11}$$

where the Matlab code is the file hat1.m so

```
function y = hat1(x,x1,x2)
% This function evaluates the hat function
    y = (x-x1)/(x2-x1);
return
```

and the other is

$$\psi_2^e(x) = \frac{x_2 - x}{x_2 - x_1} \tag{6.12}$$

where the Matlab code is the file hat2.m so

```
function y = hat2(x,x1,x2)
% This function evaluates the hat function
    y = (x2-x)/(x2-x1);
return
```

6.4.2 Define $f(x)$

```
function y = f(x)
    y = 1;  % for example
return
```

6.4.3 The Main FE Routine

```
function U = fem1d(x)

%%%%%%%%%%%%%%%%%%%%%%%%%%%%%%%%%%%%%%%%%%%%%%%%%%%%%%%%%%%%%%%%%%%
%                                                                 %
%          A simple Matlab code of 1D FE method for               %
%                                                                 %
%                  -u'' = f(x),    a <= x <= b, u(a)=u(b)=0       %
%  Input: x, Nodal points                                         %
%  Output: U, FE solution at nodal points                         %
%                                                                 %
%  Function needed: f(x).                                         %
%                                                                 %
%  Matlab functions used:                                         %
%                                                                 %
%   hat1(x,x1,x2), hat function in [x1,x2] that is 1 at x2; and   %
%   0 at x1.                                                      %
%                                                                 %
%   hat2(x,x1,x2), hat function in [x1,x2] that is 0 at x2; and   %
%   1 at x1.                                                      %
%                                                                 %
%   int_hat1_f(x1,x2): Contribution to the load vector from hat1  %
%   int_hat2_f(x1,x2): Contribution to the load vector from hat2  %
%                                                                 %
%%%%%%%%%%%%%%%%%%%%%%%%%%%%%%%%%%%%%%%%%%%%%%%%%%%%%%%%%%%%%%%%%%%

M = length(x);
for i=1:M-1,
  h(i) = x(i+1)-x(i);
end

A = sparse(M,M); F=zeros(M,1);              % Initialization
A(1,1) = 1; F(1)=0;
A(M,M) = 1; F(M)=0;
A(2,2) = 1/h(1); F(2) = int_hat1_f(x(1),x(2));

for i=2:M-2,                              % Assembling element by element
  A(i,i)      = A(i,i) + 1/h(i);
  A(i,i+1)    = A(i,i+1) - 1/h(i);
  A(i+1,i)    = A(i+1,i) - 1/h(i);
  A(i+1,i+1)  = A(i+1,i+1) + 1/h(i);
  F(i)        = F(i) + int_hat2_f(x(i),x(i+1));
  F(i+1)      = F(i+1) + int_hat1_f(x(i),x(i+1));
end

A(M-1,M-1) = A(M-1,M-1) + 1/h(M-1);
F(M-1)     = F(M-1) + int_hat2_f(x(M-1),x(M));

U = A\F;                     % Solve the linear system of equations

return
```

6.4.4 A Test Example

Let us consider the test example

$$f(x) = 1, \quad a = 0, \quad b = 1.$$

The exact solution is

$$u(x) = \frac{x(1-x)}{2}. \tag{6.13}$$

A sample Matlab drive code is listed below:

```
clear all; close all;      % Clear every thing so it won't mess up
                           % with other existing variables.

%%%%%% Generate a mesh.

x(1)=0; x(2)=0.1; x(3)=0.3; x(4)=0.333; x(5)=0.5; x(6)=0.75;x(7)=1;

U = fem1d(x);

%%%%%% Compare errors:

x2 = 0:0.05:1; k2 = length(x2);
for i=1:k2,
  u_exact(i) = soln(x2(i));
  u_fem(i) = fem_soln(x,U,x2(i));  % Compute FE solution at x2(i)
end

error = norm(u_fem-u_exact,inf)   % Compute the infinity error

plot(x2,u_fem,':', x2,u_exact)    % Solid: the exact,
                                  % dotted: FE solution
hold; plot(x,U,'o')               % Mark the solution at nodal
                                  % points
xlabel('x'); ylabel('u(x) & u_{fem}(x)');
title('Solid line: Exact solution, Dotted line: FE solution')

figure(2); plot(x2,u_fem-u_exact); title('Error plot')
xlabel('x'); ylabel('u-u_{fem}');  title('Error Plot')
```

Figure 6.4 shows the plots produced by running the code. Figure 6.4(a) shows both the true solution (the solid line) and the finite element solution (the dashed line). The little "o"s are the finite element solution values at the nodal points. Figure 6.4(b) shows the error between the true and the finite element solutions at a few selected points (zero at the nodal points in this example, although may not be so in general).

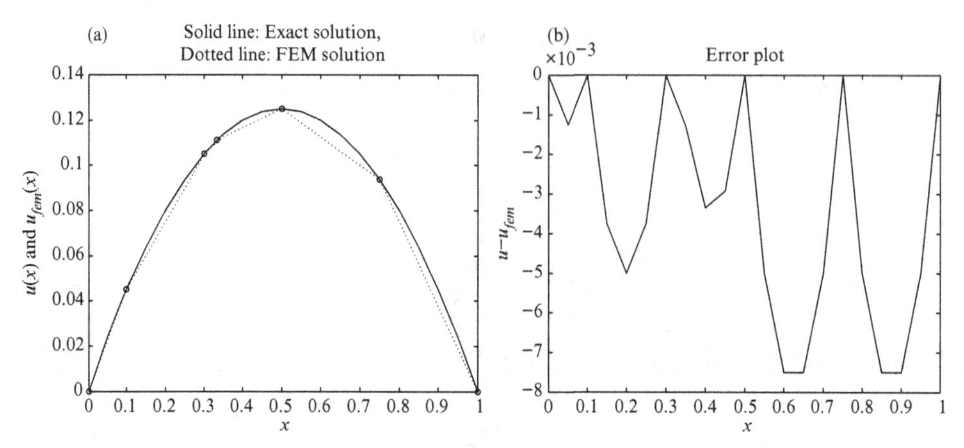

Figure 6.4. (a) Plot of the true solution (solid line) and the finite element solution (the dashed line) and (b) the error plot at some selected points.

Exercises

1. Consider the following BVP:

$$-u''(x) + u(x) = f(x), \qquad 0 < x < 1, \quad u(0) = u(1) = 0.$$

   (a) Show that the weak form (variational form) is

$$(u', v') + (u, v) = (f, v), \quad \forall\, v(x) \in H_0^1(0, 1),$$

   where

$$(u, v) = \int_0^1 u(x)v(x)dx,$$

$$H_0^1(0, 1) = \left\{ v(x), \quad v(0) = v(1) = 0, \ \int_0^1 v^2 dx < \infty \right\}.$$

   (b) Derive the linear system of the equations for the finite element approximation

$$u_h = \sum_{j=1}^{3} \alpha_j \phi_j(x),$$

   with the following information:

   - $f(x) = 1$;
   - the nodal points and the elements are indexed as

$$x_0 = 0, \quad x_2 = \frac{1}{4}, \quad x_3 = \frac{1}{2}, \quad x_1 = \frac{3}{4}, \quad x_4 = 1.$$

$$\Omega_1 = [x_3, x_1], \quad \Omega_2 = [x_1, x_4], \quad \Omega_3 = [x_2, x_3], \quad \Omega_4 = [x_0, x_2];$$

   - the basis functions are the hat functions

$$\phi_i(x_j) = \begin{cases} 1, & \text{if } i = j, \\ 0, & \text{otherwise}, \end{cases}$$

     and do not reorder the nodal points and elements; and
   - assemble the stiffness matrix and the load vector element by element.

2. (This problem involves modifying *drive.m*, *f.m* and *soln.m*.) Use the Matlab code to solve

$$-u''(x) = f(x), \qquad 0 < x < 1, \quad u(0) = u(1) = 0.$$

Try two different meshes: (a) the one given in *drive.m*; (b) the uniform mesh $x_i = ih$, $h = 1/M$, $i = 0, 1, \ldots, M$. Take $M = 10$, done in Matlab using the command: $x = 0:0.1:1$. Use the two meshes to solve the problem for the following $f(x)$ or exact $u(x)$:

(a) given $u(x) = \sin(\pi x)$, what is $f(x)$?
(b) given $f(x) = x^3$, what is $u(x)$?
(c) (extra credit) given $f(x) = \delta(x - 1/2)$, where $\delta(x)$ is the Dirac delta function, what is $u(x)$?
Hint: The Dirac delta function is defined as a distribution satisfying $\int_a^b f(x)\delta(x)dx = f(0)$ for any function $f(x) \in C(a, b)$ if $x = 0$ is in the interior of the integration.

Ensure that the errors are reasonably small.

3. (This problem involves modifying *fem1d.m*, *drive.m*, *f.m* and *soln.m*.) Assume that

$$\int_{x_i}^{x_{i+1}} \phi_i(x)\phi_{i+1}(x)dx = \frac{h}{6},$$

where $h = x_{i+1} - x_i$, and ϕ_i and ϕ_{i+1} are the hat functions centered at x_i and x_{i+1} respectively. Use the Matlab code to solve

$$-u''(x) + u(x) = f(x), \qquad 0 < x < 1, \quad u(0) = u(1) = 0.$$

Try to use the uniform grid $x = 0:0.1:1$ in Matlab, for the following exact $u(x)$:

(a) $u(x) = \sin(\pi x)$, what is $f(x)$?
(b) $u(x) = x(1 - x)/2$, what is $f(x)$?

7

Theoretical Foundations of the Finite Element Method

Using finite element methods, we need to answer these questions:

- What is the appropriate functional space V for the solution?
- What is the appropriate *weak* or *variational* form of a differential equation?
- What kind of basis functions or finite element spaces should we choose?
- How accurate is the finite element solution?

We briefly address these questions in this chapter. Recalling that finite element methods are based on *integral forms* and not on the pointwise sense as in finite difference methods, we will generalize the theory corresponding to the pointwise form to deal with integral forms.

7.1 Functional Spaces

A *functional space* is a *set* of functions with operations. For example,

$$C(\Omega) = C^0(\Omega) = \left\{ u(x), \;\; u(x) \text{ is continuous on } \Omega \right\} \tag{7.1}$$

is a linear space that contains all continuous functions on Ω, the domain where the functions are defined, *i.e.*, $\Omega = [0, \; 1]$. The space is linear because for any real numbers α and β and $u_1 \in C(\Omega)$ and $u_2 \in C(\Omega)$, we have $\alpha u_1 + \beta u_2 \in C(\Omega)$.

The functional space with first-order continuous derivatives in 1D is

$$C^1(\Omega) = \{u(x), \quad u(x), \, u'(x) \text{ are continuous on } \Omega\}, \tag{7.2}$$

and similarly

$$C^m(\Omega) = \left\{ u(x), \quad u(x), u'(x), \ldots, u^{(m)} \text{ are continuous on } \Omega \right\}. \tag{7.3}$$

Obviously,

$$C^0 \supset C^1 \supset \cdots \supset C^m \supset \cdots. \tag{7.4}$$

Then as $m \to \infty$, we define

$$C^{\infty}(\Omega) = \{u(x), \quad u(x) \text{ is indefinitely differentiable on } \Omega\}\,. \tag{7.5}$$

For example, e^x, $\sin x$, $\cos x$, and polynomials, are in $C^{\infty}(-\infty, \infty)$, but some other elementary functions such as $\log x$, $\tan x$, $\cot x$ are not if $x = 0$ is in the domain.

7.1.1 Multidimensional Spaces and Multi-index Notations

Let us now consider multidimensional functions $u(\mathbf{x}) = u(x_1, x_2, \ldots, x_n)$, $\mathbf{x} \in R^n$, and a corresponding multi-index notation that simplifies expressions for partial derivatives. We can write $\alpha = (\alpha_1, \alpha_2, \ldots, \alpha_n)$, $\alpha_i \geq 0$ for an integer vector in R^n, *e.g.*, if $n = 5$, then $\alpha = (1, 2, 0, 0, 2)$ is one of possible vectors. We can readily represent a partial derivative as

$$D^{\alpha}u(\mathbf{x}) = \frac{\partial^{|\alpha|} u}{\partial x_1^{\alpha_1} \partial x_2^{\alpha_2} \cdots \partial x_n^{\alpha_n}}, \qquad |\alpha| = \alpha_1 + \alpha_2 + \cdots + \alpha_n\,, \qquad \alpha_i \geq 0 \tag{7.6}$$

which is the so-called multi-index notation.

Example 7.1. For $n = 2$ and $u(\mathbf{x}) = u(x_1, x_2)$, all possible $D^{\alpha}u$ when $|\alpha| = 2$ are

$$\alpha = (2,0), \quad D^{\alpha}u = \frac{\partial^2 u}{\partial x_1^2},$$
$$\alpha = (1,1), \quad D^{\alpha}u = \frac{\partial^2 u}{\partial x_1 \partial x_2},$$
$$\alpha = (0,2), \quad D^{\alpha}u = \frac{\partial^2 u}{\partial x_2^2}.$$

With the multi-index notation, the C^m space in a domain $\Omega \in R^n$ can be defined as

$$C^m(\Omega) = \{u(x_1, x_2, \ldots, x_n), \quad D^{\alpha}u \text{ are continuous on } \Omega,\ |\alpha| \leq m\} \tag{7.7}$$

i.e., all possible derivatives up to order m are continuous on Ω.

Example 7.2. For $n = 2$ and $m = 3$, we have u, u_x, u_y, u_{xx}, u_{xy}, u_{yy}, u_{xxx}, u_{xxy}, u_{xyy} and u_{yyy} all continuous on Ω if $u \in C^3(\Omega)$, or simply $D^{\alpha}u \in C^3(\Omega)$ for $|\alpha| \leq 3$. Note that $C^m(\Omega)$ has infinite dimensions.

The *distance* in $C^0(\Omega)$ is defined as

$$d(u, v) = \max_{x \in \Omega} |u(x) - v(x)|$$

with the properties (1), $d(u,v) \geq 0$; (2), $d(u,v)=0$ if and only if $u \equiv v$; and (3), $d(u, v+w) \leq d(u,v) + d(u,w)$, the triangle inequality. A linear space with a distance defined is called a *metric space*.

A norm in $C^0(\Omega)$ is a nonnegative function of u that satisfies

$$\|u(x)\| = d(u,\theta) = \max_{x\in\Omega} |u(x)|, \text{ where } \theta \text{ is the zero element,}$$

with the properties (1), $\|u(x)\| \geq 0$, and $\|u(x)\| = 0$ if and only if $u \equiv 0$;

$$(2),\ \|\alpha u(x)\| = |\alpha|\|u(x)\|, \text{ where } \alpha \text{ is a number;}$$

$$(3),\ \|u(x)+v(x)\| \leq \|u(x)\| + \|v(x)\|, \text{ the triangle inequality.}$$

A linear space with a norm defined is called a *normed space*. In $C^m(\Omega)$, the distance and the norm are defined as

$$d(u,v) = \max_{0\leq|\alpha|\leq m} \max_{x\in\Omega} |D^\alpha u(x) - D^\alpha v(x)|\,, \tag{7.8}$$

$$\|u(x)\| = \max_{0\leq|\alpha|\leq m} \max_{x\in\Omega} |D^\alpha u|\,. \tag{7.9}$$

7.2 Spaces for Integral Forms, $L^2(\Omega)$ and $L^p(\Omega)$

In analogy to pointwise spaces $C^m(\Omega)$, we can define Sobolev spaces $H^m(\Omega)$ in integral forms. The square-integrable space $H^0(\Omega) = L^2(\Omega)$ is defined as

$$L^2(\Omega) = \left\{ u(x), \quad \int_\Omega u^2(x)\,dx < \infty \right\} \tag{7.10}$$

corresponding to the pointwise $C^0(\Omega)$ space. It is easy to see that

$$C(\Omega) = C^0(\Omega) \subset L^2(\Omega)\,.$$

Example 7.3. It is easy to verify that $y(x) = 1/x^{1/4} \notin C^0(0,1)$, but

$$\int_0^1 \left(\frac{1}{x^{1/4}}\right)^2 dx = \int_0^1 \frac{1}{\sqrt{x}} dx = 2 < \infty$$

so that $y(x) \in L^2(0,1)$. But it is obvious that $y(x)$ is not in $C(0,1)$ since $y(x)$ blows up as $x \to 0$ from $x>0$, see Figure 7.1 in which we also show that a piecewise constant function 0 and 1 in $(0,1)$ is in $L^2(0,1)$ but not in $C(0,1)$ since the function is discontinuous (nonremovable discontinuity) at $x=0.5$.

The *distance* in $L^2(\Omega)$ is defined as

$$d(f,g) = \left\{ \int_\Omega |f-g|^2\,dx \right\}^{1/2}, \tag{7.11}$$

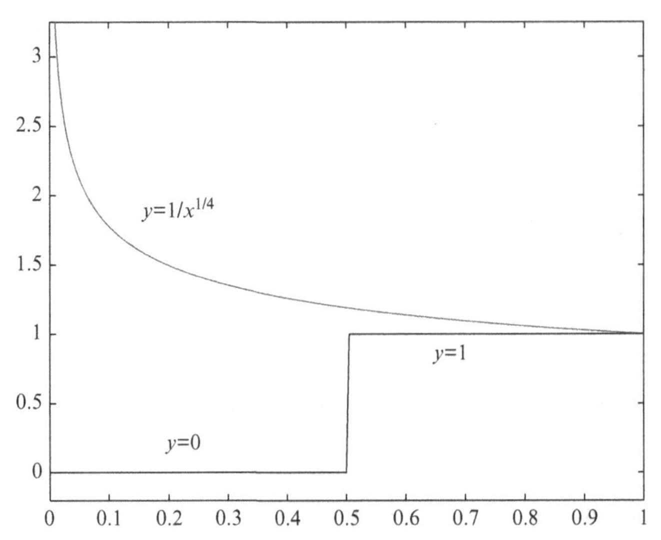

Figure 7.1. Plot of two functions that are in $L^2(0,1)$ but not in $C[0,1]$. One function is $y(x)=\frac{1}{\sqrt[4]{x}}$. The other one is $y(x)=0$ in $[0,1/2)$ and $y(x)=1$ in $[1/2,1]$.

which satisfies the three conditions of the distance definition; so $L^2(\Omega)$ is a metric space. We say that two functions f and g are identical ($f\equiv g$) in $L^2(\Omega)$ if $d(f,g)=0$. For example, the following two functions are identical in $L^2(-2,2)$:

$$f(x)=\begin{cases} 0, & \text{if } -2\leq x<0, \\ 1, & \text{if } 0\leq x\leq 2, \end{cases} \qquad g(x)=\begin{cases} 0, & \text{if } -2\leq x\leq 0, \\ 1, & \text{if } 0<x\leq 2. \end{cases}$$

The norm in the $L^2(\Omega)$ space is defined as

$$\|u\|_{L^2}=\|u\|_0=\left\{\int_\Omega |u|^2\,dx\right\}^{1/2}. \tag{7.12}$$

It is straightforward to prove that the usual properties for the distance and the norm hold.

We say that $L^2(\Omega)$ *is a complete space*, meaning that any Cauchy sequence $\{f_n(\Omega)\}$ in $L^2(\Omega)$ has a limit in $L^2(\Omega)$, *i.e.*, there is a function $f\in L^2(\Omega)$ such that

$$\lim_{n\to\infty}\|f_n-f\|_{L^2}=0, \quad \text{or} \quad \lim_{n\to\infty} f_n=f.$$

A Cauchy sequence is a sequence that satisfies the property that, for any given positive number ϵ, no matter how small it is, there is an integer N such that

$$\|f_n-f_m\|_{L^2}<\epsilon, \qquad \text{if} \quad m\geq N, \quad n\geq N.$$

A complete normed space is called a *Banach* space (a Cauchy sequence converges in terms of the norm), so $L^2(\Omega)$ is a Banach space.

7.2.1 The Inner Product in L^2

For any two vectors

$$x = \begin{bmatrix} x_1 \\ x_2 \\ \vdots \\ x_n \end{bmatrix}, \quad y = \begin{bmatrix} y_1 \\ y_2 \\ \vdots \\ y_n \end{bmatrix}$$

in R^n, we recall that the inner product is

$$(x, y) = x^T y = \sum_{i=1}^{n} x_i y_i = x_1 y_1 + x_2 y_2 + \cdots + x_n y_n .$$

Similarly, the inner product in $L^2(\Omega)$ space in R, the real number space, is defined as

$$(f, g) = \int_\Omega f(x) g(x)\, dx\,, \text{ for any } f \text{ and } g \in L^2(\Omega)\,, \tag{7.13}$$

and satisfies the familiar properties

$$\begin{aligned} (f, g) &= (g, f)\,, \\ (\alpha f, g) &= (f, \alpha g) = \alpha (f, g), \quad \forall \alpha \in R, \\ (f, g + w) &= (f, g) + (f, w) \end{aligned}$$

for any f, g, and $w \in L^2(\Omega)$. The norm, distance, and inner product in $L^2(\Omega)$ are related as follows:

$$\|u\|_0 = \|u\|_{L^2(\Omega)} = \sqrt{(u, u)} = d(u, \theta) = \left\{ \int_\Omega |u|^2\, dx \right\}^{1/2} . \tag{7.14}$$

With the $L^2(0, 1)$ inner product, for the simple model problem

$$-u'' = f, \quad 0 < x < 1, \quad u(0) = u(1) = 0,$$

we can rewrite the weak form as

$$(u', v') = (f, v)\,, \quad \forall v \in H_0^1(0, 1)\,,$$

and the minimization form as

$$\min_{v \in H_0^1(0,1)} F(v): \quad F(v) = \frac{1}{2}(v', v') - (f, v).$$

7.2.2 The Cauchy–Schwarz Inequality in $L^2(\Omega)$

For a Hilbert space with the norm $\|u\| = \sqrt{(u, u)}$, the Cauchy–Schwarz inequality is

$$|(u, v)| \leq \|u\| \|v\|. \tag{7.15}$$

Examples of the Cauchy–Schwarz inequality corresponding to inner products in R^n and in L^2 spaces are

$$\left|\sum_{i=1}^{n} x_i y_i\right| \leq \left\{\sum_{i=1}^{n} x_i^2\right\}^{1/2} \left\{\sum_{i=1}^{n} y_i^2\right\}^{1/2},$$

$$\left|\sum_{i=1}^{n} x_i\right| \leq \sqrt{n} \left\{\sum_{i=1}^{n} x_i^2\right\}^{1/2},$$

$$\left|\int_{\Omega} fg dx\right| \leq \left\{\int_{\Omega} f^2 \, dx\right\}^{1/2} \left\{\int_{\Omega} g^2 \, dx\right\}^{1/2},$$

$$\left|\int_{\Omega} f dx\right| \leq \left\{\int_{\Omega} f^2 \, dx\right\}^{1/2} \sqrt{V},$$

where V is the volume of Ω.

7.2.2.1 *A Proof of the Cauchy–Schwarz Inequality*

Noting that $(u, u) = \|u\|^2$, we construct a quadratic function of α given u and v:

$$f(\alpha) = (u + \alpha v, u + \alpha v) = (u, u) + 2\alpha(u, v) + \alpha^2(v, v) \geq 0.$$

The quadratic function is nonnegative; hence the discriminant of the quadratic form satisfies

$$\Delta = b^2 - 4ac \leq 0, \ \ i.e., \ 4(u, v)^2 - 4(u, u)(v, v) \leq 0 \text{ or } (u, v)^2 \leq (u, u)(v, v),$$

yielding the Cauchy–Schwarz inequality $|(u, v)| \leq \|u\| \|v\|$, on taking the square root of both sides.

A complete Banach space with an inner product defined is called a *Hilbert space*. Hence, $L^2(\Omega)$ is a Hilbert space (linear space, inner product, complete).

7.2.2.2 Relationships Between the Spaces

The relationships (and relevant additional properties) in the hierarchy of defined spaces may be summarized diagrammatically:

Metric Space (distance) $\Longrightarrow$ Normed Space (norm) $\Longrightarrow$ Banach space (complete) $\Longrightarrow$ Hilbert space (inner product).

7.2.3 $L^p(\Omega)$ Spaces

An $L^p(\Omega)$ $(0 < p < \infty)$ space is defined as

$$L^p(\Omega) = \left\{ u(x), \int_\Omega |u(x)|^p \, dx < \infty \right\}, \tag{7.16}$$

and the distance in $L^p(\Omega)$ is defined as

$$d(f, g) = \left\{ \int_\Omega |f - g|^p \, dx \right\}^{1/p}. \tag{7.17}$$

An $L^p(\Omega)$ space has a distance and is complete, so it is a Banach space. However, it is not a Hilbert space, because no corresponding inner product is defined unless $p = 2$.

7.3 Sobolev Spaces and Weak Derivatives

Similar to $C^m(\Omega)$ spaces, we use Sobolev spaces $H^m(\Omega)$ to define function spaces with derivatives involving integral forms. If there is no derivative, then the relevant Sobolev space is

$$H^0(\Omega) = L^2(\Omega) = \left\{ v(x), \int_\Omega |v|^2 dx < \infty \right\}. \tag{7.18}$$

7.3.1 Definition of Weak Derivatives

If $u(x) \in C^1[0, 1]$, then for any function $\phi \in C^1(0, 1)$ such that $\phi(0) = \phi(1) = 0$ we recall

$$\int_0^1 u'(x)\phi(x) = u\phi \Big|_0^1 - \int_0^1 u(x)\phi'(x) \, dx = -\int_0^1 u(x)\phi'(x) \, dx, \tag{7.19}$$

where $\phi(x)$ is a test function in $C_0^1(0, 1)$. The first-order weak derivative of $u(x) \in L^2(\Omega) = H^0(\Omega)$ is defined to be a function $v(x)$ satisfying

$$\int_\Omega v(x)\phi(x) dx = -\int_\Omega u(x)\phi'(x) dx \tag{7.20}$$

for all $\phi(x) \in C_0^1(\Omega)$ with $\phi(0) = \phi(1) = 0$. If such a function exists, then we write $v(x) = u'(x)$.

Example 7.4. Consider the following function $u(x)$

$$u(x) = \begin{cases} \frac{x}{2}, & \text{if } 0 \leq x < \frac{1}{2}, \\ \frac{1-x}{2}, & \text{if } \frac{1}{2} \leq x \leq 1. \end{cases}$$

It is obvious that $u(x) \in C[0, 1]$ but $u'(x) \notin C(0, 1)$ since the classic derivative does not exist at $x = \frac{1}{2}$. Let $\phi(x) \in C^1(0, 1)$ be any function that vanishes at two ends, *i.e.*, $\phi(0) = \phi(1) = 0$, and has first-order continuous derivative on $(0, 1)$. We carry out the following integration by parts.

$$\begin{aligned} \int_0^1 \phi' u dx &= \int_0^{\frac{1}{2}} \phi' u dx + \int_{\frac{1}{2}}^1 \phi' u dx \\ &= \phi(x)\, u(x)\Big|_0^{\frac{1}{2}} + \phi(x)\, u(x)\Big|_{\frac{1}{2}}^1 - \int_0^{\frac{1}{2}} \phi u' dx + \int_{\frac{1}{2}}^1 \phi u' dx \\ &= \phi(1/2)\Big(u(1/2-) - u(1/2+) \Big) - \int_0^{\frac{1}{2}} \frac{\phi(x)}{2} dx - \int_{\frac{1}{2}}^1 \frac{-\phi(x)}{2} dx \\ &= -\int_0^1 \psi(x)\phi(x) dx, \end{aligned}$$

where we have used the property that $\phi(0) = \phi(1) = 0$ and $\psi(x)$ is defined as

$$\psi(x) = \begin{cases} \frac{1}{2}, & \text{if } 0 < x < \frac{1}{2}, \\ -\frac{1}{2}, & \text{if } \frac{1}{2} < x < 1, \end{cases}$$

which is what we would expect. In other words, we define the weak derivative of $u(x)$ as $u'(x) = \psi(x)$ which is an $H^1(0, 1)$ but not a $C(0, 1)$ function.

Similarly, the m-th order weak derivative of $u(x) \in H^0(\Omega)$ is defined as a function $v(x)$ satisfying

$$\int_\Omega v(x)\phi(x) dx = (-1)^m \int_\Omega u(x)\phi^{(m)}(x) dx \tag{7.21}$$

for all $\phi(x) \in C_0^m(\Omega)$ with $\phi(x) = \phi'(x) = \cdots = \phi^{(m-1)}(x) = 0$ for all $x \in \partial\Omega$. If such a function exists, then we write $v(x) = u^{(m)}(x)$.

7.3.2 Definition of Sobolev Spaces $H^m(\Omega)$

The Sobolev space $H^1(\Omega)$ defined as

$$H^1(\Omega) = \left\{ v(x)\,,\; D^\alpha v \in L^2(\Omega)\,,\; |\alpha| \leq 1 \right\} \tag{7.22}$$

involves first-order derivatives, *e.g.*,

$$H^1(a,b) = \left\{ v(x)\,,\; a < x < b\,,\; \int_a^b v^2 dx < \infty\,,\; \int_a^b (v')^2 dx < \infty \right\},$$

and in two space dimensions, $H^1(\Omega)$ is defined as

$$H^1(\Omega) = \left\{ v(x,y)\,,\; v \in L^2(\Omega)\,,\; \frac{\partial v}{\partial x} \in L^2(\Omega)\,,\; \frac{\partial v}{\partial y} \in L^2(\Omega) \right\}.$$

The extension is immediate to the Sobolev space of general dimension

$$H^m(\Omega) = \left\{ v(\mathbf{x})\,,\; D^\alpha v \in L^2(\Omega)\,,\; |\alpha| \leq m \right\}. \tag{7.23}$$

7.3.3 Inner Products in $H^m(\Omega)$ Spaces

The inner product in $H^0(\Omega)$ is the same as that in $L^2(\Omega)$, *i.e.*,

$$(u,v)_{H^0(\Omega)} = (u,v)_0 = \int_\Omega uv\, dx\,.$$

The inner product in $H^1(a,b)$ is defined as

$$(u,v)_{H^1(a,b)} = (u,v)_1 = \int_a^b (uv + u'v')\; dx;$$

the inner product in $H^1(\Omega)$ of two variables is defined as

$$(u,v)_{H^1(\Omega)} = (u,v)_1 = \iint_\Omega \left(uv + \frac{\partial u}{\partial x}\frac{\partial v}{\partial x} + \frac{\partial u}{\partial y}\frac{\partial v}{\partial y} \right) dxdy\,; \quad \text{and}$$

the inner product in $H^m(\Omega)$ (of general dimension) is

$$(u,v)_{H^m(\Omega)} = (u,v)_m = \int_\Omega \sum_{|\alpha| \leq m} (D^\alpha u(\mathbf{x}))\,(D^\alpha v(\mathbf{x}))\, d\mathbf{x}. \tag{7.24}$$

The norm in $H^m(\Omega)$ (of general dimension) is

$$\|u\|_{H^m(\Omega)} = \|u\|_m = \left\{ \int_\Omega \sum_{|\alpha| \leq m} |D^\alpha u(\mathbf{x})|^2\, d\mathbf{x} \right\}^{1/2}, \tag{7.25}$$

therefore, $H^m(\Omega)$ is a Hilbert space. A norm can be defined from the inner product, *e.g.*, in $H^1(a,b)$, the norm is

$$\|u\|_1 = \left\{ \int_a^b \left(u^2 + u'^2 \right) dx \right\}^{1/2} .$$

The distance in $H^m(\Omega)$ (of general dimension) is defined as

$$d(u, v)_m = \|u - v\|_m. \tag{7.26}$$

7.3.4 Relations Between $C^m(\Omega)$ and $H^m(\Omega)$ – The Sobolev Embedding Theorem

In 1D spaces, we have

$$H^1(\Omega) \subset C^0(\Omega), \quad H^2(\Omega) \subset C^1(\Omega), \quad \ldots, H^{1+j}(\Omega) \subset C^j(\Omega).$$

Theorem 7.5. *The Sobolev embedding theorem: If $2m > n$, then*

$$H^{m+j} \subset C^j, \quad j = 0, 1, \ldots , \tag{7.27}$$

where n is the dimension of the independent variables of the elements in the Sobolev space.

Example 7.6. In 2D spaces, we have $n = 2$. The condition $2m > n$ means that $m > 1$. From the embedding theorem, we have

$$H^{2+j} \subset C^j, \quad j = 0 \Longrightarrow H^2 \subset C^0, \quad j = 1 \Longrightarrow H^3 \subset C^1, \ldots . \tag{7.28}$$

If $u(x, y) \in H^2$, which means that u, u_x, u_y, u_{xx}, u_{xy}, and u_{yy} all belong to L^2, then $u(x, y)$ is continuous, but u_x and u_y may not be continuous!

Example 7.7. In 3D spaces, we have $n = 3$ and the condition $2m > n$ means that $m > 3/2$ whose closest integer is number two, leading to the same result as in 2D:

$$H^{2+j} \subset C^j, \quad j = 0 \Longrightarrow H^2 \subset C^0, \quad j = 1 \Longrightarrow H^3 \subset C^1, \ldots . \tag{7.29}$$

We regard the *regularity* of a solution as the degree of smoothness for a class of problems measured in C^m or H^m space. Thus, for $u(x) \in H^m$ or $u(x) \in C^m$, the larger the m the smoother the function.

7.4 FE Analysis for 1D BVPs

For the simple 1D model problem

$$-u'' = f, \quad 0 < x < 1, \quad u(0) = u(1) = 0,$$

we know that the weak form is

$$\int_0^1 u'v'\,dx = \int_0^1 fv\,dx \quad \text{or} \quad (u', v') = (f, v).$$

Intuitively, because v is arbitrary we can take $v = f$ or $v = u$ to get

$$\int_0^1 u'v'\,dx = \int_0^1 u'^2\,dx, \qquad \int_0^1 fv\,dx = \int_0^1 f^2\,dx,$$

so u, u', f, v, and v' should belong to $L^2(0,1)$, *i.e.*, we have $u \in H_0^1(0,1)$ and $v \in H_0^1(0,1)$; so the solution is in the Sobolev space $H_0^1(0,1)$. We should also take v in the same space for a conforming finite element method. From the Sobolev embedding theorem, we also know that $H^1 \subset C^0$, so the solution is continuous.

7.4.1 *Conforming FE Methods*

Definition 7.8. If the finite element space is a subspace of the solution space, then the finite element space is called a *conforming finite element* space, and the finite element method is called a *conforming FE method.*

For example, the piecewise linear function over a given mesh is a conforming finite element space for the model problem. We mainly discuss conforming finite element methods in this book. On including the boundary condition, we define the solution space as

$$H_0^1(0,1) = \Big\{ v(x), \quad v(0) = v(1) = 0, \ v \in H^1(0,1) \Big\}. \tag{7.30}$$

When we look for a finite element solution in a finite-dimensional space V_h, it should be a subspace of $H_0^1(0,1)$ for a conforming finite element method. For example, given a mesh for the 1D model, we can define a finite-dimensional space using piecewise continuous linear functions over the mesh:

$$V_h = \{ v_h, \ v_h(0) = v_h(1) = 0, \ v_h \text{ is continuous piecewise linear} \}.$$

The finite element solution would be chosen from the finite-dimensional space V_h, a subspace of $H_0^1(0,1)$. If the solution of the weak form is in $H_0^1(0,1)$ but not in the V_h space; then an error is introduced on replacing the solution space with the finite-dimensional space. Nevertheless, the finite element solution is the best approximation in V_h in some norm, as discussed later.

7.4.2 FE Analysis for 1D Sturm–Liouville Problems

A 1D Sturm–Liouville problem on (x_l, x_r) with a Dirichlet boundary condition at two ends is

$$\begin{aligned} -(p(x)u'(x))' + q(x)u(x) &= f(x)\,, \quad x_l < x < x_r\,, \\ u(x_l) = 0\,, &\quad u(x_r) = 0\,, \\ p(x) \geq p_{min} > 0\,, &\quad q(x) \geq q_{min} \geq 0\,. \end{aligned} \tag{7.31}$$

The conditions on $p(x)$ and $q(x)$ guarantee the problem is well-posed, such that the weak form has a unique solution. It is convenient to assume $p(x) \in C(x_l, x_r)$ and $q(x) \in C(x_l, x_r)$. Later we will see that these conditions together with $f(x) \in L^2(x_l, x_r)$, guarantee the unique solution to the weak form of the problem. To derive the weak form, we multiply both sides of the equation by a test function $v(x)$, $v(x_l) = v(x_r) = 0$ and integrate from x_l to x_r to get

$$\begin{aligned} \int_{x_l}^{x_r} \left(-(p(x)u')' + qu\right) v\,dx &= -pu'v\Big|_{x_l}^{x_r} + \int_{x_l}^{x_r} \left(pu'v' + quv\right)\,dx \\ &= \int_{x_l}^{x_r} fv\,dx \\ \Longrightarrow \int_{x_l}^{x_r} \left(pu'v' + quv\right)\,dx &= \int_{x_l}^{x_r} fv\,dx\,, \;\forall v \in H_0^1(x_l, x_r) \quad \text{or} \quad a(u, v) = L(v). \end{aligned}$$

7.4.3 The Bilinear Form

The integral

$$a(u, v) = \int_{x_l}^{x_r} \left(pu'v' + quv\right)\,dx \tag{7.32}$$

is a bilinear form, because it is linear for both u and v from the following

$$\begin{aligned} a(\alpha u + \beta w, v) &= \int_{x_l}^{x_r} \left(p(\alpha u' + \beta w')v' + q(\alpha u + \beta w)v\right)\,dx \\ &= \alpha \int_{x_l}^{x_r} \left(pu'v' + quv\right)dx + \beta \int_{x_l}^{x_r} \left(pu'v' + qwv\right)\,dx \\ &= \alpha a(u, v) + \beta a(w, v)\,, \end{aligned}$$

where α and β are scalars; and similarly,

$$a(u, \alpha v + \beta w) = \alpha a(u, v) + \beta a(u, w)\,.$$

It is noted that this bilinear form is an inner product, usually different from the L^2 and H^1 inner products, but if $p \equiv 1$ and $q \equiv 1$ then

$$a(u, v) = (u, v).$$

Since $a(u, v)$ is an inner product, under the conditions: $p(x) \geq p_{min} > 0,\ q(x) \geq 0$, we can define the *energy norm* as

$$\|u\|_a = \sqrt{a(u, u)} = \left\{ \int_{x_l}^{x_r} \left(p(u')^2 + qu^2 \right) dx \right\}^{\frac{1}{2}}, \tag{7.33}$$

where the first term may be interpreted as the *kinetic energy* and the second term as the *potential energy*. The Cauchy–Schwarz inequality implies $|a(u, v)| \leq \|u\|_a \|v\|_a$.

The bilinear form combined with linear form often simplifies the notation for the weak and minimization forms, *e.g.*, for the above Sturm–Liouville problem the weak form becomes

$$a(u, v) = L(v)\,,\ \forall v \in H_0^1(x_l, x_r)\,, \tag{7.34}$$

and the minimization form is

$$\min_{v \in H_0^1(x_l, x_r)} F(v) = \min_{v \in H_0^1(x_l, x_r)} \left\{ \frac{1}{2} a(v, v) - L(v) \right\}. \tag{7.35}$$

Later we will see that all self-adjoint differential equations have both weak and minimization forms, and that the finite element method using the Ritz form is the same as with the Galerkin form.

7.4.4 The FE Method for 1D Sturm–Liouville Problems Using Hat Basis Functions

Consider *any* finite-dimensional space $V_h \subset H_0^1(x_l, x_r)$ with the basis

$$\phi_1(x) \in H_0^1(x_l, x_r)\,,\ \phi_2(x) \in H_0^1(x_l, x_r)\,,\ \ldots\,,\ \phi_M(x) \in H_0^1(x_l, x_r),$$

that is,

$$\begin{aligned} V_h &= span\,\{\phi_1, \phi_2, \ldots, \phi_M\} \\ &= \left\{ v_s, \quad v_s = \sum_{i=1}^{M} \alpha_i \phi_i \right\} \subset H_0^1(x_l, x_r)\,. \end{aligned}$$

The Galerkin finite method assumes the approximate solution to be

$$u_s(x) = \sum_{j=1}^{M} \alpha_j \phi_j(x), \tag{7.36}$$

and the coefficients $\{\alpha_j\}$ are chosen such that the weak form

$$a(u_s, v_s) = (f, v_s), \quad \forall\, v_s \in V_h$$

is satisfied. Thus we enforce the weak form in the finite-dimensional space V_h instead of the solution space $H_0^1(x_l, x_r)$, which introduces some error.

Since *any* element in the space is a linear combination of the basis functions, we have

$$a\,(u_s, \phi_i) = (f, \phi_i)\,, \quad i = 1, 2, \ldots, M\,,$$

or

$$a\left(\sum_{j=1}^{M} \alpha_j\, \phi_j, \phi_i\right) = (f, \phi_i)\,, \quad i = 1, 2, \ldots, M\,,$$

or

$$\sum_{j=1}^{M} a(\phi_j, \phi_i)\, \alpha_j = (f, \phi_i)\,, \quad i = 1, 2, \ldots, M\,.$$

In the matrix-vector form $AX = F$, this system of algebraic equations for the coefficients is

$$\begin{bmatrix} a(\phi_1, \phi_1) & a(\phi_1, \phi_2) & \cdots & a(\phi_1, \phi_M) \\ a(\phi_2, \phi_1) & a(\phi_2, \phi_2) & \cdots & a(\phi_2, \phi_M) \\ \vdots & \vdots & \vdots & \vdots \\ a(\phi_M, \phi_1) & a(\phi_M, \phi_2) & \cdots & a(\phi_M, \phi_M) \end{bmatrix} \begin{bmatrix} \alpha_1 \\ \alpha_2 \\ \vdots \\ \alpha_M \end{bmatrix} = \begin{bmatrix} (f, \phi_1) \\ (f, \phi_2) \\ \vdots \\ (f, \phi_M) \end{bmatrix},$$

and the system has some attractive properties.

- The coefficient matrix A is symmetric, *i.e.*, $\{a_{ij}\} = \{a_{ji}\}$ or $A = A^T$, since $a(\phi_i, \phi_j) = a(\phi_j, \phi_i)$. Note that this is only true for a self-adjoint problem such as the above, with the second-order ODE

$$-(pu')' + qu = f\,.$$

For example, the similar problem involving the ODE

$$-u'' + u' = f$$

is not self-adjoint; and the Galerkin finite element method using the corresponding weak form

$$(u', v') + (u', v) = (f, v) \;\text{ or }\; (u', v') - (u, v') = (f, v)$$

produces terms such as (ϕ_i', ϕ_j) that differ from (ϕ_j', ϕ_i), so that the coefficient matrix A is not symmetric.

- A is positive definite, *i.e.*,

$$x^T A\, x > 0 \text{ if } x \neq 0\,, \text{ and all eigenvalues of } A \text{ are positive.}$$

Proof For any $\eta \neq 0$, we show that $\eta^T A\, \eta > 0$ as follows

$$\begin{aligned}
\eta^T A\, \eta = \eta^T (A\eta) &= \sum_{i=1}^{M} \eta_i \sum_{j=1}^{M} a_{ij}\eta_j \\
&= \sum_{i=1}^{M} \eta_i \sum_{j=1}^{M} a(\phi_i, \phi_j)\eta_j \\
&= \sum_{i=1}^{M} \eta_i \sum_{j=1}^{M} a(\phi_i, \eta_j\phi_j) \\
&= \sum_{i=1}^{M} \eta_i\, a\left(\phi_i, \sum_{j=1}^{M} \eta_j\phi_j\right) \\
&= a\left(\sum_{i=1}^{M} \eta_i\phi_i, \sum_{j=1}^{M} \eta_j\phi_j\right) \\
&= a\,(v_s, v_s) = \|v_s\|_a^2 > 0\,,
\end{aligned}$$

since $v_s = \sum_{i=1}^{M} \eta_i\phi_i \neq 0$ because η is a nonzero vector and the $\{\phi_i\}$'s are linear-independent.

7.4.5 Local Stiffness Matrix and Load Vector Using the Hat Basis Functions

The local stiffness matrix using the hat basis functions is a 2×2 matrix of the following,

$$K_i^e = \begin{bmatrix} \int_{x_i}^{x_{i+1}} p(x)\,(\phi_i')^2 dx & \int_{x_i}^{x_{i+1}} p(x)\,\phi_i'\phi_{i+1}' dx \\ \int_{x_i}^{x_{i+1}} p(x)\,\phi_{i+1}'\phi_i' dx & \int_{x_i}^{x_{i+1}} p(x)\,(\phi_{i+1}')^2 dx \end{bmatrix}
+ \begin{bmatrix} \int_{x_i}^{x_{i+1}} q(x)\,\phi_i^2 dx & \int_{x_i}^{x_{i+1}} q(x)\,\phi_i\phi_{i+1} dx \\ \int_{x_i}^{x_{i+1}} q(x)\,\phi_{i+1}\phi_i dx & \int_{x_i}^{x_{i+1}} q(x)\,\phi_{i+1}^2 dx \end{bmatrix},$$

and the local load vector is

$$F_i^e = \begin{bmatrix} \int_{x_i}^{x_{i+1}} f\phi_i dx \\ \int_{x_i}^{x_{i+1}} f\phi_{i+1} dx \end{bmatrix}.$$

The global stiffness matrix and load vector can be assembled element by element.

7.5 Error Analysis of the FE Method

Error analysis for finite element methods usually includes two parts:

1. error estimates for an intermediate function in V_h, often the interpolation function; and
2. convergence analysis, a limiting process that shows the finite element solution converges to the true solution of the weak form in some norm, as the mesh size h approaches zero.

We first recall some notations and setting up:

1. Given a weak form $a(u, v) = L(v)$ and a space V, which usually has infinite dimension, the problem is to find a $u \in V$ such that the weak form is satisfied for any $v \in V$. Then u is called the solution of the weak form.
2. A *finite-dimensional* subspace of V denoted by V_h (*i.e.*, $V_h \subset V$) is adopted for a conforming finite element method and it does not have to depend on h, however.
3. The solution of the weak form in the subspace V_h is denoted by u_h, *i.e.*, we require $a(u_h, v_h) = L(v_h)$ for any $v_h \in V_h$.
4. The global error is defined by $e_h = u(x) - u_h(x)$, and we seek a sharp upper bound for $\|e_h\|$ using certain norm.

It was noted that error is introduced when the finite-dimensional space replaces the solution space, as the weak form is usually only satisfied in the subspace V_h and not in the solution space V. However, we can prove that the solution satisfying the weak form in the subspace V_h is the best approximation to the exact solution u in the finite-dimensional space in the energy norm.

Theorem 7.9. *With the notations above, we have*

1. u_h is the projection of u onto V_h through the inner product $a(u, v)$, i.e.,

$$u - u_h \perp V_h \text{ or } u - u_h \perp \phi_i, \quad i = 1, 2, \ldots, M, \tag{7.37}$$

$$a(u - u_h, v_h) = 0 \;\forall\, v_h \in V_h \text{ or } a(u - u_h, \phi_i) = 0, \quad i = 1, 2, \ldots, M, \tag{7.38}$$

where $\{\phi_i\}$'s are the basis functions.

2. *u_h is the best approximation in the energy norm, i.e.,*

$$\|u - u_h\|_a \le \|u - v_h\|_a \,,\ \forall\, v_h \in V_h.$$

Proof

$$\begin{aligned} a(u, v) &= (f, v)\,,\ \forall\, v \in V, \\ \to a(u, v_h) &= (f, v_h)\,,\ \forall\, v_h \in V_h \text{ since } V_h \subset V, \\ a(u_h, v_h) &= (f, v_h),\ \forall\, v_h \in V_h \text{ since } u_h \text{ is the solution in } V_h, \\ \text{subtract} \to a(u - u_h, v_h) &= 0 \text{ or } a(e_h, v_h) = 0\,,\ \forall\, v_h \in V_h. \end{aligned}$$

Now we prove that u_h is the best approximation in V_h.

$$\begin{aligned} \|u - v_h\|_a^2 &= a(u - v_h, u - v_h) \\ &= a(u - u_h + u_h - v_h, u - u_h + u_h - v_h) \\ &= a(u - u_h + w_h, u - u_h + w_h)\,, \text{ on letting } \quad w_h = u_h - v_h \in V_h, \\ &= a(u - u_h, u - u_h + w_h) + a(w_h, u - u_h + w_h) \\ &= a(u - u_h, u - u_h) + a(u - u_h, w_h) + a(w_h, u - u_h) + a(w_h, w_h) \\ &= \|u - u_h\|_a^2 + 0 + 0 + \|w_h\|_a^2, \quad \text{since,} \quad a(e_h, u_h) = 0 \\ &\ge \|u - u_h\|_a^2 \end{aligned}$$

i.e., $\|u - u_h\|_a \le \|u - v_h\|_a$. Figure 7.2 is a diagram to illustrate the theorem.

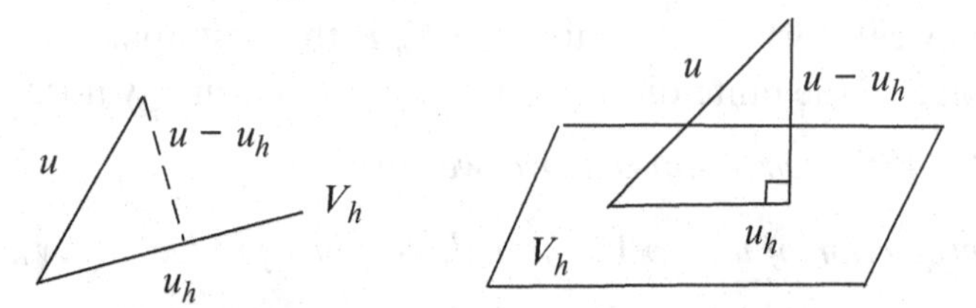

Figure 7.2. A diagram of FE approximation properties. The finite element solution is the best approximation to the solution u in the finite-dimensional space V_h in the energy norm; and the error $u - u_h$ is perpendicular to the finite-dimensional space V_h in the inner product of $a(u, v)$.

Example: For the Sturm–Liouville problem,

$$\begin{aligned}\|u-u_h\|_a^2 &= \int_a^b \left(p(x)\,(u'-u_h')^2 + q(x)\,(u-u_h)^2\right)\,dx \\ &\le p_{max}\int_a^b (u'-u_h')^2 dx + q_{max}\int_a^b (u-u_h)^2\,dx \\ &\le \max\{p_{max}, q_{max}\}\int_a^b \left((u'-u_h')^2 + (u-u_h)^2\right)dx \\ &= C\|u-u_h\|_1^2,\end{aligned}$$

where $C = \max\{p_{max}, q_{max}\}$. Thus, we obtain

$$\begin{aligned}&\|u-u_h\|_a \le \hat{C}\|u-u_h\|_1, \\ &\|u-u_h\|_a \le \|u-v_h\|_a \le \bar{C}\|u-v_h\|_1.\end{aligned}$$

7.5.1 Interpolation Functions and Error Estimates

Usually the solution is unknown; so in order to get the error estimate we choose a special $v_h^* \in V_h$, for which we can get a good error estimate. We may then use the error estimate $\|u-u_h\|_a \le \|u-v_h^*\|_a$ to get an error estimate for the finite element solution (may be overestimated). Usually we can choose a *piecewise interpolation function* for this purpose. That is another reason that we choose piecewise linear, quadratic, or cubic functions over the given mesh in finite element methods.

7.5.1.1 Linear 1D Piecewise Interpolation Function

Given a mesh $x_0, x_1, x_2, \ldots, x_M$, the linear 1D piecewise interpolation function is defined as

$$u_I(x) = \frac{x-x_i}{x_{i-1}-x_i}u(x_{i-1}) + \frac{x-x_{i-1}}{x_i-x_{i-1}}u(x_i), \quad x_{i-1} \le x \le x_i.$$

It is obvious that $u_I(x) \in V_h$, where $V_h \subset H^1$ is the set of continuous piecewise linear functions that have the first-order weak derivative, so

$$\|u-u_h\|_a \le \|u-u_I\|_a.$$

Since $u(x)$ is unknown, then so is $u_I(x)$. Nevertheless, we know the *upper error bound* of the interpolation functions.

Theorem 7.10. *Given a function $u(x) \in C^2[a,b]$ and a mesh $x_0, x_1, x_2, \ldots, x_M$, the continuous piecewise linear function u_I has the error estimates*

$$\|u - u_I\|_\infty = \max_{x\in[a,b]} |u(x) - u_I(x)| \le \frac{h^2}{8}\|u''\|_\infty\,, \tag{7.39}$$

$$\|u'(x) - u_I'(x)\|_{L^2(a,b)} \le h\sqrt{b-a}\|u''\|_\infty\,. \tag{7.40}$$

Proof If $\tilde{e_h} = u(x) - u_I(x)$, then $\tilde{e_h}(x_{i-1}) = \tilde{e_h}(x_i) = 0$. From Rolle's theorem, there must be at least one point z_i between x_{i-1} and x_i such that $\tilde{e_h}'(z_i) = 0$, hence

$$\begin{aligned}
\tilde{e_h}'(x) &= \int_{z_i}^{x} \tilde{e_h}''(t)\,dt \\
&= \int_{z_i}^{x} \left(u''(t) - u_I''(t)\right) dt \\
&= \int_{z_i}^{x} u''(t)\,dt\,.
\end{aligned}$$

Therefore, we obtain the error estimates below

$$|\tilde{e_h}'(x)| \le \int_{z_i}^{x} |u''(t)|dt \le \|u''\|_\infty \int_{z_i}^{x} dt \le \|u''\|_\infty h\,, \quad \text{and}$$

$$\|\tilde{e_h}'\|_{L^2(a,b)} = \|\tilde{e_h}'\|_0 \le \left\{ \|u''\|_\infty^2 \int_a^b h^2 dt \right\}^{\frac{1}{2}} \le \sqrt{b-a}\|u''\|_\infty h\,;$$

so we have proved the second inequality. To prove the first, assume that $x_{i-1} + h/2 \le z_i \le x_i$, otherwise we can use the other half interval. From the Taylor expansion

$$\begin{aligned}
\tilde{e_h}(x) &= \tilde{e_h}(z_i + x - z_i)\,, \text{ assuming } x_{i-1} \le x \le x_i\,, \\
&= \tilde{e_h}(z_i) + \tilde{e_h}'(z_i)(x - z_i) + \frac{1}{2}\tilde{e_h}''(\xi)(x - z_i)^2, \ x_{i-1} \le \xi \le x_i\,, \\
&= \tilde{e_h}(z_i) + \frac{1}{2}\tilde{e_h}''(\xi)(x - z_i)^2\,,
\end{aligned}$$

so at $x = x_i$ we have

$$\begin{aligned}
0 = \tilde{e_h}(x_i) &= \tilde{e_h}(z_i) + \frac{1}{2}\tilde{e_h}''(\xi)(x_i - z_i)^2\,, \\
\tilde{e_h}(z_i) &= -\frac{1}{2}\tilde{e_h}''(\xi)(x_i - z_i)^2\,, \\
|\tilde{e_h}(z_i)| &\le \frac{1}{2}\|u''\|_\infty (x_i - z_i)^2 \le \frac{h^2}{8}\|u''\|_\infty\,.
\end{aligned}$$

Note that the largest value of $\tilde{e_h}(x)$ has to be the z_i where the derivative is zero.

7.5.2 Error Estimates of the Finite Element Methods Using the Interpolation Function

Theorem 7.11. *For the 1D Sturm–Liouville problem, the following error estimates hold,*

$$\|u-u_h\|_a \le Ch\|u''\|_\infty,$$

$$\|u-u_h\|_1 \le \hat{C}h\|u''\|_\infty\,,$$

where C and $\hat{C}$ are two constants.

Proof

$$\begin{aligned}
\|u-u_h\|_a^2 &\le \|u-u_I\|_a^2 \\
&\le \int_a^b \left(p(x)\,(u'-u_I')^2 + q(x)\,(u-u_I)^2\right)\,dx \\
&\le \max\{p_{max}, q_{max}\}\int_a^b \left(\,(u'-u_I')^2 + (u-u_I)^2\right)\,dx \\
&\le \max\{p_{max}, q_{max}\}\,\|u''\|_\infty^2 \int_a^b \left(h^2 + h^4/64\right)\,dx \\
&\le Ch^2\|u''\|_\infty\,.
\end{aligned}$$

The second inequality is obtained because $\|\ \|_a$ and $\|\ \|_1$ are equivalent, so

$$c\|v\|_a \le \|v\|_1 \le C\|v\|_a\,, \quad \hat{c}\|v\|_1 \le \|v\|_a \le \hat{C}\|v\|_1\,.$$

7.5.3 Error Estimate in the Pointwise Norm

We can easily prove the following error estimate.

Theorem 7.12. *For the 1D Sturm–Liouville problem,*

$$\|u-u_h\|_\infty \le Ch\|u''\|_\infty, \tag{7.41}$$

where C is a constant. The estimate is not sharp (or optimal); or simply it is overestimated.

We note that u_h' is discontinuous at nodal points, and the infinity norm $\|u'-u_h'\|_\infty$ can only be defined for continuous functions.

Proof

$$e_h(x) = u(x) - u_h(x) = \int_a^x e_h'(t)dt$$

$$\begin{aligned}
|e_h(x)| &\le \int_a^b |e_h'(t)|dt \\
&\le \left\{\int_a^b |e_h'|^2 dt\right\}^{1/2} \left\{\int_a^b 1\,dt\right\}^{1/2} \\
&\le \sqrt{b-a}\left\{\int_a^b \frac{p}{p_{min}}|e_h'|^2 dt\right\}^{1/2} \\
&\le \sqrt{\frac{b-a}{p_{min}}}\,\|e_h\|_a \\
&\le \sqrt{\frac{b-a}{p_{min}}}\,\|\tilde{e}_h\|_a \\
&\le Ch\,\|u''\|_\infty .
\end{aligned}$$

Remark 7.13. Actually, we can prove a better inequality

$$\|u - u_h\|_\infty \le Ch^2\|u''\|_\infty,$$

so the finite element method is second-order accurate. This is an optimal (sharp) error estimate.

Exercises

1. Assuming the number of variables $n=3$, describe the Sobolev space $H^3(\Omega)$ (*i.e.*, for $m=3$) in terms of $L^2(\Omega)$, retaining all the terms but not using the multi-index notation. Then using the multi-index notation when applicable, represent the inner product, the norm, the Schwarz inequality, the distance, and the Sobolev embedding theorem in this space.
2. Consider the function $v(x)=|x|^\alpha$ on $\Omega=(-1,1)$ with $\alpha \in \mathcal{R}$. For what values of α is $v \in H^0(\Omega)$? (Consider both positive and negative α.) For what values is $v \in H^1(\Omega)$? in $H^m(\Omega)$? For what values of α is $v \in C^m(\Omega)$?

Hint: Make use of the following

$$|x|^\alpha = \begin{cases} x^\alpha & \text{if } x \ge 0 \\ (-x)^\alpha & \text{if } x < 0 , \end{cases}$$

and when k is a nonnegative integer note that

$$|x|^\alpha = \begin{cases} x^{2k} & \text{if } \alpha = 2k \\ 1 & \text{if } \alpha = 0 ; \end{cases}$$

also

$$\lim_{x \to 0} |x|^\alpha = \begin{cases} 0 & \text{if } \alpha > 0 \\ 1 & \text{if } \alpha = 0 \\ \infty & \text{if } \alpha < 0 \end{cases} \quad \text{and} \quad \int_{-1}^{1} |x|^\alpha dx = \begin{cases} \dfrac{2}{\alpha + 1} & \text{if } \alpha > -1 \\ \infty & \text{if } \alpha \le -1 . \end{cases}$$

3. Are each of the following statements true or false? Justify your answers.
   (a) If $u \in H^2(0,1)$, then u' and u'' are both continuous functions.
   (b) If $u(x,y) \in H^2(\Omega)$, then $u(x,y)$ may not have continuous partial derivatives $\partial u/\partial x$ and $\partial u/\partial y$.

   Does $u(x,y)$ have first- and second-order *weak* derivatives?
   Is $u(x,y)$ continuous in Ω?
4. Consider the Sturm–Liouville problem

$$-\left(p(x)u(x)'\right)' + q(x)u(x) = f(x) , \quad 0 < x < \pi ,$$

$$\alpha u(0) + \beta u'(0) = \gamma , \qquad u'(\pi) = u_b ,$$

where

$$0 < p_{min} \le p(x) \le p_{max} < \infty ,$$

$$0 \le q_{min} \le q(x) \le q_{max} < q_\infty .$$

   (a) Derive the weak form for the problem. Define a bilinear form $a(u, v)$ and a linear form $L(v)$ to simplify the weak form. What is the energy norm?
   (b) What kind of restrictions should we have for α, β, and γ in order that the weak form has a solution?
   (c) Determine the space where the solution resides under the weak form.
   (d) If we look for a finite element solution in a finite-dimensional space V_h using a conforming finite element method, should V_h be a subspace of C^0, or C^1, or C^2?
   (e) Given a mesh $x_0 = 0 < x_1 < x_2 < \cdots < x_{M-1} < x_M = \pi$, if the finite-dimensional space is generated by the hat functions, what kind of structure do the local and global stiffness matrix and the load vector have? Is the resulting linear system of equations formed by the global stiffness matrix and the load vector symmetric, positive definite and banded?

5. Consider the two-point BVP

$$-u''(x) = f(x), \quad a < x < b, \qquad u(a) = u(b) = 0\,.$$

Let $u_h(x)$ be the finite element solution using the piecewise linear space (in $H_0^1(a,b)$) spanned by a mesh $\{x_i\}$. Show that

$$\|u - u_h\|_\infty \le Ch^2,$$

where C is a constant.
Hint: First show $\|u_h - u_I\|_a = 0$, where $u_I(x)$ is the interpolation function in V_h.

8

Issues of the FE Method in One Space Dimension

8.1 Boundary Conditions

For a second-order two-point BVP, typical boundary conditions (BC) include one of the following at each end, say at $x = x_l$,

1. a Dirichlet condition, *e.g.*, $u(x_l) = u_l$ is given;
2. a Neumann condition, *e.g.*, $u'(x_l)$ is given; or
3. a Robin (mixed) condition, *e.g.*, $\alpha u(x_l) + \beta u'(x_l) = \gamma$ is given, where α, β, and γ are known but $u(x_l)$ and $u'(x_l)$ are both unknown.

Boundary conditions affect the bilinear and linear form, and the solution space.

Example 8.1. For example,

$$
\begin{aligned}
-u'' &= f, \qquad & 0 < x < 1, \\
u(0) &= 0, \qquad & u'(1) = 0,
\end{aligned}
$$

involves a Dirichlet BC at $x = 0$ and a Neumann BC at $x = 1$. To derive the weak form, we again follow the familiar procedure:

$$
\begin{aligned}
\int_0^1 -u'' v dx &= \int_0^1 fv\, dx, \\
-u'\, v\big|_0^1 + \int_0^1 u'v' dx &= \int_0^1 fv\, dx, \\
-u'(1)v(1) + u'(0)v(0) + \int_0^1 u'v' dx &= \int_0^1 fv\, dx\,.
\end{aligned}
$$

For a conforming finite element method, the solution function u and the test functions v should be in the same space. So it is natural to require that the test functions v satisfy the same homogeneous Dirichlet BC, *i.e.*, we require $v(0) = 0$; and the Dirichlet condition is therefore called an *essential* boundary condition. On the other hand, since $u'(1)=0$, the first term in the final expression is zero, so it does not matter what $v(1)$ is, *i.e.*, there is no constraint on $v(1)$; so the Neumann BC is called a *natural* boundary condition. It is noted that $u(1)$ is unknown. The weak form of this example is the same as before for homogeneous Dirichlet BC at both ends; but now the solution space is different:

$$(u',v') = (f,v)\,,\ \forall v \in H^1_E(0,1),$$

$$\text{where} \qquad H^1_E(0,1) = \left\{ v(x)\,,\ v(0)=0\,,\ v \in H^1(0,1) \right\},$$

where we use the subscript E in $H^1_E(0,1)$ to indicated an essential boundary condition.

8.1.1 Mixed Boundary Conditions

Consider a Sturm–Liouville problem

$$-(pu')' + qu = f, \quad x_l < x < x_r, \quad p(x) \geq p_{min} > 0\,, \quad q(x) \geq 0\,, \tag{8.1}$$

$$u(x_l) = 0, \quad \alpha u(x_r) + \beta u'(x_r) = \gamma, \quad \beta \neq 0, \quad \frac{\alpha}{\beta} \geq 0, \tag{8.2}$$

where α, β, and γ are known constants but $u(x_r)$ and $u'(x_r)$ are unknown. Integration by parts again gives

$$-p(x_r)u'(x_r)v(x_r) + p(x_l)u'(x_l)v(x_l) + \int_{x_l}^{x_r} (pu'v' + quv)\, dx = \int_{x_l}^{x_r} fv\, dx. \tag{8.3}$$

As explained earlier, we set $v(x_l)=0$ (an essential BC). Now we reexpress the mixed BC as

$$u'(x_r) = \frac{\gamma - \alpha u(x_r)}{\beta}, \tag{8.4}$$

and substitute this into (8.3) to obtain

$$-p(x_r)v(x_r)\frac{\gamma - \alpha u(x_r)}{\beta} + \int_{x_l}^{x_r} (pu'v' + quv)\; dx = \int_{x_l}^{x_r} fv\, dx$$

or equivalently

$$\int_{x_l}^{x_r} (pu'v' + quv)\; dx + \frac{\alpha}{\beta}p(x_r)u(x_r)v(x_r) = \int_{x_l}^{x_r} fv\, dx + \frac{\gamma}{\beta}p(x_r)v(x_r), \tag{8.5}$$

which is the corresponding weak (variational) form of the Sturm–Liouville problem. We define

$$a(u,v) = \int_{x_l}^{x_r} (pu'v' + quv)\,dx + \frac{\alpha}{\beta}p(x_r)u(x_r)v(x_r), \text{ the bilinear form,} \tag{8.6}$$

$$L(v) = (f,v) + \frac{\gamma}{\beta}p(x_r)v(x_r), \text{ the linear form.} \tag{8.7}$$

We can prove that:

1. $a(u,v) = a(v,u)$, *i.e.*, $a(u,v)$ is symmetric;
2. $a(u,v)$ is a bilinear form, *i.e.*,

$$a(ru+sw,v) = ra(u,v) + sa(w,v),$$
$$a(u,rv+sw) = ra(u,v) + sa(u,w),$$

for any real numbers r and s; and
3. $a(u,v)$ is an inner product, and the corresponding energy norm is

$$\|u\|_a = \sqrt{a(u,u)} = \left\{\int_{x_l}^{x_r}\left(pu'^2 + qu^2\right)dx + \frac{\alpha}{\beta}p(x_r)u(x_r)^2\right\}^{\frac{1}{2}}.$$

It is now evident why we require $\beta \neq 0$, and $\alpha/\beta \geq 0$. Using the inner product, the solution of the weak form $u(x)$ satisfies

$$a(u,v) = L(v), \quad \forall v \in H_E^1(x_l, x_r), \tag{8.8}$$

$$H_E^1(x_l,x_r) = \left\{v(x), \quad v(x_l) = 0, \ v \in H^1(x_l,x_r)\right\}, \tag{8.9}$$

and we recall that there is no restriction on $v(x_r)$. The boundary condition is essential at $x = x_l$, but natural at $x = x_r$. The solution u is also the minimizer of the functional

$$F(v) = \frac{1}{2}a(v,v) - L(v)$$

in the $H_E^1(x_l,x_r)$ space.

8.1.2 Nonhomogeneous Dirichlet Boundary Conditions

Suppose now that $u(x_l) = u_l \neq 0$ in (8.2). In this case, the solution can be decomposed as the sum of the particular solution

$$\begin{aligned} &-(pu_1')' + qu_1 = 0, \quad x_l < x < x_r, \\ &u_1(x_l) = u_l, \quad \alpha u_1(x_r) + \beta u_1'(x_r) = 0, \quad \beta \neq 0, \quad \frac{\alpha}{\beta} \geq 0, \end{aligned} \tag{8.10}$$

and

$$\begin{aligned} &-(pu_2')' + qu_2 = f, \quad x_l < x < x_r, \\ &u_2(x_l) = 0, \quad \alpha u_2(x_r) + \beta u_2'(x_r) = \gamma, \quad \beta \neq 0, \quad \frac{\alpha}{\beta} \geq 0. \end{aligned} \tag{8.11}$$

We can use the weak form to find the solution $u_2(x)$ corresponding to the homogeneous Dirichlet BC at $x = x_l$. If we can find a particular solution $u_1(x)$ of (8.10), then the solution to the original problem is $u(x) = u_1(x) + u_2(x)$.

Another simple way is to choose a function $u_0(x)$, $u_0(x) \in H^1(x_l, x_r)$ that satisfies

$$u_0(x_l) = u_l, \quad \alpha u_0(x_r) + \beta u_0'(x_r) = 0.$$

For example, the function $u_0(x) = u_l \phi_0(x)$ would be such a function, where $\phi_0(x)$ is the hat function centered at x_l if a mesh $\{x_i\}$ is given. Then $\hat{u}(x) = u(x) - u_0(x)$ would satisfy a homogeneous Dirichlet BC at x_l and the following *S-L* problem:

$$\begin{aligned} &-(p\hat{u}')' + q\hat{u} = f(x) + (pu_0')' - qu_0, \quad x_l < x < x_r, \\ &\hat{u}(x_l) = 0, \quad \alpha \hat{u}'(x_r) + \beta \hat{u}(x_r) = \gamma. \end{aligned}$$

We can apply the finite element method for $\hat{u}(x)$ previously discussed with the modified source term $f(x)$, where the weak form $u(x)$ after substituting back is the same as before:

$$a_1(\hat{u}, v) = L_1(v), \qquad \forall\, v(x) \in H_E^1(x_l, x_r),$$

where

$$\begin{aligned} a_1(\hat{u}, v) &= \int_{x_l}^{x_r} \left(p\hat{u}'v' + q\hat{u}v\right)\, dx + \frac{\alpha}{\beta} p(x_r)\hat{u}(x_r)v(x_r) \\ L_1(v) &= \int_{x_l}^{x_r} fv dx + \frac{\gamma}{\beta} p(x_r)v(x_r) + \int_{x_l}^{x_r} \left((pu_0')'v - qu_0 v\right)\, dx \\ &= \int_{x_l}^{x_r} fv dx + \frac{\gamma}{\beta} p(x_r)v(x_r) - \int_{x_l}^{x_r} \left(pu_0'v' + qu_0 v\right)\, dx \\ &\quad - \frac{\alpha}{\beta} p(x_r)u_0(x_r)v(x_r). \end{aligned}$$

If we define

$$a(u,v) = \int_{x_l}^{x_r} (pu'v' + quv)\, dx + \frac{\alpha}{\beta} p(x_r)u(x_r)v(x_r),$$

$$L(v) = \int_{x_l}^{x_r} fv dx + \frac{\gamma}{\beta} p(x_r)v(x_r),$$

as before, then we have

$$a_1(u-u_0,v) = a(u-u_0,v) = L_1(v) = L(v) - a(u_0,v), \quad \text{or} \quad a(u,v) = L(v). \tag{8.12}$$

While we still have $a(u,v) = L(v)$, the solution is not in $H_E^1(x_l,x_r)$ space due to the nonhomogeneous Dirichlet boundary condition. Nevertheless $u - u_0$ is in $H_E^1(x_l,x_r)$. The formula above is also the basis of the numerical treatment of Dirichlet boundary conditions later.

8.2 The FE Method for Sturm–Liouville Problems

Let us now consider the finite element method using the piecewise linear function over a mesh $x_1 = x_l, x_2, \ldots, x_M = x_r$ (see a diagram in Figure 8.1) for the Sturm–Liouville problem

$$-(pu')' + qu = f, \quad x_l < x < x_r,$$

$$u(x_l) = u_l, \quad \alpha u(x_r) + \beta u'(x_r) = \gamma, \quad \beta \neq 0, \quad \frac{\alpha}{\beta} \geq 0.$$

We again use the hat functions as the basis such that

$$u_h(x) = \sum_{i=0}^{M} \alpha_i \phi_i(x),$$

and now focus on the treatment of the BC.

The solution is unknown at $x = x_r$, so it is not surprising to have $\phi_M(x)$ for the natural BC. The first term $\phi_0(x)$ is the function used as $u_0(x)$, to deal with

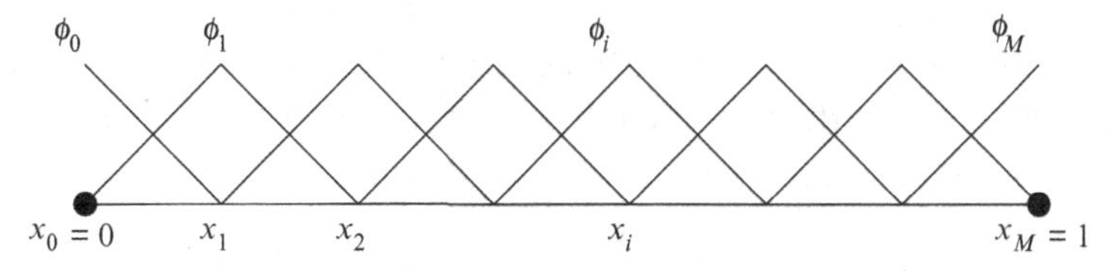

Figure 8.1. Diagram of a 1D mesh where $x_l = 0$ and $x_r = 1$.

the nonhomogeneous Dirichlet BC. The local stiffness matrix is

$$K_i^e = \begin{bmatrix} a(\phi_i,\phi_i) & a(\phi_i,\phi_{i+1}) \\ a(\phi_{i+1},\phi_i) & a(\phi_{i+1},\phi_{i+1}) \end{bmatrix}_{(x_i,x_{i+1})} = \begin{bmatrix} \int_{x_i}^{x_{i+1}} p\phi_i'^2\,dx & \int_{x_i}^{x_{i+1}} p\phi_i'\phi_{i+1}'\,dx \\ \int_{x_i}^{x_{i+1}} p\phi_{i+1}'\phi_i'\,dx & \int_{x_i}^{x_{i+1}} p{\phi_{i+1}'}^2\,dx \end{bmatrix}$$

$$+ \begin{bmatrix} \int_{x_i}^{x_{i+1}} q{\phi_i}^2 dx & \int_{x_i}^{x_{i+1}} q\phi_i\phi_{i+1}\,dx \\ \int_{x_i}^{x_{i+1}} q\phi_{i+1}\phi_i\,dx & \int_{x_i}^{x_{i+1}} q\phi_{i+1}^2\,dx \end{bmatrix}$$

$$+ \frac{\alpha}{\beta}p(x_r) \begin{bmatrix} \phi_i^2(x_r) & \phi_i(x_r)\phi_{i+1}(x_r) \\ \phi_{i+1}(x_r)\phi_i(x_r) & \phi_{i+1}^2(x_r) \end{bmatrix},$$

and the local load vector is

$$F_i^e = \begin{bmatrix} L(\phi_i) \\ L(\phi_{i+1}) \end{bmatrix} = \begin{bmatrix} \int_{x_i}^{x_{i+1}} f\phi_i dx \\ \int_{x_i}^{x_{i+1}} f\phi_{i+1} dx \end{bmatrix} + \frac{\gamma}{\beta}p(x_r) \begin{bmatrix} \phi_i(x_r) \\ \phi_{i+1}(x_r) \end{bmatrix}.$$

We can see clearly the contributions from the BC; and in particular, that the only nonzero contribution of the BC to the stiffness matrix and the load vector is from the last element (x_{M-1}, x_M) due to the compactness of the hat functions.

8.2.1 Numerical Treatments of Dirichlet BC

The finite element solution defined on the mesh can be written as

$$u_h(x) = u_l\,\phi_0(x) + \sum_{j=1}^{M} \alpha_j\phi_j(x),$$

where $\alpha_j, j = 1, 2, \ldots, M$ are the unknowns. Note that $u_l\phi_0(x)$ is an approximate particular solution in $H^1(x_l, x_r)$, and satisfies the Dirichlet boundary condition at $x = x_l$ and homogeneous Robin boundary condition at $x = x_r$. To use the finite element method to determine the coefficients, we enforce the weak form for $u_h(x) - u_a\phi_0(x)$ for the modified differential problem,

$$\begin{aligned} &-(pu')' + qu = f + u_l\,(p\,\phi_0')' - u_l q\,\phi_0, \quad x_l < x < x_r, \\ &u(x_l) = 0, \quad \alpha u(x_r) + \beta u'(x_r) = \gamma, \quad \beta \neq 0, \quad \frac{\alpha}{\beta} \geq 0. \end{aligned} \tag{8.13}$$

Thus the system of linear equations is

$$\hat{a}\,(u_h(x), \phi_i(x)) = \hat{L}(\phi_i), \quad i = 1, 2, \ldots, M,$$

where $\hat{a}(:,:)$ and $\hat{L}(:)$ are the bilinear and linear forms for the BVP above, or equivalently

$$\begin{aligned}
a(\phi_1, \phi_1)\alpha_1 + a(\phi_1, \phi_2)\alpha_2 + \cdots + a(\phi_1, \phi_M)\alpha_M &= L(\phi_1) - a(\phi_0, \phi_1)\, u_l \\
a(\phi_2, \phi_1)\alpha_1 + a(\phi_2, \phi_2)\alpha_2 + \cdots + a(\phi_2, \phi_M)\alpha_M &= L(\phi_2) - a(\phi_0, \phi_2)\, u_l \\
\cdots\cdots\cdots\cdots &= \cdots\cdots\cdots\cdots \\
a(\phi_M, \phi_1)\alpha_1 + a(\phi_M, \phi_2)\alpha_2 + \cdots + a(\phi_M, \phi_M)\alpha_M &= L(\phi_M) - a(\phi_0, \phi_M)\, u_l,
\end{aligned}$$

where the bilinear and linear forms are still

$$\begin{aligned}
a(u, v) &= \int_{x_l}^{x_r} (pu'v' + quv)\ dx + \frac{\alpha}{\beta} p(x_r) u(x_r) v(x_r); \\
L(v) &= \int_{x_l}^{x_r} fv dx + \frac{\gamma}{\beta} p(x_r) v(x_r),
\end{aligned}$$

since

$$\int_{x_l}^{x_r} \left(u_l (p\,\phi_0')' - u_l q\,\phi_0\right) \phi_i(x) dx = -u_l a(\phi_0, \phi_i).$$

After moving the $a(\phi_i, u_l\phi_0(x)) = a(\phi_i, \phi_0(x))\, u_l$ to the right-hand side, we get the matrix-vector form

$$\begin{bmatrix} 1 & 0 & \cdots & 0 \\ 0 & a(\phi_1, \phi_1) & \cdots & a(\phi_1, \phi_M) \\ \vdots & \vdots & \vdots & \vdots \\ 0 & a(\phi_M, \phi_1) & \cdots & a(\phi_M, \phi_M) \end{bmatrix} \begin{bmatrix} \alpha_0 \\ \alpha_1 \\ \vdots \\ \alpha_M \end{bmatrix} = \begin{bmatrix} u_l \\ L(\phi_1) - a(\phi_1, \phi_0) u_l \\ \vdots \\ L(\phi_M) - a(\phi_M, \phi_0) u_l \end{bmatrix}.$$

This outlines one method to deal with a nonhomogeneous Dirichlet boundary condition.

8.2.2 Contributions from Neumann or Mixed BC

The contribution of mixed boundary condition at x_r using the hat basis functions is zero until the last element $[x_{M-1}, x_M]$, where $\phi_M(x_r)$ is not zero.

The local stiffness matrix of the last element is

$$\begin{bmatrix} \int_{x_{M-1}}^{x_M} \left(p{\phi'_{M-1}}^2 + q\phi_{M-1}^2 \right) dx & \int_{x_{M-1}}^{x_M} \left(p\phi'_{M-1}\phi'_M + q\phi_{M-1}\phi_M \right) dx \\ \int_{x_{M-1}}^{x_M} \left(p\phi'_M\phi'_{M-1} + q\phi_M\phi_{M-1} \right) dx & \int_{x_{M-1}}^{x_M} \left(p{\phi'_M}^2 + \phi_M^2 \right) dx + \frac{\alpha}{\beta} p(x_r) \end{bmatrix},$$

and the local load vector is

$$F_{M-1}^e = \begin{bmatrix} \int_{x_{M-1}}^{x_M} f(x)\phi_{M-1}(x)\, dx \\ \int_{x_{M-1}}^{x_M} f(x)\phi_M(x)\, dx + \frac{\gamma}{\beta} p(x_r) \end{bmatrix}.$$

8.2.3 Pseudo-code of the FE Method for 1D Sturm–Liouville Problems Using the Hat Basis Functions

- Initialize:

```
for i=0, M
    F(i) = 0
    for j=0, M
        A(i,j) = 0
    end
end
```

- Assemble the coefficient matrix element by element:

for i=1, M

$$A(i-1,i-1) = A(i-1,i-1) + \int_{x_{i-1}}^{x_i} \left(p{\phi'_{i-1}}^2 + q\phi_{i-1}^2 \right) dx$$

$$A(i-1,i) = A(i-1,i) + \int_{x_{i-1}}^{x_i} \left(p\phi'_{i-1}\phi'_i + q\phi_{i-1}\phi_i \right) dx$$

$$A(i,i-1) = A(i-1,i)$$

$$A(i,i) = A(i,i) + \int_{x_{i-1}}^{x_i} \left(p\phi_i'^2 + q\phi_i^2 \right) dx$$

$$F(i-1) = F(i-1) + \int_{x_{i-1}}^{x_i} f(x)\phi_{i-1}(x)\, dx$$

$$F(i) = F(i) + \int_{x_{i-1}}^{x_i} f(x)\phi_i(x)\, dx$$

end.

- Deal with the Dirichlet BC:

$$
\begin{aligned}
&A(0,0)=1; \qquad F(0)=u_a \\
&\text{for } i{=}1, M \\
&\qquad F(i)=F(i) - A(i,0) * ua; \\
&\qquad A(i,0)=0; \qquad A(0,i)=0; \\
&\text{end.}
\end{aligned}
$$

- Deal with the Mixed BC at $x=b$.

$$
\begin{aligned}
A(M,M) &= A(M,M) + \frac{\alpha}{\beta} p(b); \\
F(M) &= F(M) + \frac{\gamma}{\beta} p(b).
\end{aligned}
$$

- Solve $AU=F$.
- Carry out the error analysis.

8.3 High-order Elements

To solve the Sturm–Liouville or other problems involving second-order differential equations, we can use the piecewise linear finite-dimensional space over a mesh. The error is usually $O(h)$ in the energy and H^1 norms, and $O(h^2)$ in the L^2 and L^∞ norms. If we want to improve the accuracy, we can choose to:

- refine the mesh, *i.e.*, decrease h; or
- use more accurate (high order) and larger finite-dimensional spaces, *i.e.*, the piecewise quadratic or piecewise cubic basis functions.

Let us use the Sturm–Liouville problem

$$
\begin{gathered}
-\left(p'u\right)' + qu = f, \quad x_l < x < x_r, \\
u(x_l)=0, \quad u(x_r)=0,
\end{gathered}
$$

again as the model problem for the discussion here. The other boundary conditions can be treated in a similar way, as discussed before. We assume a given mesh

$$
\begin{aligned}
&x_0=x_l,\ x_1,\ \ldots,\ x_M=x_r, \quad \text{and the elements,} \\
&\Omega_1=(x_0,x_1), \quad \Omega_2=(x_1,x_2), \ \ldots, \ \Omega_M=(x_{M-1},x_M),
\end{aligned}
$$

and consider piecewise quadratic and piecewise cubic functions, but still require the finite-dimensional spaces to be in $H_0^1(x_l,x_r)$ so that the finite element methods are conforming.

8.3.1 Piecewise Quadratic Basis Functions

Define

$$V_h = \left\{ v(x) \, , \text{ where } v(x) \text{ is continuous piecewise quadratic in } H^1 \right\},$$

over a given mesh. The piecewise linear finite-dimensional space is obviously a subspace of the space defined above, so the finite element solution is expected to be more accurate than the one obtained using the piecewise linear functions.

To use the Galerkin finite element method, we need to know *the dimension of the space* V_h of the piecewise quadratic functions in order to choose a set of basis functions. The dimension of a finite-dimensional space is sometimes called the *the degree of freedom* (DOF). Given a function $\phi(x)$ in V_h, on each element a quadratic function has the form

$$\phi(x) = a_i x^2 + b_i x + c_i, \quad x_i \le x < x_{i+1},$$

so there are three parameters to determine a quadratic function in the interval (x_i, x_{i+1}). In total, there are M elements, and so $3M$ parameters. However, they are not totally free, because they have to satisfy the continuity condition

$$\lim_{x \to x_i -} \phi(x) = \lim_{x \to x_i +} \phi(x)$$

for $x_1, x_2, \ldots, x_{M-1}$, or more precisely,

$$a_{i-1} x_i^2 + b_{i-1} x_i + c_{i-1} = a_i x_i^2 + b_i x_i + c_i, \quad i = 1, 2, \ldots, M.$$

There are $M - 1$ interior nodal points, so there are $M - 1$ constraints, and $\phi(x)$ should also satisfy the BC $\phi(x_l) = \phi(x_r) = 0$. Thus the total degree of freedom, the dimension of the finite element space, is

$$3M - (M - 1) - 2 = 2M - 1.$$

We now know that the dimension of V_h is $2M - 1$. If we can construct $2M - 1$ basis functions that are linearly independent, then all of the functions in V_h can be expressed as linear combinations of them. The desired properties are similar to those of the hat basis functions; and they should

- be continuous piecewise quadratic functions;
- have minimum support, *i.e.*, be zero almost everywhere; and
- be determined by the values at nodal points (we can choose the nodal values to be unity at one point and zero at the other nodal points).

Since the degree of freedom is $2M - 1$ and there are only $M - 1$ interior nodal points, we add M auxiliary points (not nodal points) between x_i and

x_{i+1} and define

$$z_{2i} = x_i, \quad \text{nodal points,} \tag{8.14}$$

$$z_{2i+1} = \frac{x_i + x_{i+1}}{2}, \quad \text{auxiliary points.} \tag{8.15}$$

For instance, if the nodal points are $x_0 = 0$, $x_1 = \pi/2$, $x_2 = \pi$, then $z_0 = x_0$, $z_1 = \pi/4$, $z_2 = x_1 = \pi/2$, $z_3 = 3\pi/4$, $z_4 = x_2 = \pi$. Note that in general all the basis functions should be one piece in one element (z_{2k}, z_{2k+2}), $k = 0, 1, \ldots, M-1$. Now we can define the piecewise quadratic basis functions as

$$\phi_i(z_j) = \begin{cases} 1 & \text{if } i = j, \\ 0 & \text{otherwise.} \end{cases} \tag{8.16}$$

We can derive analytic expressions of the basis functions using the properties of quadratic polynomials. As an example, let us consider how to construct the basis functions in the first element (x_0, x_1) corresponding to the interval (z_0, z_2). In this element, z_0 is the boundary point, z_2 is a nodal point, and $z_1 = (z_0 + z_2)/2$ is the mid-point (the auxiliary point). For $\phi_1(x)$, we have $\phi_1(z_0) = 0$, $\phi_1(z_1) = 1$ and $\phi_1(z_2) = 0$, $\phi_1(z_j) = 0$, $j = 3, \ldots, 2M-1$; so in the interval (z_0, z_2), $\phi_1(x)$ has the form

$$\phi_1(x) = C(x - z_0)(x - z_2),$$

because z_0 and z_2 are roots of $\phi_1(x)$. We choose C such that $\phi_1(z_1) = 1$, so

$$\phi_1(z_1) = C(z_1 - z_0)(z_1 - z_2) = 1, \quad \Longrightarrow C = \frac{1}{(z_1 - z_0)(z_1 - z_2)}$$

and the basis function $\phi_1(x)$ is

$$\phi_1(x) = \begin{cases} \dfrac{(x - z_0)(x - z_2)}{(z_1 - z_0)(z_1 - z_2)} & \text{if } z_0 \le x < z_2, \\ 0 & \text{otherwise.} \end{cases}$$

It is easy to verify that $\phi_1(x)$ is a continuous piecewise quadratic function in the domain (x_0, x_M). Similarly, we have

$$\phi_2(x) = \frac{(z - z_1)(z - z_0)}{(z_2 - z_1)(z_2 - z_0)}.$$

Generally, the three global basis functions that are nonzero in the element (x_i, x_{i+1}) have the following forms:

$$\phi_{2i}(z) = \begin{cases} 0 & \text{if } z < x_{i-1} \\ \dfrac{(z-z_{2i-1})(z-z_{2i-2})}{(z_{2i}-z_{2i-1})(z_{2i}-z_{2i-2})} & \text{if } x_{i-1} \le z < x_i \\ \dfrac{(z-z_{2i+1})(z-z_{2i+2})}{(z_{2i}-z_{2i+1})(z_{2i}-z_{2i+2})} & \text{if } x_i \le z < x_{i+1} \\ 0 & \text{if } x_{i+1} < z. \end{cases}$$

$$\phi_{2i+1}(z) = \begin{cases} 0 & \text{if } z < x_i \\ \dfrac{(z-z_{2i})(z-z_{2i+2})}{(z_{2i+1}-z_{2i})(z_{2i+1}-z_{2i+2})} & \text{if } x_i \le z < x_{i+1} \\ 0 & \text{if } x_{i+1} < z. \end{cases}$$

$$\phi_{2i+2}(z) = \begin{cases} 0 & \text{if } z < x_i \\ \dfrac{(z-z_{2i})(z-z_{2i+1})}{(z_{2i+2}-z_{2i})(z_{2i+2}-z_{2i+1})} & \text{if } x_i \le z < x_{i+1} \\ \dfrac{(z-z_{2i+3})(z-z_{2i+4})}{(z_{2i+2}-z_{2i+3})(z_{2i+2}-z_{2i+4})} & \text{if } x_{i+1} \le z < x_{i+2} \\ 0 & \text{if } x_{i+2} < z. \end{cases}$$

In Figure 8.2, we plot some quadratic basis functions in H^1. Figure 8.2(a) is the plot of the shape functions, that is, the nonzero basis functions defined in the interval $(-1, 1)$. In Figure 8.2(b), we plot all the basis functions over a three-node mesh in $(0, 1)$. In Figure 8.2(c), we plot some basis functions over the entire domain, $\phi_0(x)$, $\phi_1(x)$, $\phi_2(x)$, $\phi_3(x)$, $\phi_4(x)$, where $\phi_1(x)$ is centered at the auxiliary point z_1 and nonzero at only one element while $\phi_2(x)$, $\phi_4(x)$ are nonzero at two elements.

8.3.2 Assembling the Stiffness Matrix and the Load Vector

The finite element solution can be written as

$$u_h(x) = \sum_{i=1}^{2M-1} \alpha_i \phi_i(x).$$

The entries of the coefficient matrix are $\{a_{ij}\} = a(\phi_i, \phi_j)$ and the load vector is $F_i = L(\phi_i)$. On each element (x_i, x_{i+1}), or (z_{2i}, z_{2i+2}), there are three nonzero

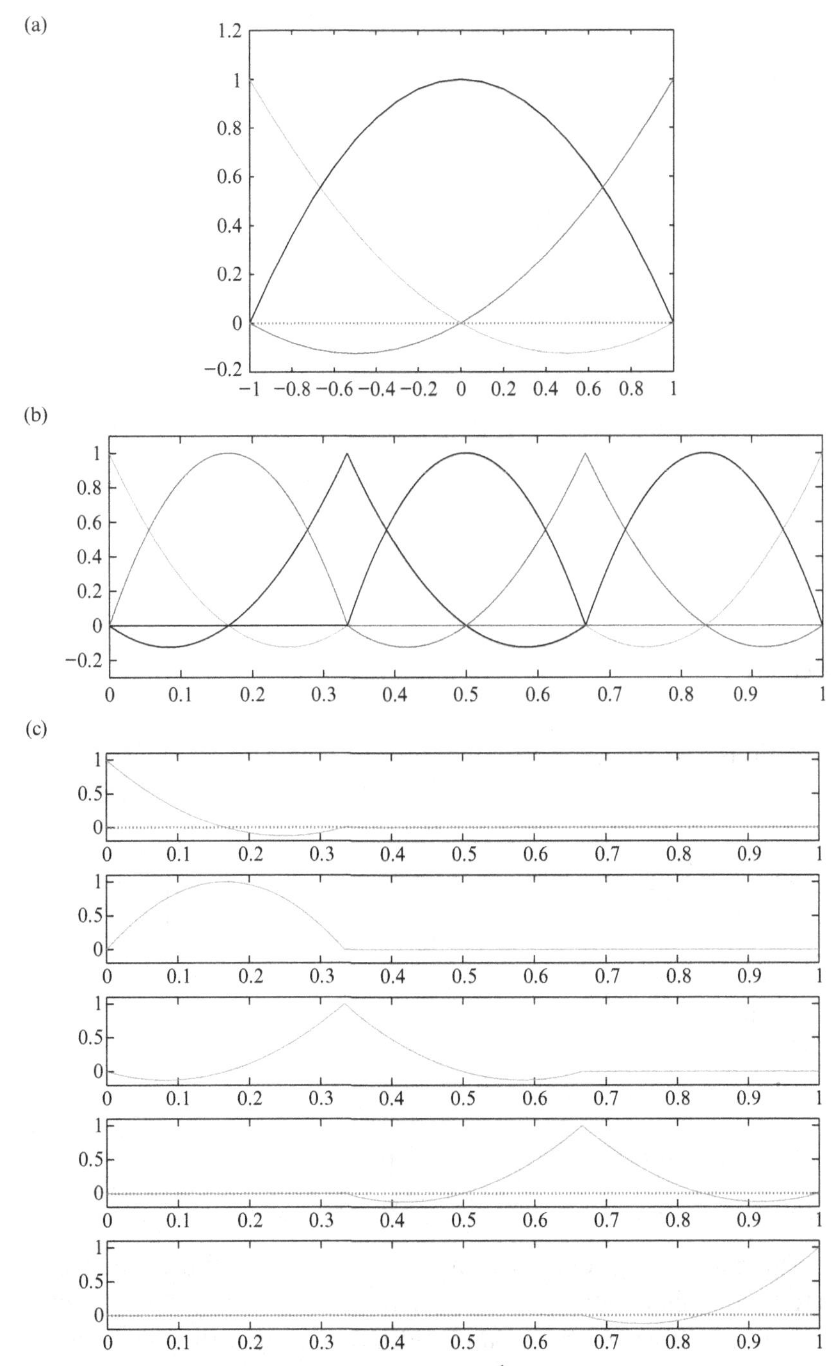

Figure 8.2. Quadratic basis functions in H^1: (a) the shape functions (basis functions in $(-1, 1)$; (b) all the basis functions over a three-node mesh in $(0, 1)$; and (c) plot of some basis functions over the entire domain, $\phi_0(x)$, $\phi_1(x)$, $\phi_2(x)$, $\phi_4(x)$, $\phi_4(x)$, where $\phi_1(x)$ is centered at the auxiliary point z_1 and nonzero at only one element, while $\phi_2(x)$, $\phi_4(x)$ are nonzero at two elements.

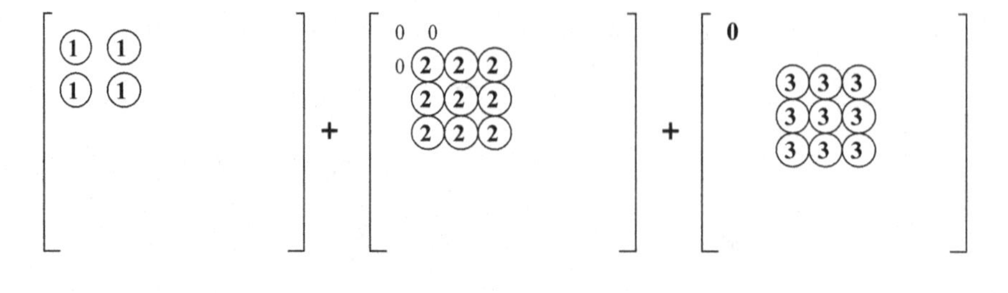

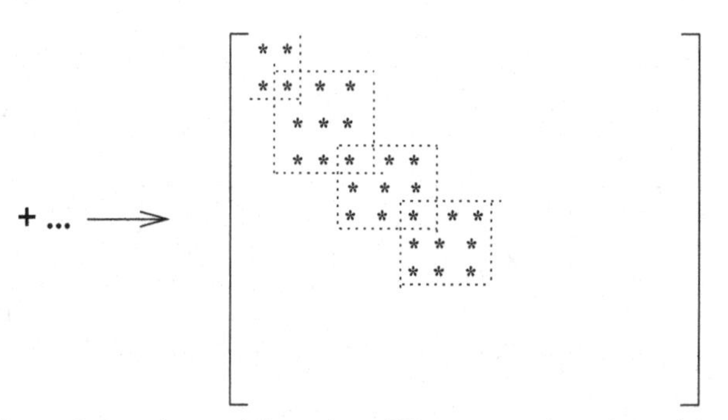

Figure 8.3. Assembling the stiffness matrix using piecewise quadratic basis functions.

basis functions: ϕ_{2i}, ϕ_{2i+1}, and ϕ_{2i+2}. Thus the local stiffness matrix is

$$K_i^e = \begin{bmatrix} a(\phi_{2i},\phi_{2i}) & a(\phi_{2i},\phi_{2i+1}) & a(\phi_{2i},\phi_{2i+2}) \\ a(\phi_{2i+1},\phi_{2i}) & a(\phi_{2i+1},\phi_{2i+1}) & a(\phi_{2i+1},\phi_{2i+2}) \\ a(\phi_{2i+2},\phi_{2i}) & a(\phi_{2i+2},\phi_{2i+1}) & a(\phi_{2i+2},\phi_{2i+2}) \end{bmatrix}_{(x_i,x_{i+1})} \tag{8.17}$$

and the local load vector is

$$L_i^e = \begin{bmatrix} L(\phi_{2i}) \\ L(\phi_{2i+1}) \\ L(\phi_{2i+2}) \end{bmatrix}_{(x_i,x_{i+1})}, \tag{8.18}$$

see the diagram in Figure 8.3 for an illustration. The stiffness matrix is still symmetric positive definite, but denser than that with the hat basis functions. It is still a banded matrix, with the band width five, a penta-diagonal matrix. The advantage in using quadratic basis functions is that the finite element solution is more accurate than that obtained on using the linear basis functions with the same mesh.

8.3.3 The Cubic Basis Functions in $H^1(x_l, x_r)$ Space

We can also construct piecewise cubic basis functions in $H^1(x_l, x_r)$. On each element (x_i, x_{i+1}), a cubic function has the form

$$\phi(x) = a_i x^3 + b_i x^2 + c_i x + d_i\,, \quad i = 0, 1, \ldots, M-1.$$

There are four parameters; and the total degree of freedom is $3M-1$ if a Dirichlet boundary condition is imposed at both ends. To construct cubic basis functions with properties similar to the piecewise linear and quadratic basis functions, we need to add two auxiliary points between x_i and x_{i+1}. The local stiffness matrix is then a 4×4 matrix. We leave the construction of the basis functions, and the application to the Sturm–Liouville BVPs as a project for students.

8.4 A 1D Matlab FE Package

A general 1D Matlab package has been written and is available at the book's depository, www4.ncsu.edu/~zhilin/FD_FEM_Book/MATLAB/1D or upon request.

- The code can be used to solve a general Sturm–Liouville problem

$$-(p(x)u')' + c(x)u' + q(x)u = f(x)\,, \quad a < x < b,$$

  with a Dirichlet, Neumann, or mixed boundary condition at $x = a$ and $x = b$.[1]
- We use conforming finite element methods.
- The mesh is

$$x_0 = a < x_1 < x_2 \cdots < x_M = b,$$

  as elaborated again later.
- The finite element spaces can be piecewise linear, quadratic, or cubic functions over the mesh.
- The integration formulas are the Gaussian quadrature of order 1, 2, 3, or 4.
- The matrix assembly is element by element.

8.4.1 Gaussian Quadrature Formulas

In a finite element method, we typically need to evaluate integrals such as $\int_a^b p(x)\phi_i'(x)\phi_j'(x)\,dx$, $\int_a^b q(x)\phi_i(x)\phi_j(x)\,dx$ and $\int_a^b f(x)\phi_i(x)\,dx$ over some

[1] In the package, $p(x)$ is expressed as $k(x)$.

intervals (a, b) such as (x_{i-1}, x_i). Although the functions involved may be arbitrary, it is usually neither practical nor necessary to find the exact integrals. A standard approach is to transfer the interval of integration to the interval $(-1, 1)$ as follows

$$\int_a^b f(x)\,dx = \int_{-1}^1 \bar{f}(\xi)\,d\xi, \tag{8.19}$$

where

$$\begin{aligned} &\xi = \frac{x-a}{b-a} + \frac{x-b}{b-a} \quad \text{or} \quad x = a + \frac{b-a}{2}\left(1+\xi\right), \\ \Longrightarrow\ &d\xi = \frac{2}{b-a}dx \quad \text{or} \quad dx = \frac{b-a}{2}d\xi. \end{aligned} \tag{8.20}$$

In this way, we have

$$\int_a^b f(x)\,dx = \frac{b-a}{2}\int_{-1}^1 f\left(a + \frac{b-a}{2}(1+\xi)\right)d\xi = \frac{b-a}{2}\int_{-1}^1 \bar{f}(\xi)\,d\xi, \tag{8.21}$$

where $\bar{f}(\xi) = f(a + \frac{b-a}{2}(1+\xi))$; and then to use a Gaussian quadrature formula to approximate the integral. The general quadrature formula can be written

$$\int_{-1}^1 g(\xi)d\xi \approx \sum_{i=1}^N w_i g(\xi_i),$$

where the Gaussian points ξ_i and weights w_i are chosen so that the quadrature is as accurate as possible. In the Newton–Cotes quadrature formulas such as the mid-point rule, the trapezoidal rule, and the Simpson methods, the ξ_i are predefined independent of w_i. In Gaussian quadrature formulas, all ξ_i's and w_i's are unknowns, and are determined simultaneously such that the quadrature formula is exact for $g(x) = 1, x, \ldots, x^{2N-1}$. The number $2N-1$ is called the *algebraic precision* of the quadrature formula.

Gaussian quadrature formulas have the following features:

- accurate with the best possible algebraic precision using the fewest points;
- open with no need to use two end points where some kind of discontinuities may occur, *e.g.*, the discontinuous derivatives of the piecewise linear functions at nodal points;
- no recursive relations for the Gaussian points ξ_i's and the weights w_i's;
- accurate enough for finite element methods, because $b - a \sim h$ is generally small and only a few points ξ_i's are needed.

We discuss some Gaussian quadrature formulas below.

8.4.1.1 Gaussian Quadrature of Order 1 (One Point):

With only one point, the Gaussian quadrature can be written as

$$\int_{-1}^{1} g(\xi)d\xi = w_1 g(\xi_1)\,.$$

We choose ξ_1 and w_1 such that the quadrature formula has the highest algebraic precision ($2N-1=1$ for $N=1$) if only one point is used. Thus we choose $g(\xi)=1$ and $g(\xi)=\xi$ to have the following,

$$\text{for } g(\xi)=1, \quad \int_{-1}^{1} g(\xi)d\xi = 2 \Longrightarrow 2 = w_1 \cdot 1\,; \quad \text{and}$$

$$\text{for } g(\xi)=\xi\,, \quad \int_{-1}^{1} g(\xi)d\xi = 0 \Longrightarrow 0 = w_1\xi_1\,.$$

Thus we get $w_1 = 2$ and $\xi_1 = 0$. The quadrature formula is simply the mid-point rule.

8.4.1.2 Gaussian Quadrature of Order 2 (Two Points):

With two points, the Gaussian quadrature can be written as

$$\int_{-1}^{1} g(\xi)d\xi = w_1 g(\xi_1) + w_2 g(\xi_2).$$

We choose ξ_1, ξ_2, and w_1, w_2 such that the quadrature formula has the highest algebraic precision ($2N-1=3$ for $N=2$) if two points are used. Thus we choose $g(\xi)=1$, $g(\xi)=\xi$, $g(\xi)=\xi^2$, and $g(\xi)=\xi^3$ to have the following,

$$\text{for } g(\xi) = 1, \quad \int_{-1}^{1} g(\xi)d\xi = 2 \Longrightarrow 2 = w_1 + w_2;$$

$$\text{for } g(\xi) = \xi, \quad \int_{-1}^{1} g(\xi)d\xi = 0 \Longrightarrow 0 = w_1\xi_1 + w_2\xi_2;$$

$$\text{for } g(\xi) = \xi^2, \quad \int_{-1}^{1} g(\xi)d\xi = \frac{2}{3} \Longrightarrow \frac{2}{3} = w_1\xi_1^2 + w_2\xi_2^2\,; \quad \text{and}$$

$$\text{for } g(\xi) = \xi^3, \quad \int_{-1}^{1} g(\xi)d\xi = 0 \Longrightarrow 0 = w_1\xi_1^3 + w_2\xi_2^3.$$

On solving the four nonlinear systems of equations by taking advantage of the symmetry, we get

$$w_1 = w_2 = 1, \quad \xi_1 = -\frac{1}{\sqrt{3}} \quad \text{and} \quad \xi_2 = \frac{1}{\sqrt{3}}.$$

So the Gaussian quadrature formula of order 2 is

$$\int_{-1}^{1} g(\xi)d\xi \simeq g\left(-\frac{1}{\sqrt{3}}\right) + g\left(\frac{1}{\sqrt{3}}\right). \tag{8.22}$$

Higher-order Gaussian quadrature formulas are likewise obtained, and for efficiency we can prestore the Gaussian points and weights in two separate matrices:

$$\xi_i \qquad\qquad\qquad\qquad w_i$$

$$\begin{bmatrix} 0 & \frac{-1}{\sqrt{3}} & \frac{-\sqrt{3}}{\sqrt{5}} & -0.8611363116 & \cdots \\ & \frac{1}{\sqrt{3}} & 0 & -0.3399810436 & \cdots \\ & & \frac{\sqrt{3}}{\sqrt{5}} & 0.3399810436 & \cdots \\ & & & 0.8611363116 & \cdots \end{bmatrix}, \quad \begin{bmatrix} 2 & 1 & \frac{5}{9} & 0.3478548451 & \cdots \\ & 1 & \frac{8}{9} & 0.6521451549 & \cdots \\ & & \frac{5}{9} & 0.6521451549 & \cdots \\ & & & 0.3478548451 & \cdots \end{bmatrix}.$$

Below is a Matlab code setint.m to store the Gaussian points and weights up to order 4.

```
function [xi,w] = setint

%%%%%%%%%%%%%%%%%%%%%%%%%%%%%%%%%%%%%%%%%%%%%%%%%%%%%%%%%%%%%%%%
%                                                              %
% Function setint provides the Gaussian points x(i), and the   %
% weights of the Gaussian quadrature formula.                  %
% Output:                                                      %
%    x(4,4):  x(:,i) is the Gaussian points of order i.        %
%    w(4,4):  w(:,i) is the weights of quadrature of order i.  %

clear x;   clear w

   xi(1,1) = 0;
   w(1,1)  = 2;                 % Gaussian quadrature of order 1

   xi(1,2) = -1/sqrt(3);
   xi(2,2) = -xi(1,2);
   w(1,2)  = 1;
   w(2,2)  = w(1,2);            % Gaussian quadrature of order 2

   xi(1,3) = -sqrt(3/5);
   xi(2,3) = 0;
   xi(3,3) = -xi(1,3);
```

```
    w(1,3)  = 5/9;
    w(2,3)  = 8/9;
    w(3,3)  = w(1,3);           % Gaussian quadrature of order 3

    xi(1,4)  = - 0.8611363116;
    xi(2,4)  = - 0.3399810436;
    xi(3,4)  = -xi(2,4);
    xi(4,4)  = -xi(1,4);
    w(1,4)  = 0.3478548451;
    w(2,4)  = 0.6521451549;
    w(3,4)  = w(2,4);
    w(4,4)  = w(1,4);           % Gaussian quadrature of order 4

    return

%----------------------- END OF SETINT -----------------------
```

8.4.2 Shape Functions

Similar to transforming an integral over some arbitrary interval to the integral over the standard interval between -1 and 1, it is easier to evaluate the basis functions and their derivatives in the standard interval $(-1,\ 1)$. Basis functions in the standard interval $(-1,\ 1)$ are called *shape functions* and often have analytic forms.

Using the transform between x and ξ in (8.19)–(8.20) for each element, on assuming $c(x)=0$ we have

$$\int_{x_i}^{x_{i+1}} \left(p(x)\phi_i'\phi_j' + q(x)\phi_i\phi_j\right)\, dx = \int_{x_i}^{x_{i+1}} f(x)\phi_i(x)\, dx$$

which is transformed to

$$\frac{x_{i+1}-x_i}{2}\int_{-1}^{1} \left(\bar{p}(\xi)\psi_i'\psi_j' + \bar{q}(\xi)\psi_i\psi_j\right)\, d\xi = \frac{x_{i+1}-x_i}{2}\int_{-1}^{1} \bar{f}(\xi)\psi_i\, d\xi$$

where

$$\bar{p}(\xi) = p\left(x_i + \frac{x_{i+1}-x_i}{2}(1+\xi)\right),$$

and so on. Here ψ_i and ψ_j are the local basis functions under the new variables, *i.e.*, the shape functions and their derivatives. For piecewise linear functions,

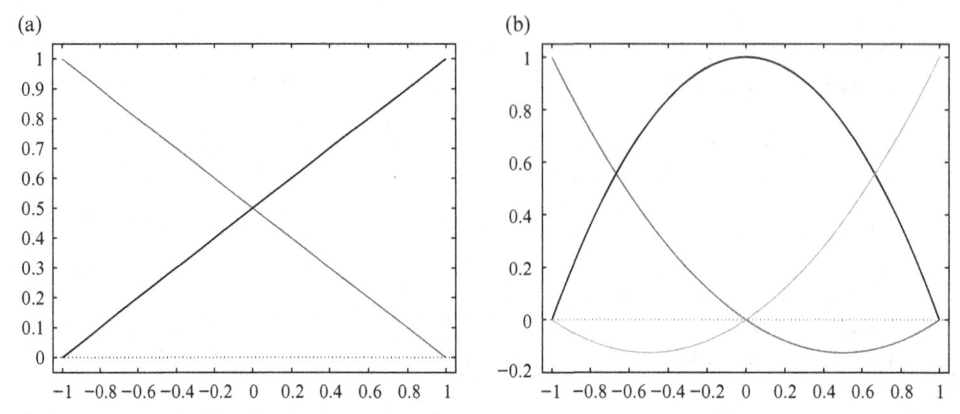

Figure 8.4. Plot of some shape functions: (a) the hat (linear) functions and (b) the quadratic functions

there are only two nonzero shape functions

$$\psi_1 = \frac{1-\xi}{2}, \qquad \psi_2 = \frac{1+\xi}{2}, \tag{8.23}$$

$$\text{with derivatives} \quad \psi_1' = -\frac{1}{2}, \qquad \psi_2' = \frac{1}{2}. \tag{8.24}$$

There are three nonzero quadratic shape functions

$$\psi_1 = \frac{\xi(\xi-1)}{2}, \quad \psi_2 = 1-\xi^2, \quad \psi_3 = \frac{\xi(\xi+1)}{2}, \tag{8.25}$$

$$\text{with derivatives} \quad \psi_1' = \xi - \frac{1}{2}, \qquad \psi_2' = -2\xi, \qquad \psi_3' = \xi + \frac{1}{2}. \tag{8.26}$$

These hat (linear) and quadratic shape functions are plotted in Figure 8.4.

It is noted that there is an extra factor in the derivatives with respect to x, due to the transform:

$$\frac{d\phi_i}{dx} = \frac{d\psi_i}{d\xi}\frac{d\xi}{dx} = \psi_i' \frac{2}{x_{i+1}-x_i}.$$

The shape functions can be defined in a Matlab function

$$[psi,\ dpsi] = shape(xi, n),$$

where $n=1$ renders the linear basis function, $n=2$ the quadratic basis function, and $n=3$ the cubic basis function values. For example, with $n=2$ the outputs are

$psi(1),\ \ psi(2),\ \ psi(3),$ three basis function values,
$dpsi(1),\ \ dpsi(2),\ \ dpsi(3),$ three derivative values.

The Matlab subroutine is as follows.

```
  function [psi,dpsi]=shape(xi,n);

%%%%%%%%%%%%%%%%%%%%%%%%%%%%%%%%%%%%%%%%%%%%%%%%%%%%%%%%%%%%%%%%%%
%                                                                %
% Function ''shape'' evaluates the values of the basis functions %
% and their derivatives at a point xi.                           %
%                                                                %
% n: The basis function. n=2, linear, n=3, quadratic, n=3, cubic. %
% xi: The point where the base function is evaluated.            %
% Output:                                                        %
% psi:  The value of the base function at xi.                    %
% dpsi: The derivative of the base function at xi.               %
%----------------------------------------------------------------%

   switch n
   case 2,
     % Linear base function
     psi(1) = (1-xi)/2;
     psi(2) = (1+xi)/2;
     dpsi(1) = -0.5;
     dpsi(2) = 0.5;
     return

   case 3,
     % quadratic base function
     psi(1) = xi*(xi-1)/2;
     psi(2) = 1-xi*xi;
     psi(3) = xi*(xi+1)/2;
     dpsi(1) =  xi-0.5;
     dpsi(2) = -2*xi;
     dpsi(3) = xi + 0.5;
     return

   case 4,
     % cubic  base function
     psi(1) = 9*(1/9-xi*xi)*(xi-1)/16;
     psi(2) = 27*(1-xi*xi)*(1/3-xi)/16;
     psi(3) = 27*(1-xi*xi)*(1/3+xi)/16;
     psi(4) = -9*(1/9-xi*xi)*(1+xi)/16;

     dpsi(1) = -9*(3*xi*xi-2*xi-1/9)/16;
     dpsi(2) = 27*(3*xi*xi-2*xi/3-1)/16;
     dpsi(3) = 27*(-3*xi*xi-2*xi/3+1)/16;
     dpsi(4) = -9*(-3*xi*xi-2*xi+1/9)/16;
     return

   end
%------------------------- END OF SHAPE --------------------------
```

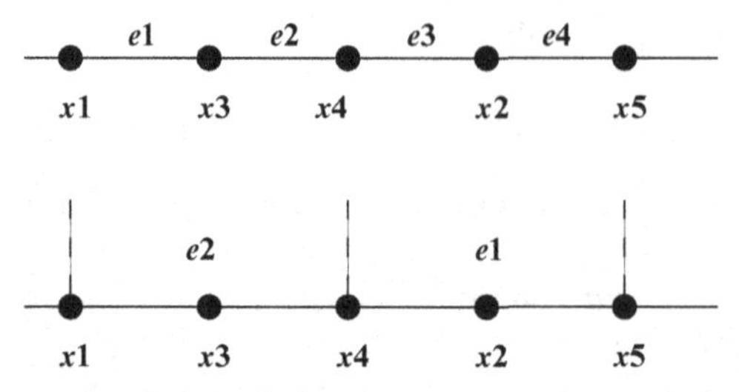

Figure 8.5. Example of the relation between nodes and elements: (a) linear basis functions and (b) quadratic basis functions

8.4.3 The Main Data Structure

In one space dimension, a mesh is a set of ordered points as described below.

- Nodal points: $x_1 = a$, x_2, ..., $x_{nnode} = b$. The number of total nodal points plus the auxiliary points is *nnode*.
- Elements: $\Omega_1, \Omega_2, \ldots, \Omega_{nelem}$. The number of elements is *nelem*.
- Connection between the nodal points and the elements: $nodes(nnode, nelem)$, where $nodes(j, i)$ is the j-th index of the nodes in the i-th element. For the linear basis function, $j = 1, 2$ since there are two nodes in an element; for the quadratic basis function, $j = 1, 2, 3$ since there are two nodes and an auxiliary point.

Example. Given the mesh and the indexing of the nodal points and the elements in Figure 8.5, for linear basis functions, we have

$$
\begin{aligned}
nodes(1,1) &= 1, & nodes(1,2) &= 3,\\
nodes(2,1) &= 3, & nodes(2,2) &= 4,\\
nodes(1,3) &= 4, & nodes(1,4) &= 2,\\
nodes(2,3) &= 2, & nodes(2,4) &= 5.
\end{aligned}
$$

Example. Given the mesh and the indexing of the nodal points and the elements in Figure 8.5, for quadratic basis functions, we have

$$
\begin{aligned}
nodes(1,1) &= 4\,, & nodes(2,1) &= 2\,, & nodes(3,1) &= 5,\\
nodes(1,2) &= 1\,, & nodes(2,2) &= 3\,, & nodes(3,2) &= 4.
\end{aligned}
$$

8.4.4 Outline of the Algorithm

```
function   [x,u]=fem1d

  global nnode nelem
  global gk gf
  global xi  w
```

```
%%% Output: x are nodal points; u is the FE solution at
%%%         nodal points.

       [xi,w] = setint;      % Get Gaussian points and weights.

%%% Input data, pre-process

       [x,kbc,ubc,kind,nint,nodes] = prospset;

  %%%  x(nnode): Nodal points,        kbc, ubc: Boundary conditions

  %%%  kind(nelen): Choice of FE spaces. kind(i)=1,2,3 indicate
  %%%  piecewise linear, quadratic, and cubic FE space over the
  %%%  triangulation.

  %%%  nint(nelen): Choice of Gaussian quadrature. nint(i)=1,2,3,4
  %%%  indicate Gaussian order 1, 2, 3, 4.

       formkf(kind,nint,nodes,x,xi,w);

  %%%  Assembling the stiffness matrix and the load vector element by
  %%%  element.

       aplyb(kbc,ubc);

  %%%  Deal with the BC.

       u = gk\gf;                  % Solve the linear system of equations

  %%%  Error analysis ...
```

8.4.5 Assembling Element by Element

The Matlab code is formkf.m

```
   function formkf(kind,nint,nodes,x,xi,w)

          ...........

    for nel = 1:nelem,
       n = kind(nel) + 1;          % Linear FE space. n = 2, quadratic n=3, ..

       i1 = nodes(1,nel);          % The first node in nel-th element.
       i2 = nodes(n,nel);          % The last node in nel-th element.
       i3 = nint(nel);             % Order of Gaussian quadrature.
       xic = xi(:,i3);             % Get Gaussian points in the column.
       wc = w(:,i3);               % Get Gaussian weights.

%%% Evaluate the local stiffness matrix ek, and the load vector ef.
       [ek,ef] = elem(x(i1),x(i2),n,i3,xic,wc);

%%% Assembling to the global stiffness matrix gk, and the load vector gf.
       assemb(ek,ef,nel,n,nodes);

    end
```

8.4.5.1 Evaluation of Local Stiffness Matrix and the Load Vector

The Matlab code is elem.m

```
   function [ek,ef] = elem(x1,x2,n,nl,xi,w)
   dx = (x2-x1)/2;

% [x1,x2] is an element [x1,x2]
% n is the choice of FE space. Linear n=2; quadratic n=3; ...

  for l=1:nl,                           % Quadrature formula that summarize.
    x = x1 + (1.0 + xi(l))*dx;          % Transform the Gaussian points.
    [xp,xc,xb,xf] = getmat(x);          % Get the coefficients at the
                                        % Gaussian points.
    [psi,dpsi] = shape(xi(l),n);        % Get the shape function and
                                        % its derivatives.
% Assembling the local stiffness matrix and the load vector.
% Notice the additional factor 1/dx in the derivatives.
    for i=1:n,
       ef(i) = ef(i) + psi(i)*xf*w(l)*dx;
       for j=1:n,
          ek(i,j)=ek(i,j)+(xp*dpsi(i)*dpsi(j)/(dx*dx) ...
               +xc*psi(i)*dpsi(j)/dx+xb*psi(i)*psi(j) )*w(l)*dx;
       end
    end
  end
```

8.4.5.2 Global Assembling

The Matlab code is assemb.m

```
    function assemb(ek,ef,nel,n,nodes)
    global gk gf

    for i=1:n,                 % Connection between nodes and the elements
      ig = nodes(i,nel);                   % Assemble global vector gf
      gf(ig) = gf(ig) + ef(i);

       for j=1:n,
         jg = nodes(j,nel);        % Assemble global stiffness matrix gk
         gk(ig,jg) = gk(ig,jg) + ek(i,j);
       end
     end
```

8.4.5.3 Input Data

The Matlab code is propset.m

```
        function [x,kbc,vbc,kind,nint,nodes] = propset
```

- The relation between the number of nodes *nnode* and the number of elements *nelem*: Linear: $nelem = nnode - 1$.
  Quadratic: $nelem = (nnode - 1)/2$.
  Cubic: $nelem = (nnode - 1)/3$.

- Nodes arranged in ascendant order. Equally spaced points are grouped together. The Matlab code is datain.m

  function [data] = datain(a,b,nnode,nelem)

  The output data has *nrec* groups

$$data(i,1) = n1, \quad \text{index of the beginning of nodes.}$$
$$data(i,2) = n2, \quad \text{number of points in this group.}$$
$$data(i,3) = x(n1), \quad \text{the first nodal point.}$$
$$data(i,4) = x(n1+n2), \quad \text{the last nodal point in this group.}$$

  The simple case is

$$data(i,1) = i\,, \quad data(i,2) = 0\,, \quad data(i,3) = x(i)\,, \quad data(i,4) = x(i)\,.$$

- The basis functions to be used in each element:

```
for i=1:nelem
   kind(i) = inf_ele = 1, or 2, or 3.
   nint(i) = 1, or 2, or 3, or 4.
   for j=1,kind(i)+1
     nodes(j,i) = j + kind(i)*(i-1);
   end
end
```

8.4.6 Input Boundary Conditions

The Matlab code aplybc.m involves an array of two elements $kbc(2)$ and a data array $vbc(2,2)$. At the left boundary

$$kbc(1) = 1\,, \quad vbc(1,1) = u_a, \quad \text{Dirichlet BC at the left end;}$$
$$kbc(1) = 2, \quad vbc(1,1) = -p(a)u'(a), \quad \text{Neumann BC at the left end;}$$
$$kbc(1) = 3, \quad vbc(1,1) = uxma, \quad vbc(2,1) = uaa, \quad \text{Mixed BC of the form:}$$
$$p(a)u'(a) = uxma(u(a) - uaa).$$

The BC will affect the stiffness matrix and the load vector and are handled in Matlab codes aplybc.m and drchlta.m.

- Dirichlet BC $u(a) = u_a = vbc(1,1)$.

```
for i=1:nnode,
    gf(i) = gf(i) - gk(i,1)*vbc(1,1);
    gk(i,1) = 0;     gk(1,i) = 0;
end
gk(1,1) = 1;     gf(1) = vbc(1,1);
```

  where *gk* is the global stiffness matrix and *gf* is the global load vector.

- Neumann BC $u'(a) = u_{xa}$. The boundary condition can be rewritten as $-p(a)u'(a) = -p(a)u_{xa} = vbc(1,1)$. We only need to change the load vector.

```
gf(1) = gf(1) + vbc(1,1);
```

- Mixed BC $\alpha u(a) + \beta u'(a) = \gamma$, $\beta \neq 0$. The BC can be rewritten as

$$p(a)u'(a) = -\frac{\alpha}{\beta}p(a)\left(u(a) - \frac{\gamma}{\alpha}\right)$$
$$= u_{xma}\left(u(a) - u_{aa}\right) = \mathbf{vbc}(1,1)\left(u(a) - \mathbf{vbc}(2,1)\right).$$

We need to change both the global stiffness matrix and the global load vector.

```
gf(1) = gf(1) + vbc(1,1)*vbc(2,1);
gk(1,1) =  gk(1,1) + vbc(1,1);
```

Examples.

1. $u(a) = 2$, we should set $kbc(1) = 1$ and $vbc(1,1) = 2$.
2. $p(x) = 2 + x^2$, $a = 2$, $u'(2) = 2$. Since $p(a) = p(2) = 6$ and $-p(a)u'(a) = -12$, we should set $kbc(1) = 2$ and $vbc(1,1) = -12$.
3. $p(x) = 2 + x^2$, $a = 2$, $2u(a) + 3u'(a) = 1$. Since

$$3u'(a) = -2u(a) + 1,$$
$$6u'(a) = -4u(a) + 2 = -4\left(u(a) - \frac{1}{2}\right),$$

we should set $kbc(1) = 3$, $vbc(1,1) = -4$ and $vbc(2,1) = 1/2$.

Similarly, at the right BC $x = b$ we should have

$kbc(2) = 1$, $vbc(1,2) = u_b$, Dirichlet BC at the right end;
$kbc(2) = 2$, $vbc(1,2) = p(b)u'(b)$, Neumann BC at the right end;
$kbc(2) = 3$, $vbc(1,2) = uxmb$, $vbc(2,2) = ubb$, Mixed BC of the form $-p(b)u'(b) = uxmb(u(b) - ubb)$.

The BC will affect the stiffness matrix and the load vector and are handled in Matlab codes aplybc.m and drchlta.m.

- Dirichlet BC $u(b) = u_b = vbc(1,2)$.

```
for i=1:nnode,
     gf(i) = gf(i) - gk(i,nnode)*vbc(1,2);
     gk(i,nnode) = 0;     gk(nnode,i) = 0;
end
gk(nnode,nnode) = 1;     gf(nnode) = vbc(1,1).
```

- Neumann BC $u'(b) = u_{xb}$ is given. The boundary condition can be rewritten as $p(b)u'(b) = p(b)u_{xb} = vbc(1,2)$. We only need to change the load vector.

```
gf(nnode) = gf(nnode) + vbc(1,2);
```

- Mixed BC $\alpha u(b) + \beta u'(b) = \gamma$, $\beta \neq 0$. The BC can be re-written as

$$
\begin{aligned}
-p(b)u'(b) &= \frac{\alpha}{\beta} p(b) \left(u(b) - \frac{\gamma}{\alpha} \right) \\
&= u_{xmb} \left(u(b) - u_{bb} \right) = \mathbf{vbc}(1,2) \left(u(b) - \mathbf{vbc}(2,2) \right).
\end{aligned}
$$

We need to change both the global stiffness matrix and the global load vector.

```
gf(nnode) = gf(nnode) + vbc(1,2)*vbc(2,2);
gk(nnode,nnode) = gk(nnode,nnode) + vbc(1,2);
```

8.4.7 A Testing Example

To check the code, we often try to compare the numerical results with some known exact solution, *e.g.*, we can choose

$$u(x) = \sin x, \qquad a \le x \le b.$$

If we set the material parameters as

$$p(x) = 1 + x, \quad c(x) = \cos x, \quad q(x) = x^2,$$

then the right-hand side can be calculated as

$$f(x) = (pu')' + cu' + qu = (1 + x) \sin x - \cos x + \cos^2 x + x^2 \sin x.$$

These functions are defined in the Matlab code getmat.m

```
function [xp,xc,xq,xf] = getmat(x);
  xp = 1+x; xc = cos(x); xq = x*x;
  xf = (1+x)*sin(x)-cos(x)+cos(x)*cos(x)+x*x*sin(x);
```

The mesh is defined in the Matlab code datain.m. All other parameters used for the finite element method are defined in the Matlab code propset.m, including the following:

- The boundary $x = a$ and $x = b$, *e.g.*, $a = 1$, $b = 4$.
- The number of nodal points, *e.g.*, $nnode = 41$.
- The choice of basis functions. If we use the same basis function, then for example we can set $inf\_ele = 2$, which is the quadratic basis function $kind(i) = inf\_ele$.
- The number of elements. If we use uniform elements, then $nelem = (nnode - 1)/inf\_ele$. We need to make it an integer.

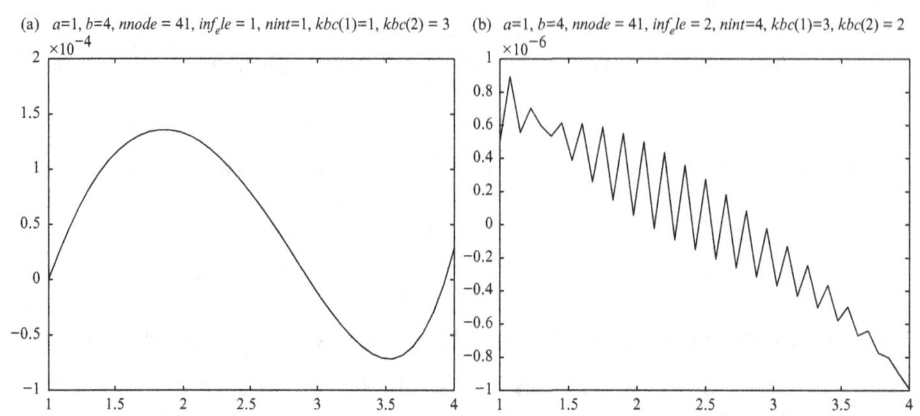

Figure 8.6. Error plots of the FE solutions at nodal points. (a) The result is obtained with piecewise linear basis function and Gaussian quadrature of order 1 in the interval $[1,4]$. Dirichlet BC at $x=a$ and mixed BC $3u(b)+4u'(b)=\gamma$ from the exact solution $u(x)=\sin x$ are used. The magnitude of the error is $O(10^{-4})$. (b) Mixed BC $3u(a)+4u'(a)=\gamma$ at $x=a$ and the Neumann BC at $x=b$ from the exact solution are used. The result is obtained with piecewise quadratic basis functions and Gaussian quadrature of order 4. The magnitude of the error is $O(10^{-6})$.

- The choice of Gaussian quadrature formula, *e.g.*, $nint(i)=4$. The order of the Gaussian quadrature formula should be the same or higher than the order of the basis functions, for otherwise it may not converge! For example, if we use linear elements (*i.e.*, *inf_ele* = 1), then we can choose $nint(i)=1$ or $nint(i)=2$, *etc.*
- Determine the BCs $kbc(1)$ and $kbc(2)$, and $vbc(i,j)$, $i,j=1,2$. Note that the linear system of equations is singular if both BCs are Neumann, for the solution either does not exist or is not unique.

To run the program, simply type the following into the Matlab:

```
[x,u]= fem1d;
```

To find out the detailed usage of the finite element code, read README carefully. Figure 8.6 gives the error plots for two different boundary conditions.

8.5 The FE Method for Fourth-Order BVPs in 1D

Let us now discuss how to solve fourth-order differential equations using the finite element method. An important fourth-order differential equation is the biharmonic equation, such as in the model problem

$$u'''' + q(x)u = f(x), \; 0 < x < 1, \text{ subject to the BC}$$

$$\begin{aligned} &I: u(0) = u'(0) = 0, \; u(1) = u'(1) = 0; \text{ or} \\ &II: u(0) = u'(0) = 0, \; u(1) = 0, \; u''(1) = 0; \text{ or} \\ &III: u(0) = u'(0) = 0, \; u''(1) = 0, \; u'''(1) = 0. \end{aligned}$$

Note that there is no negative sign in the highest derivative term. To derive the weak form, we again multiply by a test function $v(x) \in V$ and integrate by parts to get

$$\begin{aligned} \int_0^1 (u'''' + q(x)u)v \, dx &= \int_0^1 fv \, dx, \\ u'''v\Big|_0^1 - \int_0^1 u'''v' \, dx + \int_0^1 quv \, dx &= \int_0^1 fv \, dx, \\ u'''v\Big|_0^1 - u''v'\Big|_0^1 + \int_0^1 \left(u''v'' + quv\right) dx &= \int_0^1 fv \, dx, \\ u'''(1)v(1) - u'''(0)v(0) - u''(1)v'(1) + u''(0)v'(0) + \int_0^1 \left(u''v'' + quv\right) dx & \\ = \int_0^1 fv \, dx. & \end{aligned}$$

For $u(0) = u'(0) = 0$, $u(1) = u'(1) = 0$, they are *essential boundary conditions*, thus we set

$$v(0) = v'(0) = v(1) = v'(1) = 0. \tag{8.27}$$

The weak form is

$$a(u, v) = f(v), \tag{8.28}$$

where the bilinear form and the linear form are

$$a(u, v) = \int_0^1 \left(u''v'' + quv\right) dx, \tag{8.29}$$

$$L(v) = \int_0^1 fv \, dx. \tag{8.30}$$

Since the weak form involves second-order derivatives, the solution space is

$$H_0^2(0, 1) = \left\{ v(x), \; v(0) = v'(0) = v(1) = v'(1) = 0, \quad v, v' \text{ and } v'' \in L^2 \right\}, \tag{8.31}$$

and from the Sobolev embedding theorem we know that $H^2 \subset C^1$.

For the boundary conditions $u(1) = u''(1) = 0$, we still have $v(1) = 0$, but there is no restriction on $v'(1)$ and the solution space is

$$H_E^2 = \left\{ v(x), \quad v(0) = v'(0) = v(1) = 0, \quad v \in H^2(0,1) \right\}. \tag{8.32}$$

For the boundary conditions $u''(1) = u'''(1) = 0$, there are no restrictions on both $v(1)$ and $v'(1)$ and the solution space is

$$H_E^2 = \left\{ v(x), \quad v(0) = v'(0) = 0, \quad v \in H^2(0,1) \right\}. \tag{8.33}$$

For nonhomogeneous *natural or mixed* boundary conditions, the weak form and the linear form may be different. For homogeneous *essential BC*, the weak form and the linear form will be the same. We often need to do something to adjust the essential boundary conditions.

8.5.1 The Finite Element Discretization

Given a mesh

$$0 = x_0 < x_1 < x_2 < \cdots < x_M = 1,$$

we want to construct a finite-dimensional space V_h. For conforming finite element methods we have $V_h \in H^2(0,1)$, therefore we cannot use the piecewise linear functions since they are in the Sobolev space $H^1(0,1)$ but not in $H^2(0,1)$.

For piecewise quadratic functions, theoretically we can find a finite-dimensional space that is a subset of $H^2(0,1)$; but this is not practical as the basis functions would have large support and involve at least six nodes. The most practical conforming finite-dimensional space in 1D is the piecewise cubic functions over the mesh

$$V_h = \left\{ v(x), \ v(x) \text{ is a continuous piecewise cubic function}, \ v \in H_0^2(0,1) \right\}. \tag{8.34}$$

The degree of freedom. On each element, we need four parameters to determine a cubic function. For essential boundary conditions at both $x = a$ and $x = b$, there are $4M$ parameters for M elements; and at each interior nodal point, the cubic and its derivative are continuous and there are four boundary conditions, so the dimension of the finite element space is

$$4M - 2(M-1) - 4 = 2(M-1).$$

8.5.1.1 Construct the Basis Functions in H^2 in 1D

Since the derivative has to be continuous, we can use the piecewise Hermite interpolation and construct the basis functions in two categories. The first category is

$$\phi_i(x_j) = \begin{cases} 1 & \text{if } i=j, \\ 0 & \text{otherwise,} \end{cases} \qquad (8.35)$$
$$\text{and} \quad \phi_i'(x_j) = 0, \quad \text{for any } x_j,$$

i.e., the basis functions in this group have unity at one node and are zero at other nodes, and the derivatives are zero at all nodes. To construct the local basis function in the element (x_i, x_{i+1}), we can set

$$\phi_i(x) = \frac{(x - x_{i+1})^2 \, (a(x - x_i) + 1)}{(x_i - x_{i+1})^2}.$$

It is obvious that $\phi_i(x_i) = 1$ and $\phi_i(x_{i+1}) = \phi_i'(x_{i+1}) = 0$, *i.e.*, x_{i+1} is a double root of the polynomial. We use $\phi'(x_i) = 0$ to find the coefficient a, to finally obtain

$$\phi_i(x) = \frac{(x - x_{i+1})^2 \left(\dfrac{2(x - x_i)}{(x_{i+1} - x_i)} + 1 \right)}{(x_i - x_{i+1})^2}. \qquad (8.36)$$

The global basis function can thus be written as

$$\phi_i(x) = \begin{cases} 0 & \text{if } x \le x_{i-1}, \\ \dfrac{(x - x_{i-1})^2 \left(\dfrac{2(x - x_i)}{(x_{i-1} - x_i)} + 1 \right)}{(x_i - x_{i-1})^2} & \text{if } x_{i-1} \le x \le x_i, \\ \dfrac{(x - x_{i+1})^2 \left(\dfrac{2(x - x_i)}{(x_{i+1} - x_i)} + 1 \right)}{(x_i - x_{i+1})^2}, & \text{if } x_i \le x \le x_{i+1}, \\ 0 & \text{if } x_{i+1} \le x. \end{cases} \qquad (8.37)$$

There are $M - 1$ such basis functions. The second group of basis functions satisfy

$$\bar{\phi}_i'(x_j) = \begin{cases} 1 & \text{if } i=j, \\ 0 & \text{otherwise,} \end{cases} \qquad (8.38)$$
$$\text{and} \quad \bar{\phi}_i(x_j) = 0, \quad \text{for any } x_j,$$

i.e., the basis functions in this group are zero at all nodes, and the derivatives are unity at one node and zero at other nodes. To construct the local basis functions in an element (x_i, x_{i+1}), we can set

$$\bar{\phi}_i(x) = C(x - x_i)(x - x_{i+1})^2,$$

since x_i and x_{i+1} are zeros of the cubic and x_{i+1} is a double root of the cubic. The constant C is chosen such that $\psi_i'(x_i) = 1$, so we finally obtain

$$\bar{\phi}_i(x) = \frac{(x - x_i)(x - x_{i+1})^2}{(x_i - x_{i+1})^2}. \tag{8.39}$$

The global basis function for this category is thus

$$\bar{\phi}_i(x) = \begin{cases} 0 & \text{if } x \leq x_{i-1}, \\ \dfrac{(x - x_i)(x - x_{i-1})^2}{(x_i - x_{i-1})^2} & \text{if } x_{i-1} \leq x \leq x_i, \\ \dfrac{(x - x_i)(x - x_{i+1})^2}{(x_{i+1} - x_i)^2} & \text{if } x_i \leq x \leq x_{i+1}, \\ 0 & \text{if } x_{i+1} \leq x. \end{cases} \tag{8.40}$$

8.5.2 The Shape Functions

There are four shape functions in the interval $(-1, 1)$, namely,

$$\psi_1(\xi) = \frac{(\xi - 1)^2(\xi + 2)}{4},$$

$$\psi_2(\xi) = \frac{(\xi + 1)^2(-\xi + 2)}{4},$$

$$\psi_3(\xi) = \frac{(\xi - 1)^2(\xi + 1)}{4},$$

$$\psi_4(\xi) = \frac{(\xi + 1)^2(\xi - 1)}{4}.$$

In Figure 8.7, we show the shape functions and some global basis functions from the Hermite cubic interpolation. It is noted that there are two basis functions *centered* at each node, so-called a *double node*. The finite element solution can be written as

$$u_h(x) = \sum_{j=1}^{M-1} \alpha_j \phi_j(x) + \sum_{j=1}^{M-1} \beta_j \bar{\phi}_j(x), \tag{8.41}$$

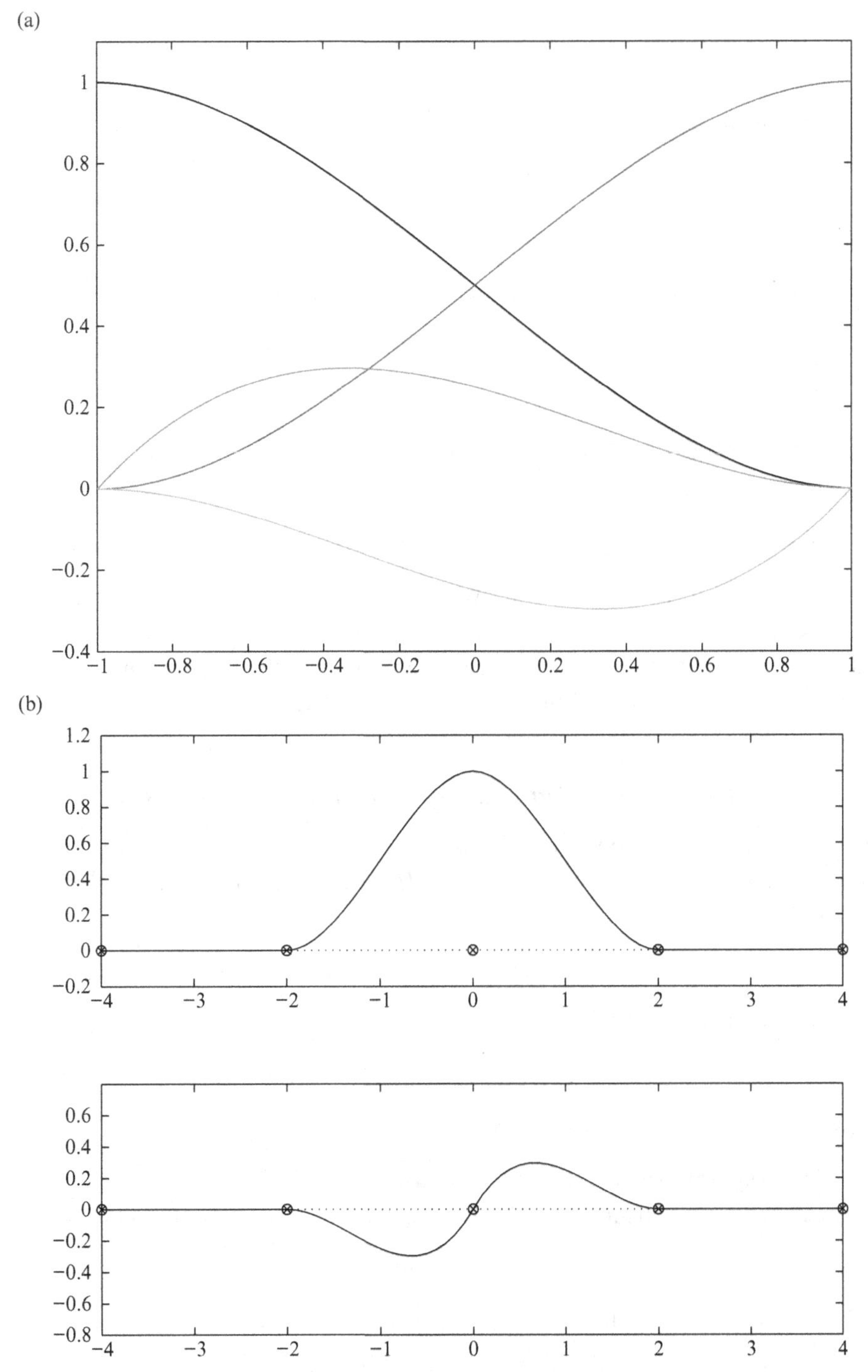

Figure 8.7. (a) Hermite cubic shape functions on a master element and (b) corresponding global functions at a node i in a mesh.

and after the coefficients α_j and β_j are found we have

$$u_h(x_j) = \alpha_j, \qquad u_h'(x_j) = \beta_j.$$

There are four nonzero basis functions on each element (x_i, x_{i+1}); and on adopting the order $\phi_1, \phi_2, \psi_1, \psi_2, \phi_3, \phi_3, \psi_3, \ldots$, the local stiffness matrix has the form

$$\begin{bmatrix} a(\phi_i, \phi_i) & a(\phi_i, \phi_{i+1}) & a(\phi_i, \bar{\phi}_i) & a(\phi_i, \bar{\phi}_{i+1}) \\ a(\phi_{i+1}, \phi_i) & a(\phi_{i+1}, \phi_{i+1}) & a(\phi_{i+1}, \bar{\phi}_i) & a(\phi_{i+1}, \bar{\phi}_{i+1}) \\ a(\bar{\phi}_i, \phi_i) & a(\bar{\phi}_i, \phi_{i+1}) & a(\bar{\phi}_i, \bar{\phi}_i) & a(\bar{\phi}_i, \bar{\phi}_{i+1}) \\ a(\bar{\phi}_{i+1}, \phi_i) & a(\bar{\phi}_{i+1}, \phi_{i+1}) & a(\bar{\phi}_{i+1}, \bar{\phi}_i) & a(\bar{\phi}_{i+1}, \bar{\phi}_{i+1}) \end{bmatrix}_{(x_i, x_{i+1})}.$$

This global stiffness matrix is still banded, and has band width six.

8.6 The Lax–Milgram Lemma and the Existence of FE Solutions

One of the most important issues is whether the weak form has a solution, and if so under what assumptions. Further, if the solution does exist, is it unique, and how close is it to the solution of the original differential equations? Answers to these questions are based on the Lax–Milgram Lemma.

8.6.1 General Settings: Assumptions and Conditions

Let V be a Hilbert space with an inner product $(u, v)_V$ and the norm $\|u\|_V = \sqrt{(u, u)_V}$, *e.g.*, C^m, the Sobolev spaces H^1 and H^2, *etc.* Assume there is a *bilinear* form

$$a(u, v), \qquad V \times V \longmapsto R,$$

and a linear form

$$L(v), \qquad V \longmapsto R,$$

that satisfy the following conditions:

1. $a(u, v)$ is symmetric, *i.e.*, $a(u, v) = a(v, u)$;
2. $a(u, v)$ is continuous in both u and v, *i.e.*, there is a constant γ such that

   $$|a(u, v)| \le \gamma \|u\|_V \|v\|_V,$$

   for any u and $v \in V$; the norm of the operator $a(u, v)$[2];

[2] If this condition is true, then $a(u, v)$ is called a bounded operator and the largest lower bound of such a $\gamma > 0$ is called the norm of $a(u, v)$.

3. $a(u, v)$ is V-elliptic, *i.e.*, there is a constant α such that

$$a(v, v) \geq \alpha \|v\|_V^2$$

   for any $v \in V$ (alternative terms are *coercive*, or inf–sup condition); and
4. L is continuous, *i.e.*, there is a constant Λ such that

$$|L(v)| \leq \Lambda \|v\|_V,$$

   for any $v \in V$.

8.6.2 The Lax–Milgram Lemma

Theorem 8.2. *Under the above conditions 2 to 4, there exists a unique element $u \in V$ such that*

$$a(u, v) = L(v), \quad \forall\, v \in V.$$

Furthermore, if the condition 1 is also true, i.e., $a(u, v)$ is symmetric, then

1. $\|u\|_V \leq \dfrac{\Lambda}{\alpha}$ and
2. u is the unique global minimizer of

$$F(v) = \frac{1}{2} a(v, v) - L(v)\,.$$

Sketch of the proof. The proof exploits the Riesz representation theorem from functional analysis. Since $L(v)$ is a bounded linear operator in the Hilbert space V with the inner product $a(u, v)$, there is unique element u^* in V such that

$$L(v) = a(u^*, v), \quad \forall\, v \in V.$$

Next we show that the a-norm is equivalent to V norm. From the continuity condition of $a(u, v)$, we get

$$\|u\|_a = \sqrt{a(u,u)} \leq \sqrt{\gamma \|u\|_V^2} = \sqrt{\gamma}\, \|u\|_V.$$

From the V-elliptic condition, we have

$$\|u\|_a = \sqrt{a(u,u)} \geq \sqrt{\alpha \|u\|_V^2} = \sqrt{\alpha}\, \|u\|_V,$$

therefore

$$\sqrt{\alpha}\, \|u\|_V \leq \|u\|_a \leq \sqrt{\gamma}\, \|u\|_V,$$

or

$$\frac{1}{\sqrt{\gamma}}\|u\|_a \le \|u\|_V \le \frac{1}{\sqrt{\alpha}}\|u\|_a.$$

Often $\|u\|_a$ is called the energy norm.

Now we show that $F(u^*)$ the global minimizer. For any $v \in V$, if $a(u,v) = a(v,u)$, then

$$\begin{aligned}
F(v) &= F(u^* + v - u^*) = F(u^* + w) = \frac{1}{2}a(u^* + w, u^* + w) - L(u^* + w) \\
&= \frac{1}{2}\left(a(u^* + w, u^*) + a(u^* + w, w)\right) - L(u^*) - L(w) \\
&= \frac{1}{2}\left(a(u^*, u^*) + a(w, u^*) + a(u^*, w) + a(w, w)\right) - L(u^*) - L(w) \\
&= \frac{1}{2}a(u^*, u^*) - L(u^*) + \frac{1}{2}a(w, w) + a(u^*, w) - L(w) \\
&= F(u^*) + \frac{1}{2}a(w, w) - 0 \\
&\ge F(u^*).
\end{aligned}$$

Finally we show the proof of the stability. We have

$$\alpha\|u^*\|_V^2 \le a(u^*, u^*) = L(u^*) \le \Lambda\|u^*\|_V,$$

therefore

$$\alpha\|u^*\|_V^2 \le \Lambda\|u^*\|_V \Longrightarrow \|u^*\|_V \le \frac{\Lambda}{\alpha}.$$

Remark: The Lax–Milgram Lemma is often used to prove the existence and uniqueness of the solutions of ODEs/PDEs.

8.6.3 An Example using the Lax–Milgram Lemma

Let us consider the 1D Sturm–Liouville problem once again:

$$-(pu')' + qu = f, \quad a < x < b,$$

$$u(a) = 0, \quad \tilde{\alpha}u(b) + \tilde{\beta}u'(b) = \tilde{\gamma}, \quad \tilde{\beta} \ne 0, \quad \frac{\tilde{\alpha}}{\tilde{\beta}} \ge 0.$$

The bilinear form is

$$a(u, v) = \int_a^b (pu'v' + quv)\, dx + \frac{\tilde{\alpha}}{\tilde{\beta}}p(b)u(b)v(b),$$

and the linear form is

$$L(v) = (f, v) + \frac{\tilde{\gamma}}{\tilde{\beta}}p(b)v(b).$$

The space is $V = H_E^1(a,b)$. To consider the conditions of the Lax–Milgram theorem, we need the *Poincaré inequality*:

Theorem 8.3. *If $v(x) \in H^1$ and $v(a) = 0$, then*

$$\int_a^b v^2\,dx \le (b-a)^2 \int_a^b |v'(x)|^2\,dx \quad \textit{or} \quad \int_a^b |v'(x)|^2\,dx \ge \frac{1}{(b-a)^2}\int_a^b v^2\,dx. \tag{8.42}$$

Proof We have

$$\begin{aligned} v(x) &= \int_a^x v'(t)\,dt \\ \Longrightarrow |v(x)| &\le \int_a^x |v'(t)|\,dt \le \left\{\int_a^x |v'(t)|^2 dt\right\}^{1/2} \left\{\int_a^x dt\right\}^{1/2} \\ &\le \sqrt{b-a}\left\{\int_a^b |v'(t)|^2\right\}^{1/2}, \end{aligned}$$

so that

$$\begin{aligned} v^2(x) &\le (b-a)\int_a^b |v'(t)|^2\,dt \\ \Longrightarrow \int_a^b v^2(x)\,dx &\le (b-a)\int_a^b |v'(t)|^2 dt \int_a^b dx \le (b-a)^2 \int_a^b |v'(x)|\,dx. \end{aligned}$$

This completes the proof.

We now verify the Lax–Milgram Lemma conditions for the Sturm–Liouville problem.

- Obviously $a(u,v) = a(v,u)$.
- The bilinear form is continuous:

$$\begin{aligned} |a(u,v)| &= \left| \int_a^b (pu'v' + quv)\,dx + \frac{\tilde{\alpha}}{\tilde{\beta}} p(b)u(b)v(b) \right| \\ &\le \max\{p_{max}, q_{max}\} \left(\int_a^b (|u'v'| + |uv|)\,dx + \frac{\tilde{\alpha}}{\tilde{\beta}}|u(b)v(b)| \right) \\ &\le \max\{p_{max}, q_{max}\} \left(\int_a^b |u'v'|dx + \int_a^b |uv|\,dx + \frac{\tilde{\alpha}}{\tilde{\beta}}|u(b)v(b)| \right) \\ &\le \max\{p_{max}, q_{max}\} \left(2\|u\|_1\|v\|_1 + \frac{\tilde{\alpha}}{\tilde{\beta}}|u(b)v(b)| \right). \end{aligned}$$

From the inequality

$$
\begin{aligned}
|u(b)v(b)| &= \left| \int_a^b u'(x)\,dx \int_a^b v'(x)\,dx \right| \\
&\le (b-a)\sqrt{\int_a^b |u'(x)|^2\,dx}\sqrt{\int_a^b |v'(x)|^2\,dx} \\
&\le (b-a)\sqrt{\int_a^b \left(|u'(x)|^2 + |u(x)|^2\right)\,dx}\sqrt{\int_a^b \left(|v'(x)|^2 + |v(x)|^2\right)\,dx} \\
&\le (b-a)\|u\|_1\|v\|_1,
\end{aligned}
$$

we get

$$
|a(u,v)| \le \max\{p_{max}, q_{max}\}\left(2 + \frac{\tilde{\alpha}}{\tilde{\beta}}(b-a)\right)\|u\|_1\|v\|_1
$$

i.e., the constant γ can be determined as

$$
\gamma = \max\{p_{max}, q_{max}\}\left(2 + \frac{\tilde{\alpha}}{\tilde{\beta}}(b-a)\right).
$$

- $a(v,v)$ is V-elliptic. We have

$$
\begin{aligned}
a(v,v) &= \int_a^b \left(p(v')^2 + qv^2\right)\,dx + \frac{\tilde{\alpha}}{\tilde{\beta}}p(b)v(b)^2 \\
&\ge \int_a^b p(v')^2\,dx \\
&\ge p_{min}\int_a^b (v')^2\,dx \\
&= p_{min}\left(\frac{1}{2}\int_a^b (v')^2\,dx + \frac{1}{2}\int_a^b (v')^2\,dx\right) \\
&\ge p_{min}\left(\frac{1}{2}\frac{1}{(b-a)^2}\int_a^b v^2\,dx + \frac{1}{2}\int_a^b (v')^2\,dx\right) \\
&= p_{min}\min\left\{\frac{1}{2(b-a)^2}, \frac{1}{2}\right\}\|v\|_1^2,
\end{aligned}
$$

i.e., the constant α can be determined as

$$
\alpha = p_{min}\min\left\{\frac{1}{2(b-a)^2}, \frac{1}{2}\right\}.
$$

- $L(v)$ is continuous because

$$
\begin{aligned}
L(v) &= \int_a^b f(x)v(x)\,dx + \frac{\tilde{\gamma}_1}{\tilde{\beta}} p(b)v(b) \\
|L(v)| &\le (|f|,|v|)_0 + \left|\frac{\tilde{\gamma}_1}{\tilde{\beta}}\right| p(b)\sqrt{b-a}\,\|v\|_1 \\
&\le \|f\|_0\|v\|_1 + \left|\frac{\tilde{\gamma}_1}{\tilde{\beta}}\right| p(b)\sqrt{b-a}\,\|v\|_1 \\
&\le \left(\|f\|_0 + \left|\frac{\tilde{\gamma}_1}{\tilde{\beta}}\right| p(b)\sqrt{b-a}\right)\|v\|_1,
\end{aligned}
$$

i.e., the constant Λ can be determined as

$$\Lambda = \|f\|_0 + \left|\frac{\tilde{\gamma}_1}{\tilde{\beta}}\right| p(b)\,\sqrt{b-a}.$$

Thus we have verified the conditions of the Lax–Milgram lemma under certain assumptions such as $p(x) \ge p_{min} > 0$, $q(x) \ge 0$, *etc.*, and hence conclude that there is the unique solution in $H_e^1(a,b)$ to the original differential equation. The solution also satisfies

$$\|u\|_1 \le \frac{\|f\|_0 + \left|\tilde{\gamma}/\tilde{\beta}\right| p(b)\sqrt{b-a}}{p_{min}\min\left\{\frac{1}{2(b-a)^2}, \frac{1}{2}\right\}}.$$

8.6.4 Abstract FE Methods

In the same setting, let us assume that V_h is a finite-dimensional subspace of V and that $\{\phi_1, \phi_2, \ldots, \phi_M\}$ is a basis for V_h. We can formulate the following abstract finite element method using the finite-dimensional subspace V_h. We seek $u_h \in V_h$ such that

$$a(u_h, v) = L(v), \quad \forall v \in V_h, \tag{8.43}$$

or equivalently

$$F(u_h) \le F(v), \quad \forall v \in V_h. \tag{8.44}$$

We apply the weak form in the finite-dimensional V_h:

$$a(u_h, \phi_i) = L(\phi_i), \qquad i = 1, \ldots, M. \tag{8.45}$$

Let the finite element solution u_h be

$$u_h = \sum_{j=1}^{M} \alpha_j \phi_j.$$

Then from the weak form in V_h we get

$$a\left(\sum_{j=1}^{M} \alpha_j \phi_j, \phi_i\right) = \sum_{j=1}^{M} \alpha_j a(\phi_j, \phi_i) = L(\phi_i), \qquad i = 1, \ldots, M,$$

which in the matrix-vector form is

$$AU = F,$$

where $U \in R^M, F \in R^M$ with $F(i) = L(\phi_i)$ and A is an $M \times M$ matrix with entries $a_{ij} = a(\phi_j, \phi_i)$. Since any element in V_h can be written as

$$v = \sum_{i=1}^{M} \eta_i \phi_i,$$

we have

$$a(v, v) = a\left(\sum_{i=1}^{M} \eta_i \phi_i, \sum_{j=1}^{M} \eta_j \phi_j\right) = \sum_{i,j=1}^{M} \eta_i a(\phi_i, \phi_j) \eta_j = \eta^T A \eta > 0$$

provided $\eta^T = \{\eta_1, \ldots, \eta_M\} \neq 0$. Consequently, A is symmetric positive definite. The minimization form using V_h is

$$\frac{1}{2} U^T A U - F^T \mathbf{U} = \min_{\eta \in R^M} \left(\frac{1}{2} \eta^T A \eta - F^T \eta\right). \tag{8.46}$$

The existence and uniqueness of the abstract FE method.
Since the matrix A is symmetric positive definite and it is invertible, so there is a unique solution to the discrete weak form. Also from the conditions of Lax–Milgram lemma, we have

$$\alpha \|u_h\|_V^2 \leq a(u_h, u_h) = L(u_h) \leq \Lambda \|u_h\|_V,$$

whence

$$\|u_h\|_V \leq \frac{\Lambda}{\alpha}.$$

Error estimates. If $e_h = u - u_h$ is the error, then:

- $a(e_h, v_h) = (e_h, v_h)_a = 0$, $\forall v_h \in V_h$;
- $\|u - u_h\|_a = \sqrt{a(e_h, e_h)} \le \|u - v_h\|_a$, $\forall v_h \in V_h$, *i.e.*, u_h is the best approximation to u in the energy norm; and
- $\|u - u_h\|_V \le \frac{\gamma}{\alpha}\|u - v_h\|_V$, $\forall v_h \in V_h$, which gives the error estimates in the V norm.

Sketch of the proof: From the weak form, we have

$$a(u, v_h) = L(v_h), \quad a(u_h, v_h) = L(v_h) \quad \Longrightarrow \quad a(u - u_h, v_h) = 0.$$

This means the finite element solution is the projection of u onto the space V_h. It is the best solution in V_h in the energy norm, because

$$\begin{aligned}
\|u - v_h\|_a^2 &= a(u - v_h, u - v_h) = a(u - u_h + w_h, u - u_h + w_h) \\
&= a(u - u_h, u - u_h) + a(u - u_h, w_h) + a(w_h, u - u_h) + a(w_h, w_h) \\
&= a(u - u_h, u - u_h) + a(w_h, w_h) \\
&\ge \|u - u_h\|_a^2,
\end{aligned}$$

where $w_h = u_h - v_h \in V_h$. Finally, from the condition 3, we have

$$\begin{aligned}
\alpha\|u - u_h\|_V^2 &\le a(u - u_h, u - u_h) = a(u - u_h, u - u_h) + a(u - u_h, w_h) \\
&= a(u - u_h, u - u_h + w_h) = a(u - u_h, u - v_h) \\
&\le \gamma\|u - u_h\|_V\|u - v_h\|_V.
\end{aligned}$$

The last inequality is obtained from condition 2.

8.7 *1D IFEM for Discontinuous Coefficients

Now we revisit the 1D interface problems discussed in Section 2.10

$$-(pu')' = f(x), \qquad 0 < x < 1, \qquad u(0) = 0, \quad u(1) = 0, \tag{8.47}$$

and consider the case in which the coefficient has a finite jump,

$$p(x) = \begin{cases} \beta^-(x) & \text{if } 0 < x < \alpha, \\ \beta^+(x) & \text{if } \alpha < x < 1. \end{cases} \tag{8.48}$$

The theoretical analysis about the solution still holds if the natural jump conditions

$$[u]_\alpha = 0, \qquad [\beta u']_\alpha = 0, \tag{8.49}$$

are satisfied, where $[u]_\alpha$ means the jump defined at α.

Given a uniform mesh x_i, $i=0,1,\ldots,n$, $x_{i+1}-x_i=h$. Unless the interface α in (8.47) itself is a node, the solution obtained from the standard finite element method using the linear basis functions is only first-order accurate in the maximum norm. In Li (1998), modified basis functions that are defined below,

$$\phi_i(x_k)=\begin{cases}1, & \text{if } k=i,\\ 0, & \text{otherwise,}\end{cases} \tag{8.50}$$

$$[\phi_i]_\alpha=0, \qquad [\beta\phi_i']_\alpha=0, \tag{8.51}$$

are proposed. Obviously, if $x_j\le\alpha<x_{j+1}$, then only ϕ_j and ϕ_{j+1} need to be changed to satisfy the second jump condition. Using the method of undetermined coefficients, that is, we look for the basis function $\phi_j(x)$ in the interval (x_j,x_{j+1}) as

$$\phi_j(x)=\begin{cases}a_0+a_1x & \text{if } x_j\le x<\alpha,\\ b_0+b_1x & \text{if } \alpha\le x\le x_{j+1},\end{cases} \tag{8.52}$$

which should satisfy $\phi_j(x_j)=1$, $\phi_j(x_{j+1})=0$, $\phi_j(\alpha-)=\phi_j(\alpha+)$, and $\beta^-\phi_j'(\alpha-)=\beta^+\phi_j'(\alpha+)$. There are four unknowns and four conditions. It has been proved in Li (1998) that the coefficients are unique determined and have the following closed form if β is a piecewise constant and $\beta^-\beta^+>0$,

$$\phi_j(x)=\begin{cases}0, & 0\le x<x_{j-1},\\ \dfrac{x-x_{j-1}}{h}, & x_{j-1}\le x<x_j,\\ \dfrac{x_j-x}{D}+1, & x_j\le x<\alpha,\\ \dfrac{\rho\,(x_{j+1}-x)}{D}, & \alpha\le x<x_{j+1},\\ 0, & x_{j+1}\le x\le 1,\end{cases} \qquad \phi_{j+1}(x)=\begin{cases}0, & 0\le x<x_j,\\ \dfrac{x-x_j}{D}, & x_j\le x<\alpha,\\ \dfrac{\rho\,(x-x_{j+1})}{D}+1, & \alpha\le x<x_{j+1},\\ \dfrac{x_{j+2}-x}{h}, & x_{j+1}\le x<x_{j+2},\\ 0, & x_{j+2}\le x\le 1.\end{cases}$$

where

$$\rho=\frac{\beta^-}{\beta^+}, \qquad D=h-\frac{\beta^+-\beta^-}{\beta^+}\,(x_{j+1}-\alpha).$$

Figure 8.8 shows several plots of the modified basis functions $\phi_j(x)$, $\phi_{j+1}(x)$, and some neighboring basis functions, which are the standard hat functions. At the interface α, we can clearly see kinks in the basis functions which reflect the natural jump condition.

Using the modified basis functions, it has been shown in Li (1998) that the finite element solution obtained from the Galerkin finite method with the new basis functions is second-order accurate in the maximum norm.

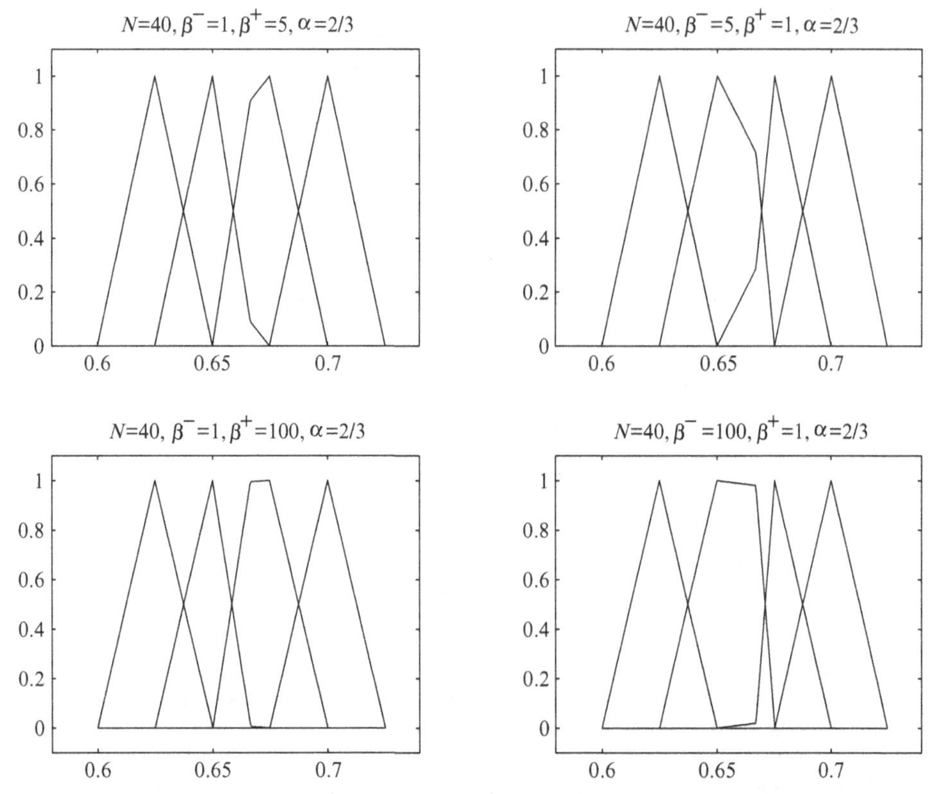

Figure 8.8. Plot of some basis function near the interface with different β^- and β^+. The interface is at $\alpha = \frac{2}{3}$.

For 1D interface problems, the finite difference and finite element methods are not much different. The finite element method likely performs better for self-adjoint problems, while the finite difference method is more flexible for general elliptic interface problems.

Exercises

1. (Purpose: Review abstract FE methods.) Consider the Sturm–Liouville problem

$$-u'' + u = f, \quad 0 < x < \pi,$$
$$u(0) = 0, \quad u(\pi) + u'(\pi) = 1.$$

Let V_f be the finite-dimensional space

$$V_f = span\{\, x,\ \sin(x), \sin(2x)\,\}.$$

Find the best approximation to the solution of the weak form from V_f in the energy norm $(\,\|:\|_a = \sqrt{a(:,\,:)}\,)$. You can use either analytic derivation or computer software packages (*e.g.*, Maple, Matlab, SAS, *etc.*). Take $f = 1$ for the computation. Compare this approach with the finite element method using three hat basis functions. Find the true solution, and plot the solution and the error of the finite element solution.

2. Consider the Sturm–Liouville problem

$$-((1+x^2)u')' + xu = f, \quad 0 < x < 1,$$
$$u(1) = 2.$$

Transform the problem to a problem with homogeneous Dirichlet boundary condition at $x = 1$. Write down the weak form for each of the following case:

(a) $u(0) = 3$. **Hint:** Construct a function $u_0(x) \in H^1$ such that $u_0(0) = 3$ and $u_0(1) = 2$.
(b) $u'(0) = 3$. **Hint:** Construct a function $u_0(x) \in H^1$ such that $u_0(1) = 2$ and $u_0'(0) = 0$.
(c) $u(0) + u'(0) = 3$. **Hint:** Construct a function $u_0(x) \in H^1$ such that $u_0(1) = 2$ and $u_0(0) + u_0'(0) = 0$.

3. Consider the Sturm–Liouville problem

$$-(pu')' + qu = f, \quad a < x < b,$$
$$u(a) = 0, \qquad u(b) = 0.$$

Consider a mesh $a = x_0 < x_1 \cdots < x_M = b$ and the finite element space

$$V_h = \left\{ v(x) \in H_0^1(a,b),\, v(x) \text{ is piecewise cubic function over the mesh} \right\}.$$

(a) Find the dimension of V_h.
(b) Find all nonzero shape functions $\psi_i(\xi)$ where $-1 \le \xi \le 1$, and plot them.
(c) What is the size of the local stiffness matrix and load vector? Sketch the assembling process.
(d) List some advantages and disadvantages of this finite element space, compared with the piecewise continuous linear finite-dimensional space (the hat functions).

4. Download the files of the 1D finite element Matlab package. Consider the following analytic solution and parameters,

$$u(x) = e^x \sin x, \quad p(x) = 1 + x^2, \quad q(x) = e^{-x}, \quad c(x) = 1,$$

and $f(x)$ determined from the differential equation

$$-(pu')' + c(x)u' + qu = f, \quad a < x < b.$$

Use this example to become familiar with the 1D finite element Matlab package, by trying the following boundary conditions:

(a) Dirichlet BC at $x = a$ and $x = b$, where $a = -1, b = 2$;
(b) Neumann BC at $x = a$ and Dirichlet BC at $x = b$, where $a = -1$ and $b = 2$;
(c) Mixed BC $\gamma = 3u(a) - 5u'(a)$ at $x = a = -1$, and Neumann BC at $x = b = 2$.

Using linear, quadratic, and cubic basis functions, tabulate the errors in the infinity norm

$$e_M = \max_{0 \le i \le M} |u(x_i) - U_i|$$

at the nodes and auxiliary points as follows:

M	Basis	Gaussian	error	e_M/e_{2M}

for different $M=4, 8, 16, 32, 64$ (nnode= $M+1$), or the **closest integers** if necessary. What are the respective convergence orders?
(*Note*: The method is second-, third-, or fourth-order convergent if the ratio e_M/e_{2M} approaches 4, 8, or 16, respectively.)
For the **last case**:
(1) Print out the stiffness matrix for the *linear basis function* with $M=5$. Is it symmetric?
(2) Plot the computed solution against the exact one, and the error plot for the case of the linear basis function. Take enough points to plot the exact solution to see the whole picture.
(3) Plot the error versus $h=1/M$ in log–log scale for the three different bases.
The slope of such a plot is the convergence order of the method employed. For this problem, you will only produce *five* plots for the last case.
Find the energy norm, H^1 norm and L^2 norm of the error and do the grid refinement analysis.

5. Use the Lax–Milgram Lemma to show whether the following two-point value problem has a unique solution:

$$\begin{aligned} -u'' + q(x)u &= f, \quad 0<x<1, \\ u'(0) = u'(1) &= 0, \end{aligned} \tag{8.53}$$

where $q(x) \in C(0,1)$, $q(x) \geq q_{min} > 0$. What happens if we relax the condition to $q(x) \geq 0$? Give counterexamples if necessary.

6. Consider the general fourth-order two-point BVP

$$a_4 u'''' + a_3 u''' + a_2 u'' + a_1 u' + a_0 u = f, \quad a<x<b,$$

with the mixed BC

$$2u'''(a) - u''(a) + \gamma_1 u'(a) + \rho_1 u(a) = \delta_1, \tag{8.54}$$

$$u'''(a) + u''(a) + \gamma_2 u'(a) + \rho_2 u(a) = \delta_2, \tag{8.55}$$

$$u(b) = 0,$$

$$u'(b) = 0.$$

Derive the weak form for this problem.
Hint: Solve for $u'''(a)$ and $u''(a)$ from (8.54) and (8.55). The weak form should only involve up to second-order derivatives.

7. (An eigenvalue problem) Consider

$$-(pu')' + qu - \lambda u = 0, \quad 0<x<\pi, \tag{8.56}$$

$$u(0) = 0, \quad u(\pi) = 0. \tag{8.57}$$

(a) Find the weak form of the problem.
(b) Check whether the conditions of the Lax–Milgram Lemma are satisfied. Which condition is violated? Is the solution unique for arbitrary λ?
Note: It is obvious that $u=0$ is a solution. For some λ, we can find nontrivial solutions $u(x) \neq 0$. Such a λ is an eigenvalue of the system, and the nonzero solution is an eigenfunction corresponding to that eigenvalue. The problem to find the eigenvalues and the eigenfunctions is called an eigenvalue problem.
(c) Find all the eigenvalues and eigenfunctions when $p(x)=1$ and $q(x)=0$.
Hint: $\lambda_1 = 1$ and $u(x) = \sin(x)$ is one pair of the solutions.

8. Use the 1D finite element package with linear basis functions and a uniform grid to solve the eigenvalue problem

$$-(pu')' + qu - \lambda u = 0, \quad 0 < x < \pi,$$

$$u(0) = 0,\ u'(\pi) + \alpha u(\pi) = 0,$$

$$\text{where} \quad p(x) \geq p_{min} > 0,\ q(x) \geq 0,\ \alpha \geq 0$$

in each of the following two cases:

(a) $p(x) = 1,\ q(x) = 1,\ \alpha = 1.$
(b) $p(x) = 1 + x^2,\ q(x) = x,\ \alpha = 3.$

Try to solve the eigenvalue problem with $M = 5$ and $M = 20$. Print out the eigenvalues but not the eigenfunctions. Plot all the eigenfunctions in a single plot for $M = 5$, and plot two typical eigenfunctions for $M = 20$ (6 plots in total).
Hint: The approximate eigenvalues $\lambda_1,\ \lambda_2,\ \ldots,\ \lambda_M$ and the eigenfunction $u_{\lambda_i}(x)$ are the generalized eigenvalues of

$$Ax = \lambda Bx,$$

where A is the stiffness matrix and $B = \{b_{ij}\}$ with $b_{ij} = \int_0^{\pi} \phi_i(x)\phi_j(x)dx$. You can generate the matrix B either numerically or analytically; and in Matlab you can use $[V, D] = EIG(A, B)$ to find the generalized eigenvalues and the corresponding eigenvectors. For a computed eigenvalue λ_i, the corresponding eigenfunction is

$$u_{\lambda_i}(x) = \sum_{j=1}^{M} \alpha_{i,j}\phi_j(x),$$

where $[\alpha_{i,1}, \alpha_{i,2}, \ldots, \alpha_{i,M}]^T$ is the eigenvector corresponding to the generalized eigenvalue.
Note: if we can find the eigenvalues and corresponding eigenfunctions, the solution to the differential equation can be expanded in terms of the eigenfunctions, similar to Fourier series.

9. (An application.) Consider a nuclear fuel element of spherical form, consisting of a sphere of "fissionable" material surrounded by a spherical shell of aluminum "cladding" as shown in the figure. We wish to determine the temperature distribution in the nuclear fuel element and the aluminum cladding. The governing equations for the two regions are the same, except that there is no heat source term for the aluminum cladding. Thus

$$-\frac{1}{r^2}\frac{d}{dr}r^2 k_1 \frac{dT_1}{dr} = q, \quad 0 \leq r \leq R_F,$$

$$-\frac{1}{r^2}\frac{d}{dr}r^2 k_2 \frac{dT_2}{dr} = 0, \quad R_F \leq r \leq R_C,$$

where the subscripts 1 and 2 refer to the nuclear fuel element and the cladding, respectively. The heat generation in the nuclear fuel element is assumed to be of the form

$$q_1 = q_0 \left[1 + c\left(\frac{r}{R_F}\right)^2\right],$$

where q_0 and c are constants depending on the nuclear material. The BC are

$$kr^2\frac{dT_1}{dr} = 0 \text{ at } r = 0 \quad \text{(natural BC)},$$

$$T_2 = T_0 \text{ at } r = R_C,$$

where T_0 is a constant. Note the temperature at $r = R_F$ is continuous.

- Derive a weak form for this problem. (**Hint:** First multiply both sides by r^2.)
- Use two linear elements $[0, R_F]$ and $[R_F, R_C]$ to determine the finite element solution.
- Compare the nodal temperatures $T(0)$ and $T(R_F)$ with the values from the exact solution

$$T_1 = T_0 + \frac{q_0 R_F^2}{6k_1}\left\{\left[1-\left(\frac{r}{R_F}\right)^2\right] + \frac{3}{10}c\left[1-\left(\frac{r}{R_F}\right)^4\right]\right\}$$
$$+ \frac{q_0 R_F^2}{3k_2}\left(1+\frac{3}{5}c\right)\left(1-\frac{R_F}{R_C}\right),$$
$$T_2 = T_0 + \frac{q_0 R_F^2}{3k_2}\left(1+\frac{3}{5}c\right)\left(\frac{R_F}{r}-\frac{R_F}{R_C}\right).$$

Take $T_0 = 80$, $q_0 = 5$, $k_1 = 1$, $k_2 = 50$, $R_F = 0.5$, $R_C = 1$, $c = 1$ for plotting and comparison.

9

The Finite Element Method for 2D Elliptic PDEs

The procedure of the finite element method to solve 2D problems is the same as that for 1D problems, as the flow chart below demonstrates.

$$\text{PDE} \longrightarrow \text{Integration by parts} \longrightarrow \text{weak form in a space } V : a(u,v) = L(v)$$
$$\text{or } \min_{v \in V} F(v) \longrightarrow V_h \text{ (finite-dimensional space and basis functions)}$$
$$\longrightarrow a(u_h, v_h) = L(v_h) \longrightarrow u_h \text{ and error analysis.}$$

9.1 The Second Green's Theorem and Integration by Parts in 2D

Let us first recall the 2D version of the well-known divergence theorem in Cartesian coordinates.

Theorem 9.1. *If* $\mathbf{F} \in H^1(\Omega) \times H^1(\Omega)$ *is a vector in 2D, then*

$$\iint_{\Omega} \nabla \cdot \mathbf{F}\, dxdy = \int_{\partial\Omega} \mathbf{F} \cdot \mathbf{n}\, ds, \tag{9.1}$$

where $\mathbf{n}$ *is the unit normal direction pointing outward at the boundary* $\partial\Omega$ *with line element ds, and* ∇ *is the gradient operator,* $\nabla = \left[\frac{\partial}{\partial x}, \frac{\partial}{\partial y}\right]^T$.

The second Green's theorem is a corollary of the divergence theorem if we set $\mathbf{F} = v\nabla u = \left[v\dfrac{\partial u}{\partial x}, v\dfrac{\partial u}{\partial y}\right]^T$. Thus since

$$\begin{aligned}
\nabla \cdot \mathbf{F} &= \frac{\partial}{\partial x}\left(v\frac{\partial u}{\partial x}\right) + \frac{\partial}{\partial y}\left(v\frac{\partial u}{\partial y}\right) \\
&= \frac{\partial u}{\partial x}\frac{\partial v}{\partial x} + v\frac{\partial^2 u}{\partial x^2} + \frac{\partial u}{\partial y}\frac{\partial v}{\partial y} + v\frac{\partial^2 u}{\partial y^2} \\
&= \nabla u \cdot \nabla v + v\,\Delta u,
\end{aligned}$$

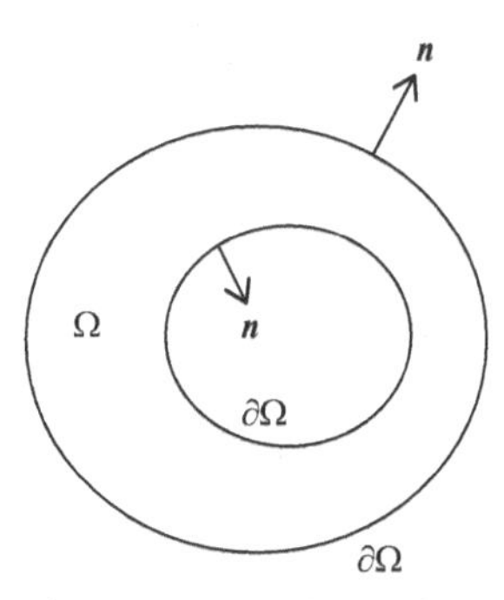

Figure 9.1. A diagram of a 2D domain Ω, its boundary ∂Ω and its unit normal direction.

where $\Delta u = \nabla \cdot \nabla u = u_{xx} + u_{yy}$, we obtain

$$\begin{aligned}\iint_\Omega \nabla \cdot \mathbf{F}\, dxdy &= \iint_\Omega (\nabla u \cdot \nabla v + v\,\Delta u)\, dxdy \\ &= \int_{\partial\Omega} \mathbf{F} \cdot \mathbf{n}\, ds \\ &= \int_{\partial\Omega} v\,\nabla u \cdot \mathbf{n}\, ds = \int_{\partial\Omega} v \frac{\partial u}{\partial n}\, ds,\end{aligned}$$

where $\mathbf{n} = (n_x, n_y)$ $(n_x^2 + n_y^2 = 1)$ is the unit normal direction, and $\frac{\partial u}{\partial n} = \nabla u \cdot \mathbf{n} = n_x \frac{\partial u}{\partial x} + n_y \frac{\partial u}{\partial y}$, the normal derivative of u, see Figure 9.1 for an illustration. This result immediately yields the formula for integration by parts in 2D.

Theorem 9.2. *If $u(x,y) \in H^2(\Omega)$ and $v(x,y) \in H^1(\Omega)$ where Ω is a bounded domain, then*

$$\iint_\Omega v\,\Delta u\, dxdy = \int_{\partial\Omega} v \frac{\partial u}{\partial n}\, ds - \iint_\Omega \nabla u \cdot \nabla v\, dxdy. \tag{9.2}$$

Note: The normal derivative $\partial u/\partial n$ is sometimes written more concisely as u_n.

Some important elliptic PDEs in 2D Cartesian coordinates are:

$$\begin{aligned} u_{xx} + u_{yy} &= 0, && \text{Laplace equation,} \\ -u_{xx} - u_{yy} &= f(x,y), && \text{Poisson equation,} \\ -u_{xx} - u_{yy} + \lambda u &= f, && \text{generalized Helmholtz equation,} \\ u_{xxxx} + 2u_{xxyy} + u_{yyyy} &= 0, && \text{Biharmonic equation.} \end{aligned}$$

When $\lambda > 0$, the generalized Helmholtz equation is easier to solve than when $\lambda < 0$. Incidentally, the expressions involved in these PDEs may also be abbreviated using the gradient operator ∇, *e.g.*, $u_{xx} + u_{yy} = \nabla \cdot \nabla u = \Delta u$ as mentioned before. We also recall that a general linear second-order elliptic PDE has the form

$$a(x,y)u_{xx} + 2b(x,y)u_{xy} + c(x,y)u_{yy} + d(x,y)u_x \\ + e(x,y)u_y + g(x,y)u = f(x,y)$$

with discriminant $b^2 - ac < 0$. A second-order self-adjoint elliptic PDE has the form

$$-\nabla \cdot (p(x,y)\nabla u) + q(x,y)u = f(x,y). \tag{9.3}$$

9.1.1 Boundary Conditions

In 2D, the domain boundary $\partial\Omega$ is one or several curves. We consider the following various linear boundary conditions.

- Dirichlet boundary condition on the entire boundary, *i.e.*, $u(x,y)|_{\partial\Omega} = u_0(x,y)$ is given.
- Neumann boundary condition on the entire boundary, *i.e.*, $\partial u/\partial n|_{\partial\Omega} = g(x,y)$ is given.
  In this case, the solution to a Poisson equation may not be unique or even exist, depending upon whether a compatibility condition is satisfied. Integrating the Poisson equation over the domain, we have

$$\iint_\Omega f\,dxdy = -\iint_\Omega \Delta u\,dxdy = -\iint_\Omega \nabla \cdot \nabla u\,dxdy \\ = -\int_{\partial\Omega} u_n\,ds = -\int_{\partial\Omega} g(x,y)\,ds, \tag{9.4}$$

  which is the compatibility condition to be satisfied for the solution to exist. If a solution does exist, it is not unique as it is determined within an arbitrary constant.
- Mixed boundary condition on the entire boundary, *i.e.*,

$$\alpha(x,y)u(x,y) + \beta(x,y)\frac{\partial u}{\partial n} = \gamma(x,y)$$

  is given, where $\alpha(x,y)$, $\beta(x,y)$, and $\gamma(x,y)$ are known functions.
- Dirichlet, Neumann, and Mixed boundary conditions on some parts of the boundary.

9.2 Weak Form of Second-Order Self-Adjoint Elliptic PDEs

Now we derive the weak form of the self-adjoint PDE (9.3) with a homogeneous Dirichlet boundary condition on part of the boundary $\partial\Omega_D$, $u|_{\partial\Omega_D} = 0$ and a homogeneous Neumann boundary condition on the rest of boundary $\partial\Omega_N = \partial\Omega - \partial\Omega_D$, $\frac{\partial u}{\partial n}|_{\partial\Omega_N} = 0$. Multiplying the equation (9.3) by a test function $v(x, y) \in H^1(\Omega)$, we have

$$\iint_\Omega \Big\{ -\nabla \cdot (p(x, y)\nabla u) + q(x, y)\, u \Big\}\, v\, dxdy = \iint_\Omega fv\, dxdy\,;$$

and on using the formula for integration by parts the left-hand side becomes

$$\iint_\Omega \Big(p\nabla u \cdot \nabla v + quv \Big)\, dxdy - \int_{\partial\Omega} pvu_n\, ds\,,$$

so the weak form is

$$\begin{aligned}\iint_\Omega (p\nabla u \cdot \nabla v + quv)\, dxdy &= \iint_\Omega fv\, dxdy \\ &+ \int_{\partial\Omega_N} pg(x, y)v(x, y)\, ds \qquad \forall v(x, y) \in H^1(\Omega)\,.\end{aligned} \tag{9.5}$$

Here $\partial\Omega_N$ is the part of boundary where a Neumann boundary condition is applied; and the solution space resides in

$$V = \Big\{ v(x, y)\,,\ v(x, y) = 0\,,\ (x, y) \in \partial\Omega_D\,,\ v(x, y) \in H^1(\Omega) \Big\}, \tag{9.6}$$

where $\partial\Omega_D$ is the part of boundary where a Dirichlet boundary condition is applied.

9.2.1 Verification of Conditions of the Lax–Milgram Lemma

The bilinear form for (9.3) is

$$a(u, v) = \iint_\Omega (p\nabla u \cdot \nabla v + quv)\, dxdy\,, \tag{9.7}$$

and the linear form is

$$L(v) = \iint_\Omega fv\, dxdy \tag{9.8}$$

for a Dirichlet BC on the entire boundary. As before, we assume that

$$0 < p_{min} \le p(x, y) \le p_{max}\,,\ 0 \le q(x) \le q_{max}\,,\ p \in C(\Omega)\,,\ q \in C(\Omega)\,.$$

We need the Poincaré inequality to prove the V-elliptic condition.

Theorem 9.3. *If $v(x,y) \in H_0^1(\Omega)$, $\Omega \subset R^2$, i.e., $v(x,y) \in H^1(\Omega)$ and vanishes at the boundary $\partial\Omega$ (can be relaxed to a point on the boundary), then*

$$\iint_\Omega v^2 dxdy \le C \iint_\Omega |\nabla v|^2 \, dxdy \,, \tag{9.9}$$

where C is a constant.

Now we are ready to check the conditions of the Lax–Milgram Lemma.

1. It is obvious that $a(u, v) = a(v, u)$.
2. It is easy to see that

$$\begin{aligned} |a(u,v)| &\le \max\{p_{max}, q_{max}\} \left| \iint_\Omega (|\nabla u \cdot \nabla v| + |uv|)\, dxdy \right| \\ &= \max\{p_{max}, q_{max}\} \left|(|u|, |v|)_1\right| \\ &\le \max\{p_{max}, q_{max}\} \|u\|_1 \|v\|_1 \,, \end{aligned}$$

   so $a(u, v)$ is a continuous and bounded bilinear operator.
3. From the Poincaré inequality

$$\begin{aligned} |a(v,v)| &= \left| \iint_\Omega p\left(|\nabla v|^2 + qv^2\right) dxdy \right| \\ &\ge p_{min} \iint_\Omega |\nabla v|^2 \, dxdy \\ &= \frac{1}{2} p_{min} \iint_\Omega |\nabla v|^2 \, dxdy + \frac{1}{2} p_{min} \iint_\Omega |\nabla v|^2 \, dxdy \\ &\ge \frac{1}{2} p_{min} \iint_\Omega |\nabla v|^2 \, dxdy + \frac{p_{min}}{2C} \iint_\Omega |v|^2 \, dxdy \\ &\ge \frac{1}{2} p_{min} \min\left\{1, \frac{1}{C}\right\} \|v\|_1^2 \,, \end{aligned}$$

   therefore $a(u, v)$ is V-elliptic.
4. Finally, we show that $L(v)$ is continuous:

$$|L(v)| = |(f, v)_0| \le \|f\|_0 \|v\|_0 \le \|f\|_0 \|v\|_1 \,.$$

Consequently, the solutions to the weak form and the minimization form are unique and bounded in $H_0^1(\Omega)$.

9.3 Triangulation and Basis Functions

The general procedure of the finite element method is the same for any dimension, and the Galerkin finite element method involves the following main steps.

- Generate a triangulation over the domain. Usually the triangulation is composed of either triangles or rectangles. There are a number of mesh generation software packages available, *e.g.*, the Matlab PDE toolbox from Mathworks, Triangle from Carnegie Mellon University, *etc.* Some are available through the Internet.
- Construct basis functions over the triangulation. We mainly consider the conforming finite element method in this book.
- Assemble the stiffness matrix and the load vector element by element, using either the Galerkin finite method (the weak form) or the Ritz finite element method (the minimization form).
- Solve the system of equations.
- Do the error analysis.

In Figure 9.2, we show a diagram of a simple mesh generation process. The circular domain is approximated by a polygon with five vertices (selected points on the boundary). We then connect the five vertices and an interior point to get an initial five triangles (solid line) to obtain an initial coarse mesh. We can refine the mesh using the so-called middle point rule by connecting all the middle points of all triangles in the initial mesh to obtain a finer mesh (solid and dashed lines).

9.3.1 Triangulation and Mesh Parameters

Given a general domain, we can approximate the domain by a polygon and then generate a triangulation over the polygon, and we can refine the triangulation if

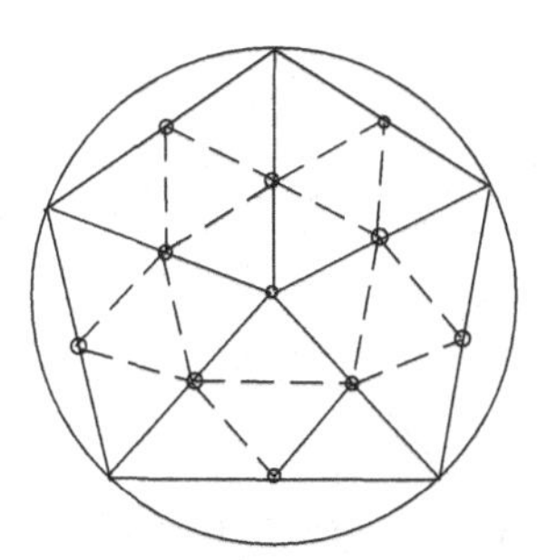

Figure 9.2. A diagram of a simple generation process and the middle point rule.

necessary. A simple approach is the mid-point rule by connecting all the middle points of three sides of existing triangles to get a refined mesh.

A triangulation usually has the mesh parameters

$$
\begin{aligned}
\Omega_p &: \quad \text{polygonal region} = K_1 \cup K_2 \cup K_3 \cdots \cup K_{nelem},\\
K_j &: \quad \text{are nonoverlapping triangles, } j = 1, 2, \ldots, nelem,\\
N_i &: \quad \text{are nodal points, } i = 1, 2, \ldots, nnode,\\
h_j &: \quad \text{the longest side of } K_j,\\
\rho_j &: \quad \text{the diameter of the circle inscribed in } K_j \text{ (encircle)},\\
h &: \quad \text{the largest of all } h_j, \quad h = \max\{h_j\},\\
\rho &: \quad \text{the smallest of all } \rho_j, \quad \rho = \min\{\rho_j\},
\end{aligned}
$$

with

$$1 \ge \frac{\rho_j}{h_j} \ge \beta > 0,$$

where the constant β is a measurement of the triangulation quality (see Figure 9.7 for an illustration of such ρ's and h's). The larger the β, the better the quality of the triangulation. Given a triangulation, a node is also the vertex of all adjacent triangles. We do not discuss hanging nodes here.

9.3.2 *The FE Space of Piecewise Linear Functions over a Triangulation*

For linear second-order elliptic PDEs, we know that the solution space is in the $H^1(\Omega)$. Unlike the 1D case, an element $v(x, y)$ in $H^1(\Omega)$ may not be continuous

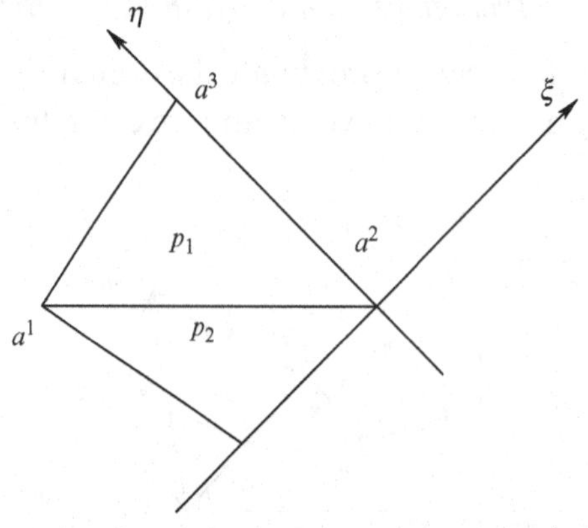

Figure 9.3. A diagram of a triangle with three vertices a^1, a^2, and a^3; an adjacent triangle with a common side; and the local coordinate system in which a^2 is the origin and a^2a^3 is the η-axis.

under the Sobolev embedding theorem. However, in practice most solutions are indeed continuous, especially for second-order PDEs with certain regularities. Thus, we still look for a solution in the continuous function space $C^0(\Omega)$. Let us first consider how to construct piecewise linear functions over a triangulation with the Dirichlet BC

$$u(x,y)|_{\partial\Omega} = 0.$$

Given a triangulation, we define

$$V_h = \Big\{ v(x,y) \text{ is continuous in } \Omega \text{ and piecewise linear over each } K_j, \\ v(x,y)|_{\partial\Omega} = 0 \Big\}. \tag{9.10}$$

We need to determine the dimension of this space and construct a set of basis functions. On each triangle, a linear function has the form

$$v_h(x,y) = \alpha + \beta x + \gamma y\,, \tag{9.11}$$

where α, β and γ are constants (three free parameters). Let

$$P_k = \{ p(x,y)\,,\ \text{a polynomial of degree of } k \}\,. \tag{9.12}$$

We have the following theorem.

Theorem 9.4.

1. *A linear function $p_1(x,y) = \alpha + \beta x + \gamma y$ defined on a triangle is uniquely determined by its values at the three vertices.*
2. *If $p_1(x,y) \in P_1$ and $p_2(x,y) \in P_1$ are such that $p_1(A) = p_2(A)$ and $p_1(B) = p_2(B)$, where A and B are two points in the xy-plane, then $p_1(x,y) \equiv p_2(x,y)$, $\forall (x,y) \in I_{AB}$, where I_{AB} is the line segment between A and B.*

Proof Assume the vertices of the triangle are (x_i, y_i), $i = 1,2,3$. The linear function takes the value v_i at the vertices, *i.e.*,

$$p(x_i, y_i) = v_i,$$

so we have the three equations

$$\begin{aligned} \alpha + \beta x_1 + \gamma y_1 &= v_1, \\ \alpha + \beta x_2 + \gamma y_2 &= v_2, \\ \alpha + \beta x_3 + \gamma y_3 &= v_3. \end{aligned}$$

The determinant of this linear algebraic system is

$$det \begin{bmatrix} 1 & x_1 & y_1 \\ 1 & x_2 & y_2 \\ 1 & x_3 & y_3 \end{bmatrix} = \pm 2 \,\text{area of the triangle} \neq 0 \,\text{ since } \frac{\rho_j}{h_j} \geq \beta > 0, \qquad (9.13)$$

hence the linear system of equations has a unique solution.

Now let us prove the second part of the theorem. Suppose that the equation of the line segment is

$$l_1 x + l_2 y + l_3 = 0, \quad l_1^2 + l_2^2 \neq 0.$$

We can solve for x or for y:

$$x = -\frac{l_2 y + l_3}{l_1} \quad \text{if} \quad l_1 \neq 0,$$

$$\text{or} \quad y = -\frac{l_1 x + l_3}{l_2} \quad \text{if} \quad l_2 \neq 0.$$

Without loss of generality, let us assume $l_2 \neq 0$ such that

$$\begin{aligned} p_1(x,y) &= \alpha + \beta x + \gamma y \\ &= \alpha + \beta x - \frac{l_1 x + l_3}{l_2}\gamma \\ &= \left(\alpha - \frac{l_3}{l_2}\gamma\right) + \left(\beta - \frac{l_1}{l_2}\gamma\right) x \\ &= \alpha_1 + \beta_1 x. \end{aligned}$$

Similarly, we have

$$p_2(x,y) = \bar{\alpha}_1 + \bar{\beta}_1 x.$$

Since $p_1(A) = p_2(A)$ and $p_1(B) = p_2(B)$,

$$\begin{aligned} \alpha_1 + \beta_1 x_1 &= p(A), & \bar{\alpha}_1 + \bar{\beta}_1 x_1 &= p(A), \\ \alpha_1 + \beta_1 x_2 &= p(B), & \bar{\alpha}_1 + \bar{\beta}_1 x_2 &= p(B), \end{aligned}$$

where both of the linear systems of algebraic equations have the same coefficient matrix

$$\begin{bmatrix} 1 & x_1 \\ 1 & x_2 \end{bmatrix}$$

that is nonsingular since $x_1 \neq x_2$ (because points A and B are distinct). Thus we conclude that $\alpha_1 = \bar{\alpha}_1$ and $\beta_1 = \bar{\beta}_1$, so the two linear functions have the same expression along the line segment, *i.e.*, they are identical along the line segment.

Corollary 9.5. A piecewise linear function in $C^0(\Omega) \cap H^1(\Omega)$ over a triangulation (a set of nonoverlapping triangles) is uniquely determined by its values at the vertices.

Theorem 9.6. *The dimension of the finite-dimensional space composed of piecewise linear functions in $C^0(\Omega) \cap H^1(\Omega)$ over a triangulation for (9.3) is the number of interior nodal points plus the number of nodal points on the boundary where the natural BC are imposed (Neumann and mixed boundary conditions).*

Example 9.7. Given the triangulation shown in Figure 9.4, a piecewise continuous function $v_h(x, y)$ is determined by its values on the vertices of all triangles, more precisely, $v_h(x, y)$ is determined from

$$
\begin{aligned}
&(0, 0, v(N_1)), \quad (x, y) \in K_1, && (0, v(N_2), v(N_1)), \quad (x, y) \in K_2, \\
&(0, 0, v(N_2)), \quad (x, y) \in K_3, && (0, 0, v(N_2)), \quad (x, y) \in K_4, \\
&(0, v(N_3), v(N_2)), \quad (x, y) \in K_5, && (0, 0, v(N_3)), \quad (x, y) \in K_6, \\
&(0, v(N_1), v(N_3)), \quad (x, y) \in K_7, && (v(N_1), v(N_2), v(N_3)), \quad (x, y) \in K_8 .
\end{aligned}
$$

Note that although three values of the vertices are the same, like the values for K_3 and K_4, the geometries are different, hence, the functions will likely have different expressions on different triangles.

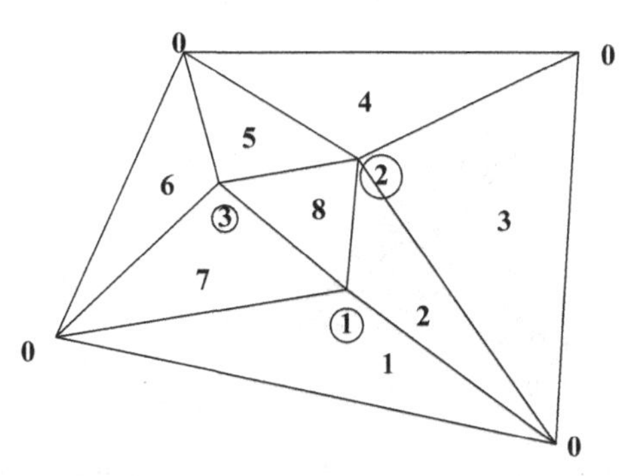

Figure 9.4. A diagram of a simple triangulation with a homogeneous boundary condition.

9.3.3 Global Basis Functions

A global basis function of the piecewise linear functions in $C^0(\Omega) \cap H^1(\Omega)$ can be defined as

$$\phi_i(N_j) = \begin{cases} 1 & \text{if } i = j, \\ 0 & \text{otherwise,} \end{cases} \tag{9.14}$$

where N_j are nodal points. The shape (mesh plot) of $\phi_i(N_j)$ looks like a "tent" without a door; and its support of $\phi_i(N_j)$ is the union of the triangles surrounding the node N_i (*cf.* Figure 9.5, where Figure 9.5(a) is the mesh plot of the global basis function and Figure 9.5(b) is the plot of a triangulation and the contour plot of the global basis function centered at a node). The basis function is piecewise linear and it is supported only in the surrounding triangles.

It is almost impossible to give a closed form of a global basis function except for some very special geometries (*cf.* the example in the next section). However, it is much easier to write down the shape function.

Example 9.8. Let us consider a Poisson equation and a uniform mesh, as an example to demonstrate the piecewise linear basis functions and the finite element method:

$$-(u_{xx} + u_{yy}) = f(x, y), \quad (x, y) \in (a, b) \times (c, d),$$

$$u(x, y)|_{\partial\Omega} = 0.$$

We know how to use the standard central finite difference scheme with the five-point stencil to solve the Poisson equation. With some manipulations, the

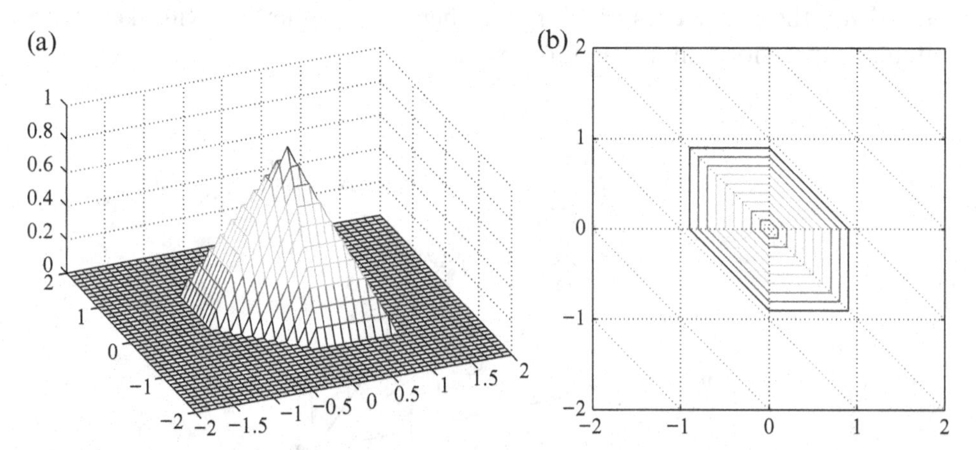

Figure 9.5. A global basis function ϕ_j: (a) the mesh plot of the global function and (b) the triangulation and the contour plot of the global basis function.

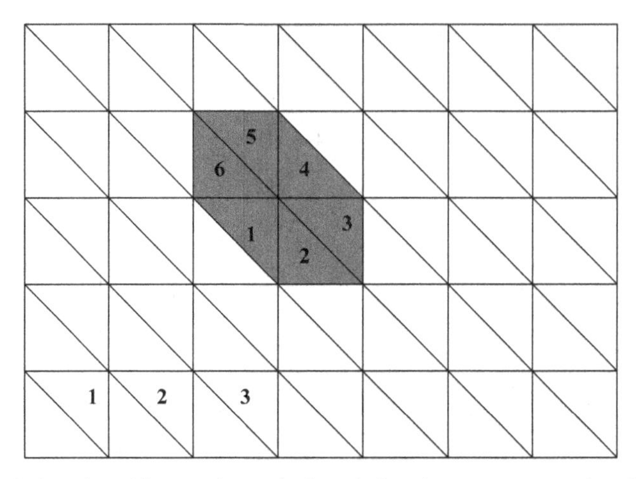

Figure 9.6. A uniform triangulation defined on a rectangular domain.

linear system of equations on using the finite element method with a uniform triangulation (*cf.* Figure 9.6) proves to be the same as that obtained from the finite difference method.

Given a uniform triangulation as shown in Figure 9.6, if we use row-wise natural ordering for the nodal points

$$(x_i, y_j), \quad x_i = ih, \quad y_j = jh, \quad h = \frac{1}{n}, \quad i = 1, 2, \ldots, m-1, \; j = 1, 2, \ldots, n-1,$$

then the global basis function defined at $(x_i, y_j) = (ih, jh)$ is

$$\phi_{j(n-1)+i} = \begin{cases} \dfrac{x-(i-1)h+y-(j-1)h}{h} - 1 & \text{Region 1} \\ \dfrac{y-(j-1)h}{h} & \text{Region 2} \\ \dfrac{h-(x-ih)}{h} & \text{Region 3} \\ 1 - \dfrac{x-ih+y-jh}{h} & \text{Region 4} \\ \dfrac{h-(y-jh)}{h} & \text{Region 5} \\ \dfrac{x-(i-1)h}{h} & \text{Region 6} \\ 0 & \text{otherwise}. \end{cases}$$

If $m = n = 3$, there are nine interior nodal points such that the stiffness matrix is a 9×9 matrix:

$$A = \begin{bmatrix} * & * & 0 & * & 0 & 0 & 0 & 0 & 0 \\ * & * & * & o & * & 0 & 0 & 0 & 0 \\ 0 & * & * & 0 & o & * & 0 & 0 & 0 \\ * & o & 0 & * & * & 0 & * & 0 & 0 \\ 0 & * & o & * & * & * & o & * & 0 \\ 0 & 0 & * & 0 & * & * & 0 & o & * \\ 0 & 0 & 0 & * & o & 0 & * & * & 0 \\ 0 & 0 & 0 & 0 & * & o & * & * & * \\ 0 & 0 & 0 & 0 & 0 & * & 0 & * & * \end{bmatrix},$$

where "$*$" stands for the nonzero entries and "o" happens to be zero for Poisson equations. Generally, the stiffness matrix is block tridiagonal:

$$A = \begin{bmatrix} B & -I & 0 & & & \\ -I & B & -I & & & \\ & & \cdots & \cdots & & \\ & & & \cdots & \cdots & \\ & & & -I & B & -I \\ & & & & -I & B \end{bmatrix}, \text{ where } B = \begin{bmatrix} 4 & -1 & 0 & & & \\ -1 & 4 & -1 & & & \\ & & \cdots & \cdots & & \\ & & & \cdots & \cdots & \\ & & & -1 & 4 & -1 \\ & & & & -1 & 4 \end{bmatrix}$$

and I is the identity matrix. The component of the load vector F_i can be approximated as

$$\int\!\!\int_D f(x,y)\phi_i dxdy \simeq f_{ij} \int\!\!\int_D \phi_i \, dxdy = h^2 f_{ij},$$

so after dividing by h^2 we get the same system of equations as in the finite difference scheme, namely,

$$-\frac{U_{i-1,j} + U_{i+1,j} + U_{i,j-1} + U_{i,j+1} - 4U_{ij}}{h^2} = f_{ij},$$

with the same ordering.

9.3.4 The Interpolation Function and Error Analysis

We know that the finite element solution u_h is the best solution in terms of the energy norm in the finite-dimensional space V_h, *i.e.*, $\|u - u_h\|_a \le \|u - v_h\|_a$, assuming that u is the solution to the weak form. However, this does not give a quantitative estimate for the finite element solution, and we may wish to have a more precise error estimate in terms of the solution information and the mesh size h. This can be done through the interpolation function, for which an error estimate is often available from the approximation theory. Note that the solution information appears as part of the error constants in the error estimates, even though the solution is unknown. We will use the mesh parameters defined on page 6 in the discussion here.

Definition 9.9. Given a triangulation of T_h, let $K \in T_h$ be a triangle with vertices a^i, $i = 1, 2, 3$. The interpolation function for a function $v(x, y)$ on the triangle is defined as

$$v_I(x,y) = \sum_{i=1}^{3} v(a^i)\phi_i(x,y), \qquad (x,y) \in K, \tag{9.15}$$

where $\phi_i(x, y)$ is the piecewise linear function that satisfies $\phi_i(a^j) = \delta_i^j$ (with δ_i^j being the Kronecker delta). A global interpolation function is defined as

$$v_I(x,y) = \sum_{i=1}^{nnode} v(a^i)\phi_i(x,y), \qquad (x,y) \in T_h\,, \tag{9.16}$$

where a^i's are all nodal points and $\phi_i(x, y)$ is the global basis function centered at a^i.

Theorem 9.10. *If $v(x, y) \in C^2(K)$, then we have an error estimate for the interpolation function on a triangle K,*

$$\|v - v_I\|_\infty \le 2h^2 \max_{|\alpha|=2} \|D^\alpha v\|_\infty\,, \tag{9.17}$$

where h is the longest side. Furthermore, we have

$$\max_{|\alpha|=1} \|D^\alpha\,(v - v_I)\,\|_\infty \le \frac{8h^2}{\rho} \max_{|\alpha|=2} \|D^\alpha v\|_\infty. \tag{9.18}$$

Proof From the definition of the interpolation function and the Taylor expansion of $v(a^i)$ at (x,y), we have

$$\begin{aligned}
v_I(x,y) &= \sum_{i=1}^{3} v(a^i)\phi_i(x,y) \\
&= \sum_{i=1}^{3} \phi_i(x,y)\left(v(x,y) + \frac{\partial v}{\partial x}(x,y)(x_i - x) + \frac{\partial v}{\partial y}(x,y)(y_i - y) \right. \\
&\quad \left. + \frac{1}{2}\frac{\partial^2 v}{\partial x^2}(\xi,\eta)(x_i - x)^2 + \frac{\partial^2 v}{\partial x \partial y}(\xi,\eta)(x_i - x)(y_i - y) + \frac{1}{2}\frac{\partial^2 v}{\partial y^2}(\xi,\eta)(y_i - y)^2 \right) \\
&= \sum_{i=1}^{3} \phi_i(x,y)v(x,y) + \sum_{i=1}^{3} \phi_i(x,y)\left(\frac{\partial v}{\partial x}(x,y)(x_i - x) + \frac{\partial v}{\partial y}(x,y)(y_i - y) \right) \\
&\quad + R(x,y),
\end{aligned}$$

where (ξ,η) is a point in the triangle K. It is easy to show that

$$|R(x,y)| \le 2h^2 \max_{|\alpha|=2} \|D^\alpha v\|_\infty \sum_{i=1}^{3} |\phi_i(x,y)| = 2h^2 \max_{|\alpha|=2} \|D^\alpha v\|_\infty ,$$

since $\phi(x,y) \ge 0$ and $\sum_{i=1}^{3} \phi_i(x,y) = 1$. If we take $v(x,y) = 1$, which is a linear function, then $\partial v/\partial x = \partial v/\partial y = 0$ and $\max_{|\alpha|=2} \|D^\alpha v\|_\infty = 0$. The interpolation is simply the function itself, since it is uniquely determined by the values at the vertices of T, hence

$$v_I(x,y) = v(x,y) = \sum_{i=1}^{3} v(a^i)\phi_i(x,y) = \sum_{i=1}^{3} \phi_i(x,y) = 1 . \tag{9.19}$$

If we take $v(x,y) = d_1 x + d_2 y$, which is also a linear function, then $\partial v/\partial x = d_1$, $\partial v/\partial y = d_2$, and $\max_{|\alpha|=2} \|D^\alpha v\|_\infty = 0$. The interpolation is again simply the function itself, since it is uniquely determined by the values at the vertices of K. Thus from the previous Taylor expansion and the identity $\sum_{i=1}^{3} \phi_i(x,y) = 1$, we have

$$v_I(x,y) = v(x,y) = v(x,y) + \sum_{i=1}^{3} \phi_i(x,y)\left(d_1(x_i - x) + d_2(y_i - y)\right) = v(x,y), \tag{9.20}$$

hence $\sum_{i=1}^{3} \phi_i(x,y)\left(d_1(x_i - x) + d_2(y_i - y)\right) = 0$ for any d_1 and d_2, *i.e.*, the linear part in the expansion is the interpolation function. Consequently, for a general function $v(x,y) \in C^2(K)$ we have

$$v_I(x,y) = v(x,y) + R(x,y), \qquad \|v - v_I\|_\infty \le 2h^2 \max_{|\alpha|=2} \|D^\alpha v\|_\infty,$$

which completes the proof of the first part of the theorem.

To prove the second part concerning the error estimate for the gradient, choose a point (x_0, y_0) inside the triangle K and apply the Taylor expansion at (x_0, y_0) to get

$$\begin{aligned} v(x,y) &= v(x_0,y_0) + \frac{\partial v}{\partial x}(x_0,y_0)(x-x_0) + \frac{\partial v}{\partial y}(x_0,y_0)(y-y_0) + R_2(x,y), \\ &= p_1(x,y) + R_2(x,y), \qquad |R_2(x,y)| \le 2h^2 \max_{|\alpha|=2} \|D^\alpha v\|_\infty . \end{aligned}$$

Rewriting the interpolation function $v_I(x,y)$ as

$$v_I(x,y) = v(x_0,y_0) + \frac{\partial v}{\partial x}(x_0,y_0)(x-x_0) + \frac{\partial v}{\partial y}(x_0,y_0)(y-y_0) + R_1(x,y),$$

where $R_1(x,y)$ is a linear function of x and y, we have

$$v_I(a^i) = p_1(a^i) + R_1(a^i), \; i = 1,2,3,$$

from the definition above. On the other hand, $v_I(x,y)$ is the interpolation function, such that also

$$v_I(a^i) = v(a^i) = p_1(a^i) + R_2(a^i), \qquad i = 1,2,3.$$

Since $p_1(a^i) + R_1(a^i) = p_1(a^i) + R_2(a^i)$, it follows that $R_1(a^i) = R_2(a^i)$, *i.e.*, $R_1(x,y)$ is the interpolation function of $R_2(x,y)$ in the triangle K, and we have

$$R_1(x,y) = \sum_{i=1}^{3} R_2(a^i)\phi_i(x,y) .$$

With this equality and on differentiating

$$v_I(x,y) = v(x_0,y_0) + \frac{\partial v}{\partial x}(x_0,y_0)(x-x_0) + \frac{\partial v}{\partial y}(x_0,y_0)(y-y_0) + R_1(x,y)$$

with respect to x, we get

$$\frac{\partial v_I}{\partial x}(x,y) = \frac{\partial v}{\partial x}(x_0,y_0) + \frac{\partial R_1}{\partial x}(x,y) = \frac{\partial v}{\partial x}(x_0,y_0) + \sum_{i=1}^{3} R_2(a^i)\frac{\partial \phi_i}{\partial x}(x,y) .$$

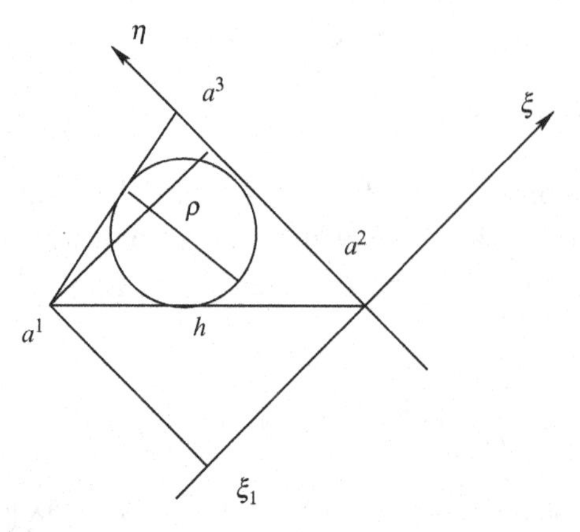

Figure 9.7. A diagram used to prove Theorem 9.10.

Applying the Taylor expansion for $\partial v(x,y)/\partial x$ at (x_0, y_0) gives

$$\frac{\partial v}{\partial x}(x,y) = \frac{\partial v}{\partial x}(x_0,y_0) + \frac{\partial^2 v}{\partial x^2}(\bar{x},\bar{y})(x-x_0) + \frac{\partial^2 v}{\partial x \partial y}(\bar{x},\bar{y})(y-y_0)\,,$$

where $(\bar{x},\bar{y})$ is a point in the triangle K. From the last two equalities, we obtain

$$\begin{aligned}\left|\frac{\partial v}{\partial x} - \frac{\partial v_I}{\partial x}\right| &= \left|\frac{\partial^2 v}{\partial x^2}(\bar{x},\bar{y})(x-x_0) + \frac{\partial^2 v}{\partial x \partial y}(\bar{x},\bar{y})(y-y_0) - \sum_{i=1}^{3} R_2(a^i)\frac{\partial \phi_i}{\partial x}\right| \\ &\le \max_{|\alpha|=2} \|D^\alpha v\|_\infty \left(2h + 2h^2 \sum_{i=1}^{3}\left|\frac{\partial \phi_i}{\partial x}\right|\right).\end{aligned}$$

It remains to prove that $|\partial\phi_i/\partial x| \le 1/\rho$, $i = 1,2,3$. We take $i=1$ as an illustration, and use a shift and rotation coordinate transform such that a^2a^3 is the η axis and a^2 is the origin (*cf.* Figure 9.7):

$$\begin{aligned}\xi &= (x-x_2)\cos\theta + (y-y_2)\sin\theta\,, \\ \eta &= -(x-x_2)\sin\theta + (y-y_2)\cos\theta\,.\end{aligned}$$

Then $\phi_1(x,y) = \phi_1(\xi,\eta) = C\xi = \xi/\xi_1$, where ξ_1 is the ξ coordinate in the (ξ,η) coordinate system, such that

$$\left|\frac{\partial \phi_1}{\partial x}\right| = \left|\frac{\partial \phi_1}{\partial \xi}\cos\theta - \frac{\partial \phi_1}{\partial \eta}\sin\theta\right| \le \left|\frac{1}{\xi_1}\cos\theta\right| \le \frac{1}{|\xi_1|} \le \frac{1}{\rho}\,.$$

The same estimate applies to $\partial\phi_i/\partial x$, $i = 2,3$, so finally we have

$$\left|\frac{\partial v}{\partial x} - \frac{\partial v_I}{\partial x}\right| \le \max_{|\alpha|=2}\|D^\alpha v\|_\infty \left(2h + \frac{6h^2}{\rho}\right) \le \frac{8h^2}{\rho}\max_{|\alpha|=2}\|D^\alpha v\|_\infty\,,$$

from the fact that $\rho \leq h$. Similarly, we may obtain the same error estimate for $\partial v_I/\partial y$. □

Corollary 9.11. Given a triangulation of T_h, we have the following error estimates for the interpolation function:

$$\|v - v_I\|_{L^2(T_h)} \leq C_1 h^2 \|v\|_{H^2(T_h)}, \qquad \|v - v_I\|_{H^1(T_h)} \leq C_2 h \|v\|_{H^2(T_h)}, \tag{9.21}$$

where C_1 and C_2 are constants.

9.3.5 Error Estimates of the FE Solution

Let us now recall the 2D Sturm–Liouville problem in a bounded domain Ω:

$$-\nabla \cdot (p(x,y)\nabla u(x,y)) + q(x,y)u(x,y) = f(x,y), \quad (x,y) \in \Omega,$$

$$u(x,y)_{\partial\Omega} = u_0(x,y),$$

where $u_0(x,y)$ is a given function, *i.e.*, a Dirichlet BC is prescribed. If we assume that $p, q \in C(\Omega)$, $p(x,y) \geq p_0 > 0$, $q(x,y) \geq 0$, $f \in L^2(\Omega)$ and the boundary $\partial\Omega$ is smooth (in C^1), then we know that the weak form has a unique solution and the energy norm $\|v\|_a$ is equivalent to the H^1 norm $\|v\|_1$. Furthermore, we know that the solution $u(x,y) \in H^2(\Omega)$. Given a triangulation T_h with a polygonal approximation to the outer boundary $\partial\Omega$, let V_h be the piecewise linear function space over the triangulation T_h, and u_h be the finite element solution. With those assumptions, we have the following theorem for the error estimates.

Theorem 9.12.

$$\|u - u_h\|_a \leq C_1 h \|u\|_{H^2(T_h)}, \qquad \|u - u_h\|_{H^1(T_h)} \leq C_2 h \|u\|_{H^2(T_h)}, \tag{9.22}$$

$$\|u - u_h\|_{L^2(T_h)} \leq C_3 h^2 \|u\|_{H^2(T_h)}, \qquad \|u - u_h\|_\infty \leq C_4 h^2 \|u\|_{H^2(T_h)}, \tag{9.23}$$

where C_i are constants.

Sketch of the proof. Since the finite element solution is the best solution in the energy norm, we have

$$\|u - u_h\|_a \leq \|u - u_I\|_a \leq \bar{C}_1 \|u - u_I\|_{H^1(T_h)} \leq \bar{C}_1 \bar{C}_2 h \|u\|_{H^2(T_h)},$$

because the energy norm is equivalent to the H^1 norm. Furthermore, because of the equivalence we get the estimate for the H^1 norm as well. The error estimates for the L^2 and L^∞ norm are not trivial in 2D, and the reader may care to consult other advanced textbooks on finite element methods.

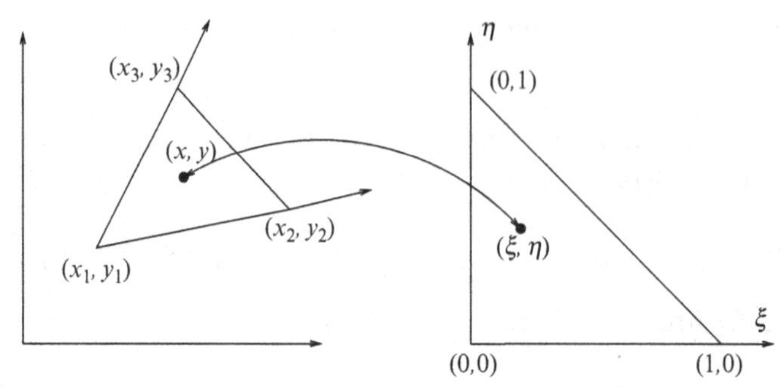

Figure 9.8. The linear transform from an arbitrary triangle to the standard triangle (master element) and the inverse map.

9.4 Transforms, Shape Functions, and Quadrature Formulas

Any triangle with nonzero area can be transformed to the right-isosceles master triangle, or standard triangle $\triangle$ (*cf.* Figure 9.8). There are three nonzero basis functions over this standard triangle $\triangle$, namely,

$$\psi_1(\xi,\eta) = 1 - \xi - \eta, \tag{9.24}$$

$$\psi_2(\xi,\eta) = \xi, \tag{9.25}$$

$$\psi_3(\xi,\eta) = \eta. \tag{9.26}$$

The linear transform from a triangle with vertices (x_1, y_1), (x_2, y_2), and (x_3, y_3) arranged in the counterclockwise direction to the master triangle $\triangle$ is

$$x = \sum_{j=1}^{3} x_j \psi_j(\xi,\eta), \qquad y = \sum_{j=1}^{3} y_j \psi_j(\xi,\eta), \tag{9.27}$$

or

$$\xi = \frac{1}{2A_e}\Big((y_3 - y_1)(x - x_1) - (x_3 - x_1)(y - y_1)\Big), \tag{9.28}$$

$$\eta = \frac{1}{2A_e}\Big(-(y_2 - y_1)(x - x_1) + (x_2 - x_1)(y - y_1)\Big), \tag{9.29}$$

where A_e is the area of the triangle that can be calculated using the formula in (9.13).

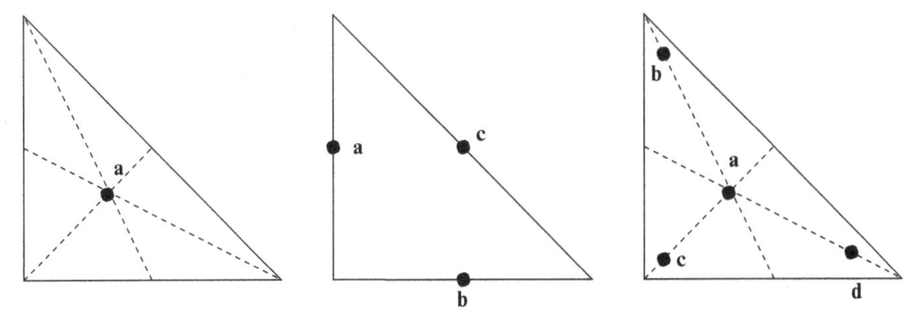

Figure 9.9. A diagram of the quadrature formulas in 2D with one, three, and four quadrature points, respectively.

Table 9.1. Quadrature points and weights corresponding to the geometry in Figure 9.9

L	Points	(ξ_k, η_k)	w_k
1	a	$\left(\frac{1}{3}, \frac{1}{3}\right)$	$\frac{1}{2}$
3	a	$\left(0, \frac{1}{2}\right)$	$\frac{1}{6}$
	b	$\left(\frac{1}{2}, 0\right)$	$\frac{1}{6}$
	c	$\left(\frac{1}{2}, \frac{1}{2}\right)$	$\frac{1}{6}$
4	a	$\left(\frac{1}{3}, \frac{1}{3}\right)$	$-\frac{27}{96}$
	b	$\left(\frac{2}{15}, \frac{11}{15}\right)$	$\frac{25}{96}$
	c	$\left(\frac{2}{15}, \frac{2}{15}\right)$	$\frac{25}{96}$
	d	$\left(\frac{11}{15}, \frac{2}{15}\right)$	$\frac{25}{96}$

9.4.1 Quadrature Formulas

In the assembling process, we need to evaluate the double integrals

$$\iint_{\Omega_e} q(x,y)\phi_i(x,y)\phi_j(x,y)\,dxdy = \iint_{\triangle} q(\xi,\eta)\,\psi_i(\xi,\eta)\psi_j(\xi,\eta)\left|\frac{\partial(x,y)}{(\partial\xi,\eta)}\right|\,d\xi d\eta,$$

$$\iint_{\Omega_e} f(x,y)\phi_j(x,y)\,dxdy = \iint_{\triangle} f(\xi,\eta)\,\psi_j(\xi,\eta)\left|\frac{\partial(x,y)}{(\partial\xi,\eta)}\right| d\xi d\eta,$$

$$\iint_{\Omega_e} p(x,y)\nabla\phi_i \cdot \nabla\phi_j\,dxdy = \iint_{\triangle} p(\xi,\eta)\,\nabla_{(x,y)}\psi_i \cdot \nabla_{(x,y)}\psi_j \left|\frac{(\partial(x,y)}{\partial\xi,\eta)}\right| d\xi d\eta$$

in which, for example, $q(\xi,\eta)$ should really be $q(x(\xi,\eta),y(\xi,\eta))=\bar{q}(\xi,\eta)$ and so on. For simplification of the notations, we omit the bar symbol.

A quadrature formula has the form

$$\iint_{S_\triangle} g(\xi,\eta)d\xi d\eta = \sum_{k=1}^{L} w_k\, g(\xi_k,\eta_k), \tag{9.30}$$

where $S_\triangle$ is the standard right triangle and L is the number of points involved in the quadrature. In Table 9.1 we list some commonly used quadrature formulas in 2D using one, three, and four points. The geometry of the points is illustrated in Figure 9.9, and the coordinates of the points and the weights are given in Table 9.1. It is noted that only the three-point quadrature formula is closed, since the three points are on the boundary of the triangle, and the other quadrature formulas are open.

9.5 Some Implementation Details

The procedure is essentially the same as in the 1D case, but some details are slightly different.

9.5.1 Description of a Triangulation

A triangulation is determined by its elements and nodal points. We use the following notation:

- Nodal points: N_i, $(x_1,y_1),(x_2,y_2),\ldots,(x_{nnode},y_{nnode})$, *i.e.*, we assume there are *nnode* nodal points.
- Elements: K_i, $K_1,K_2,\ldots,K_{nelem}$, *i.e.*, we assume there are *nelem* elements.
- A 2D array *nodes* is used to describe the relation between the nodal points and the elements: *nodes*$(3, nelem)$. The first index is the index of nodal point in an element, usually in the counterclockwise direction, and the second index is the index of the element.

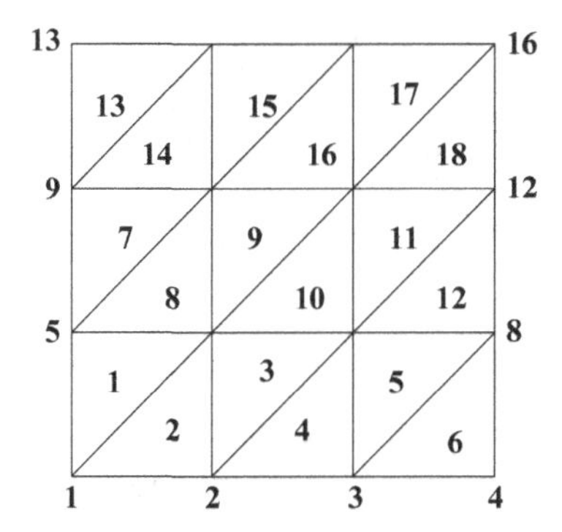

Figure 9.10. A simple triangulation with the row-wise natural ordering.

Example 9.13. Below we show the relation between the index of the nodal points and elements, and its relations, *cf.* also Figure 9.10.

$$nodes(1,1)=5, \quad (x_5,y_5)=(0,h),$$
$$nodes(2,1)=1, \quad (x_1,y_1)=(0,0)\,,$$
$$nodes(3,1)=6, \quad (x_6,y_6)=(h,h)\,,$$

$$nodes(1,10)=7, \quad (x_7,y_7)=(2h,h),$$
$$nodes(2,10)=11, \quad (x_{11},y_{11})=(2h,2h)\,,$$
$$nodes(3,10)=6\,, \quad (x_6,y_6)=(h,h).$$

9.5.2 Outline of the FE Algorithm using the Piecewise Linear Basis Functions

The main assembling process is the following loop.

```
for nel = 1:nelem
   i1 = nodes(1,nel);        % (x(i1),y(i1)), get nodal points
   i2 = nodes(2,nel);        % (x(i2),y(i2))
   i3 = nodes(3,nel);        % (x(i3),y(i3))
        ..............
```

- Computing the local stiffness matrix and the load vector.

```
ef=zeros(3,1);
ek = zeros(3,3);
for l=1:nq                    % nq is the number of quadrature points.
  [xi_x(l),eta_y(l)] = getint,     % Get a quadrature point.
  [psi,dpsi] = shape(xi_x(l),eta_y(l));
  [x_l,y_l]  = transform,  % Get (x,y) from (\xi_x(l), \eta_y(l))
```

```
  [xk,xq,xf] = getmat(x_1,y_1);      % Get the material
                                   %coefficients at the quadrature point.
  for i= 1:3
     ef(i) = ef(i) + psi(i)*xf*w(1)*J;    % J is the Jacobian
     for j=1:3
        ek(i,j)=ek(i,j)+ (T + xq*psi(i)*psi(j) )*J    % see below
     end
  end
end
```

Note that *psi* has three values corresponding to three nonzero basis functions; *dpsi* is a 3×2 matrix which contains the partial derivatives $\partial\psi_i/\partial\xi$ and $\partial\psi_i/\partial\eta$. The evaluation of T is

$$\iint_{\Omega_e} p(x,y)\,\nabla\phi_i\cdot\nabla\phi_j\,dx\,dy = \iint_{\Omega_e} p(\xi,\eta)\left(\frac{\partial\psi_i}{\partial x}\frac{\partial\psi_j}{\partial x}+\frac{\partial\psi_i}{\partial y}\frac{\partial\psi_j}{\partial y}\right)|J|\,d\xi\,d\eta\,,$$

where $J = \frac{\partial(x,y)}{\partial(\xi,\eta)}$ is the Jacobian of the transform. We need to calculate $\partial\psi_i/\partial x$ and $\partial\psi_i/\partial y$ in terms of ξ and η. Notice that

$$\frac{\partial\psi_i}{\partial x} = \frac{\partial\psi_i}{\partial\xi}\frac{\partial\xi}{\partial x} + \frac{\partial\psi_i}{\partial\eta}\frac{\partial\eta}{\partial x}\,,$$

$$\frac{\partial\psi_i}{\partial y} = \frac{\partial\psi_i}{\partial\xi}\frac{\partial\xi}{\partial y} + \frac{\partial\psi_i}{\partial\eta}\frac{\partial\eta}{\partial y}\,.$$

Since we know that

$$\xi = \frac{1}{2A_e}\Big((y_3-y_1)(x-x_1)-(x_3-x_1)(y-y_1)\Big)\,,$$

$$\eta = \frac{1}{2A_e}\Big(-(y_2-y_1)(x-x_1)+(x_2-x_1)(y-y_1)\Big)\,,$$

we obtain those partial derivatives below,

$$\frac{\partial\xi}{\partial x} = \frac{1}{2A_e}(y_3-y_1), \qquad \frac{\partial\xi}{\partial y} = -\frac{1}{2A_e}(x_3-x_1)\,,$$

$$\frac{\partial\eta}{\partial x} = -\frac{1}{2A_e}(y_2-y_1), \qquad \frac{\partial\eta}{\partial y} = \frac{1}{2A_e}(x_2-x_1)\,.$$

- Add to the global stiffness matrix and the load vector.

```
for i= 1:3
   ig = nodes(i,nel);
   gf(ig) = gf(ig) + ef(i);
   for j=1:3
     jg  = nodes(j,nel);
     gk(ig,jg) = gk(ig,jg) + ek(i,j);
   end
end
```

- Solve the system of equations gk U= gf.
  - Direct method, *e.g.*, Gaussian elimination.
  - Sparse matrix technique, *e.g.*, $A = sparse(M, M)$.
  - Iterative method plus preconditioning, *e.g.*, Jacobi, Gauss–Seidel, SOR(ω), conjugate gradient methods, *etc.*
- Error analysis.
  - Construct interpolation functions.
  - Error estimates for interpolation functions.
  - Finite element solution is the best approximation in the finite element space in the energy norm.

9.6 Simplification of the FE Method for Poisson Equations

With constant coefficients, there is a closed form for the local stiffness matrix, in terms of the coordinates of the nodal points; so the finite element algorithm can be simplified. We now introduce the simplified finite element algorithm. A good reference is White (1985): *An introduction to the finite element method with applications to nonlinear problems* by R.E. White, John Wiley & Sons.

Let us consider the Poisson equation below

$$\begin{aligned} -\Delta u &= f(x,y),\ (x,y)\in\Omega, \\ u(x,y) &= g(x,y)\,,\ (x,y)\in\partial\Omega_1, \\ \frac{\partial u}{\partial n} &= 0\,,\ (x,y)\in\partial\Omega_2, \end{aligned}$$

where Ω is an arbitrary but bounded domain. We can use Matlab PDE Toolbox to generate a triangulation for the domain Ω.

The weak form is

$$\iint_\Omega \nabla u\cdot\nabla v\,dxdy = \iint_\Omega fv\,dxdy.$$

With the piecewise linear basis functions defined on a triangulation on Ω, we can derive analytic expressions for the basis functions and the entries of the local stiffness matrix.

Theorem 9.14. *Consider a triangle determined by (x_1, y_1), (x_2, y_2) and (x_3, y_3). Let*

$$a_i = x_j y_m - x_m y_j, \tag{9.31}$$

$$b_i = y_j - y_m, \tag{9.32}$$

$$c_i = x_m - x_j, \tag{9.33}$$

where i, j, m is a positive permutation of 1, 2, 3, e.g., $i=1, j=2$ and $m=3$; $i=2$, $j=3$ and $m=1$; and $i=3, j=1$ and $m=2$. Then the corresponding three nonzero basis functions are

$$\psi_i(x,y) = \frac{a_i + b_i x + c_i y}{2\Delta}, \quad i=1,2,3, \tag{9.34}$$

where $\psi_i(x_i, y_i) = 1$, $\psi_i(x_j, y_j) = 0$ if $i \neq j$, and

$$\Delta = \frac{1}{2} \, det \begin{bmatrix} 1 & x_1 & y_1 \\ 1 & x_2 & y_2 \\ 1 & x_3 & y_3 \end{bmatrix} = \pm \text{ area of the triangle.} \tag{9.35}$$

We prove the theorem for $\psi_1(x,y)$. Substitute a_1, b_1, and c_1 in terms of x_i and y_i in the definition of ψ_1, we have,

$$\begin{aligned} \psi_1(x,y) &= \frac{a_1 + b_1 x + c_1 y}{2\Delta}, \\ &= \frac{(x_2 y_3 - x_3 y_2) + (y_2 - y_3)x + (x_3 - x_2)y}{2\Delta}, \end{aligned}$$

so
$$\begin{aligned} \psi_1(x_2,y_2) &= \frac{(x_2 y_3 - x_3 y_2) + (y_2 - y_3)x_2 + (x_3 - x_2)y_2}{2\Delta} = 0, \\ \psi_1(x_3,y_3) &= \frac{(x_2 y_3 - x_3 y_2) + (y_2 - y_3)x_3 + (x_3 - x_2)y_3}{2\Delta} = 0, \\ \psi_1(x_1,y_1) &= \frac{(x_2 y_3 - x_3 y_2) + (y_2 - y_3)x_1 + (x_3 - x_2)y_1}{2\Delta} = \frac{2\Delta}{2\Delta} = 1. \end{aligned}$$

We can prove the same feature for ψ_2 and ψ_3.

We also have the following theorem, which is essential for the simplified finite element method.

Theorem 9.15. *With the same notations as in Theorem 9.14, we have*

$$\iint_{\Omega_e} (\psi_1)^m (\psi_2)^n (\psi_3)^l \, dxdy = \frac{m!\, n!\, l!}{(m+n+l+2)\,!} 2\Delta, \tag{9.36}$$

$$\iint_{\Omega_e} \nabla\psi_i \cdot \nabla\psi_j \, dxdy = \frac{b_i b_j + c_i c_j}{4\Delta},$$

$$F_1^e = \iint_{\Omega_e} \psi_1 f(x,y)\, dxdy \simeq f_1 \frac{\Delta}{6} + f_2 \frac{\Delta}{12} + f_3 \frac{\Delta}{12},$$

$$F_2^e = \iint_{\Omega_e} \psi_2 f(x,y)\, dxdy \simeq f_1 \frac{\Delta}{12} + f_2 \frac{\Delta}{6} + f_3 \frac{\Delta}{12},$$

$$F_3^e = \iint_{\Omega_e} \psi_3 f(x,y)\, dxdy \simeq f_1 \frac{\Delta}{12} + f_2 \frac{\Delta}{12} + f_3 \frac{\Delta}{6},$$

where $f_i = f(x_i, y_i)$.

The proof is straightforward since we have the analytic form for ψ_i. We approximate $f(x,y)$ using

$$f(x,y) \simeq f_1\psi_1 + f_2\psi_2 + f_3\psi_3, \tag{9.37}$$

and therefore

$$\begin{aligned} F_1^e &\simeq \iint_{\Omega_e} \psi_1 f(x,y)\, dxdy \\ &= f_1 \iint_{\Omega_e} \psi_1^2 dxdy + f_2 \iint_{\Omega_e} \psi_1\psi_2\, dxdy + f_3 \iint_{\Omega_e} \psi_1\psi_3\, dxdy. \end{aligned} \tag{9.38}$$

Note that the integrals in the last expression can be obtained from the formula (9.36). There is a negligible error from approximating $f(x,y)$ compared with the error from the finite element approximation when we seek approximate solution only in V_h space instead of $H^1(\Omega)$ space. Similarly we can get approximation F_2^e and F_3^e.

9.6.1 A Pseudo-code of the Simplified FE Method

Assume that we have a triangulation, *e.g.*, a triangulation generated from Matlab by saving the mesh. Then we have

$p(1,1), p(1,2), \ldots, p(1, nnode)$ as x coordinates of the nodal points,
$p(2,1), p(2,2), \ldots, p(2, nnode)$ as y coordinates of the nodal points;

and the array t (the nodes in our earlier notation)

$t(1,1),\ t(1,2),\ \ldots,\ t(1, nele)$ as the index of the first node of an element,
$t(2,1),\ t(2,2),\ \ldots,\ t(2, nele)$ as the index of the second node of the element,
$t(3,1),\ t(3,2),\ \ldots,\ t(3, nele)$ as the index of the third node of the element;

and the array e to describe the nodal points on the boundary

$e(1,1),\ e(1,2),\ \ldots,\ e(1, nbc)$ as the index of the beginning node of a boundary edge,
$e(2,1),\ e(2,2),\ \ldots,\ e(2, nbc)$ as the index of the end node of the boundary edge.

A Matlab code for the simplified finite element method is listed below.

```
% Set-up: assume we have a triangulation p,e,t from Matlab PDE tool box
% already.

    [ijunk,nelem] = size(t);
    [ijunk,nnode] = size(p);

    for i=1:nelem
      nodes(1,i)=t(1,i);
      nodes(2,i)=t(2,i);
      nodes(3,i)=t(3,i);
    end

    gk=zeros(nnode,nnode);
    gf = zeros(nnode,1);

    for nel = 1:nelem,    % Begin to assemble by element.

      for j=1:3,            % The coordinates of the nodes in the
        jj = nodes(j,nel);    % element.
        xx(j) = p(1,jj);
        yy(j) = p(2,jj);
      end

    for nel = 1:nelem,    % Begin to assemble by element.

      for j=1:3,            % The coordinates of the nodes in the
        jj = nodes(j,nel);    % element.
        xx(j) = p(1,jj);
        yy(j) = p(2,jj);
      end

      for i=1:3,
```

```
          j = i+1 - fix((i+1)/3)*3;
          if j == 0
             j = 3;
          end
          m = i+2 - fix((i+2)/3)*3;
          if m  == 0
             m = 3;
          end

          a(i) = xx(j)*yy(m) - xx(m)*yy(j);
          b(i) = yy(j) - yy(m);
          c(i) = xx(m) - xx(j);
        end

        delta = ( c(3)*b(2) - c(2)*b(3) )/2.0;    % Area.

        for ir = 1:3,
          ii = nodes(ir,nel);
          for ic=1:3,
            ak = (b(ir)*b(ic) + c(ir)*c(ic))/(4*delta);
            jj = nodes(ic,nel);
            gk(ii,jj) = gk(ii,jj) + ak;
          end
            j = ir+1 - fix((ir+1)/3)*3;
               if j == 0
                  j = 3;
               end
            m = ir+2 - fix((ir+2)/3)*3;
               if m == 0
                  m = 3;
               end
            gf(ii) = gf(ii)+( f(xx(ir),yy(ir))*2.0 + f(xx(j),yy(j)) ...
                                   + f(xx(m),yy(m)) )*delta/12.0;
        end

    end                      % End assembling by element.

%-------------------------------------------------------
% Now deal with the Dirichlet BC

    [ijunk,npres] = size(e);
    for i=1:npres,
       xb = p(1,e(1,i));  yb=p(2,e(1,i));
       g1(i) = uexact(xb,yb);
    end

    for i=1:npres,
      nod = e(1,i);
      for k=1:nnode,
         gf(k) = gf(k) - gk(k,nod)*g1(i);
         gk(nod,k) = 0;
         gk(k,nod) = 0;
```

```
        end
           gk(nod,nod) = 1;
           gf(nod) = g1(i);
    end

    u=gk\gf;              % Solve the linear system.
    pdemesh(p,e,t,u)      % Plot the solution.

% End.
```

Example 9.16. We test the simplified finite element method to solve a Poisson equation using the following example:

- Domain: Unit square with a hole (*cf.* Figure 9.11).
- Exact solution: $u(x,y) = x^2 + y^2$, for $f(x,y) = -4$.
- BC: Dirichlet condition on the whole boundary.
- Use Matlab PDE Toolbox to generate initial mesh and then *export it.*

Figure 9.11 shows the domain and the mesh generated by the Matlab PDE Toolbox. Figure 9.12(a) is the mesh plot for the finite element solution, and the Figure 9.12(b) is the error plot (the magnitude of the error is $O(h^2)$).

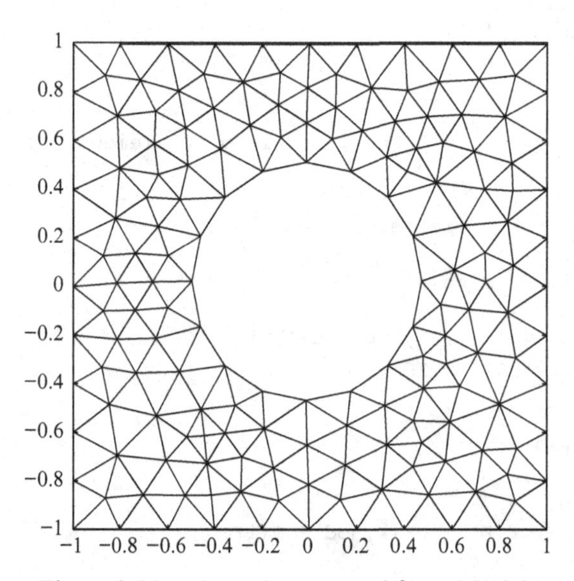

Figure 9.11. A mesh generated from Matlab.

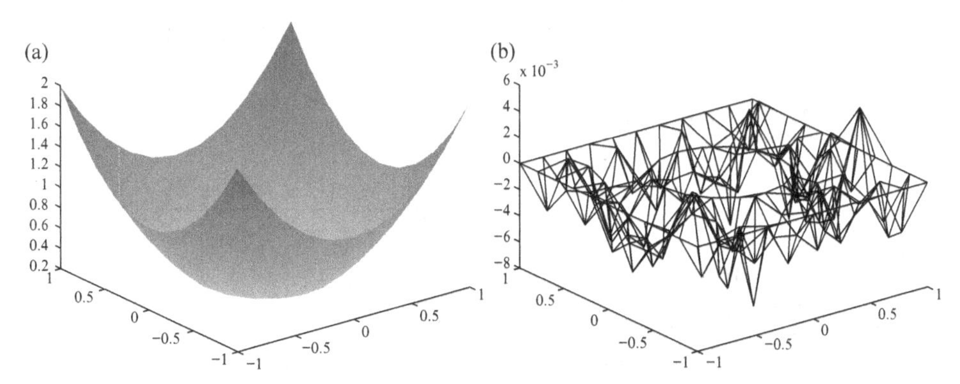

Figure 9.12. (a) A plot of the finite element solution when $f(x, y) = -4$ and (b) the corresponding error plot.

9.7 Some FE Spaces in $H^1(\Omega)$ and $H^2(\Omega)$

Given a triangulation (triangles, rectangles, quadrilaterals, *etc.*), let us construct different finite element spaces with finite dimensions. There are several reasons to do so, including:

- better accuracy of the finite element solution, with piecewise higher-order polynomial basis functions, and
- to allow for higher-order derivatives in higher-order PDEs, *e.g.*, in solving the biharmonic equation in H^2 space.

As previously mentioned, we consider conforming piecewise polynomial finite element spaces. A set of polynomials of degree k is denoted by

$$P_k = \left\{ v(x, y), \quad v(x, y) = \sum_{i,j=0}^{i+j \le k} a_{ij}\, x^i x^j \right\},$$

in the xy-plane. Below we list some examples,

$$\begin{aligned}
P_1 &= \{\, v(x, y), \quad v(x, y) = a_{00} + a_{10}x + a_{01}y \,\}, \\
P_2 &= \left\{\, v(x, y), \quad v(x, y) = a_{00} + a_{10}x + a_{01}y + a_{20}x^2 + a_{11}xy + a_{02}y^2 \,\right\}, \\
P_3 &= P_2 + \left\{ a_{30}x^3 + a_{21}x^2y + a_{12}xy^2 + a_{03}y^3 \right\}, \\
&\cdots.
\end{aligned}$$

Degree of freedom of P_k. For any fixed x^i, all the possible y^j terms in a $p_k(x, y) \in P_k$ are $y^0, y^1, \ldots, y^{k-i}$, *i.e.*, j ranges from 0 to $k - i$. Thus there are

$k-i+1$ parameters for a given x^i, and the total degree of freedom is

$$\sum_{i=0}^{k}(k-i+1)=\sum_{i=0}^{k}(k+1)-\sum_{i=0}^{k} i$$
$$=(k+1)^2-\frac{k(k+1)}{2}=\frac{(k+1)(k+2)}{2}.$$

Some degrees of freedom for different k's are:

- 3 when $k=1$, the linear function space P_1;
- 6 when $k=2$, the quadratic function space P_2;
- 10 when $k=3$, the cubic function space P_3;
- 15 when $k=4$, the fourth-order polynomials space P_4; and
- 21 when $k=5$, the fifth-order polynomials space P_5.

Regularity requirements: Generally, we cannot conclude that $v(x,y)\in C^0$ if $v(x,y)\in H^1$. However, if V_h is a finite-dimensional space of piecewise polynomials, then that is indeed true. Similarly, if $v(x,y)\in H^2$ and $v(x,y)|_{K_i}\in P_k$, $\forall K_i\in T_h$, then $v(x,y)\in C^1$. The regularity requirements are important for the construction of finite element spaces.

As is quite well known, there are two ways to improve the accuracy. One way is to decrease the mesh size h, and the other is to use high-order polynomial spaces P_k. If we use a P_k space on a given triangulation T_h for a linear second-order elliptic PDE, the error estimates for the finite element solution u_h are

$$\|u-u_h\|_{H^1(\Omega)}\le C_1h^k\|u\|_{H^{k+1}(\Omega)},\quad \|u-u_h\|_{L^2(\Omega)}\le C_2h^{k+1}\|u\|_{H^{k+1}(\Omega)}. \tag{9.39}$$

9.7.1 A Piecewise Quadratic Function Space

The degree of freedom of a quadratic function on a triangle is six, so we may add three auxiliary middle points along the three sides of the triangle.

Theorem 9.17. *Consider a triangle $K=(a^1,a^2,a^3)$, as shown in Figure 9.13. A function $v(x,y)\in P_2(K)$ is uniquely determined by its values at*

$$v(a^i),\ i=1,2,3,\ \textit{and the three middle points}\ v(a^{12}),\ v(a^{23}),\ v(a^{31}).$$

As there are six parameters and six conditions, we expect to be able to determine the quadratic function uniquely. Highlights of the proof are as follows.

- We just need to prove the homogeneous case $v(a^i)=0$, $v(a^{ij})=0$, since the right-hand side does not affect the existence and uniqueness.

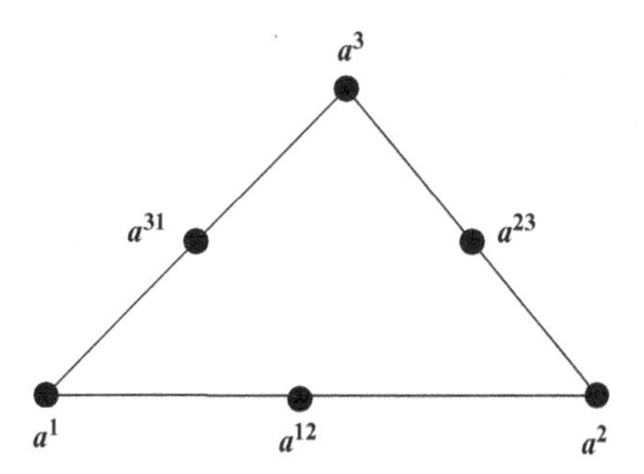

Figure 9.13. A diagram of six points in a triangle to determine a quadratic function.

- We can represent a quadratic function as a product of two linear functions, *i.e.*, $v(\mathbf{x}) = \psi_1(\mathbf{x})\omega(\mathbf{x}) = \psi_1(\mathbf{x})\psi_2(\mathbf{x})\omega_0$, with $\psi_i(\mathbf{x})$ denoting the local linear basis function such that $\psi_i(a^i) = 1$ and $\psi_i(a^j) = 0$ if $i \neq j$. Note that here we use $\mathbf{x} = (x, y)$ notation for convenience.
- It is easier to introduce a coordinate axis aligned with one of the three sides.

Proof We introduce the new coordinates (*cf.* Figure 9.7)

$$\xi = (x - x_2)\cos\alpha + (y - y_2)\sin\alpha\,,$$
$$\eta = -(x - x_2)\sin\alpha + (y - y_2)\cos\alpha\,,$$

such that a^2 is the origin and a^2a^3 is the η- axis. Then $v(x, y)$ can be written as

$$v(x,y) = v(x(\xi,\eta), y(\xi,\eta)) = \bar{v}(\xi,\eta) = \bar{a}_{00} + \bar{a}_{10}\xi + \bar{a}_{01}\eta + \bar{a}_{20}\xi^2 + \bar{a}_{11}\xi\eta + \bar{a}_{02}\eta^2.$$

Furthermore, under the new coordinates, we have

$$\psi_1(\xi,\eta) = \sigma + \beta\xi + \gamma\eta = \beta\xi\,, \quad \beta \neq 0,$$

since $\psi_1(a^2) = \psi_1(a^3) = 0$. Along the η-axis ($\xi = 0$), $\bar{v}(\xi,\eta)$ has the following form

$$\bar{v}(0,\eta) = \bar{a}_{00} + \bar{a}_{01}\eta + \bar{a}_{02}\eta^2.$$

Since $\bar{v}(a^2) = \bar{v}(a^3) = \bar{v}(a^{23}) = 0$, we get $\bar{a}_{00} = 0$, $\bar{a}_{01} = 0$ and $\bar{a}_{02} = 0$, therefore,

$$\begin{aligned}\bar{v}(\xi,\eta) &= \bar{a}_{10}\xi + \bar{a}_{11}\xi\eta + \bar{a}_{20}\xi^2 = \xi\,(\bar{a}_{10} + \bar{a}_{11}\eta + \bar{a}_{20}\xi)\\ &= \beta\xi\left(\frac{\bar{a}_{10}}{\beta} + \frac{\bar{a}_{20}}{\beta}\xi + \frac{\bar{a}_{11}}{\beta}\eta\right)\\ &= \psi_1(\xi,\eta)\omega(\xi,\eta).\end{aligned}$$

Similarly, along the edge a^1a^3, we have

$$v(a^{13}) = \psi_1(a^{13})\,\omega(a^{13}) = \frac{1}{2}\,\omega(a^{13}) = 0,$$
$$v(a^1) = \psi_1(a^1)\omega(a^1) = \omega(a^1) = 0,$$

i.e.,

$$\omega(a^{13}) = 0, \quad \omega(a^1) = 0.$$

By similar arguments, we conclude that

$$\omega(x,y) = \psi_2(x,y)\,\omega_0,$$

and hence

$$v(x,y) = \psi_1(x,y)\psi_2(x,y)\omega_0.$$

Using the zero value of v at a^{12}, we have

$$v(a^{12}) = \psi_1(a^{12})\,\psi_2(a^{12})\,\omega_0 = \frac{1}{2}\frac{1}{2}\omega_0 = 0,$$

so we must have $\omega_0 = 0$ and hence $v(x,y) \equiv 0$.

9.7.1.1 Continuity Along the Edges

Along each edge, a quadratic function $v(x,y)$ can be written as a quadratic function of one variable. For example, if the edge is represented as

$$y = ax + b \quad \text{or} \quad x = ay + b,$$

then

$$v(x,y) = v(x, ax+b) \quad \text{or} \quad v(x,y) = v(ay+b, y).$$

Thus, the piecewise quadratic functions defined on two triangles with a common side are identical on the entire side if they have the same values at the two end points and at the mid-point of the side.

9.7.1.2 Representing Quadratic Basis Functions using Linear Functions

To define quadratic basis functions with minimum compact support, we can determine the six nonzero functions using the values at three vertices and the mid-points $\mathbf{v} = (v(a^1), v(a^2), v(a^3), v(a^{12}), v(a^{23}), v(a^{13})) \in \mathbf{R}^6$. We can either take $\mathbf{v} = \mathbf{e}_i \in \mathbf{R}^6$, $i = 1, 2, \ldots, 6$, respectively, or determine a quadratic function on the triangle using the linear basis functions as stated in the following theorem.

Theorem 9.18. *A quadratic function on a triangle can be represented by*

$$v(x,y) = \sum_{i=1}^{3} v(a^i)\phi_i(x,y)\Big(2\phi_i(x,y) - 1\Big) + \sum_{i,j=1,i<j}^{3} 4\,v(a^{ij})\,\phi_i(x,y)\,\phi_j(x,y), \tag{9.40}$$

where $\phi_i(x,y)$, $i=1,2,3$, is one of the three linear basis functions centered at one of the vertices a^i.

Proof It is easy to verify the vertices if we substitute a^j into the right-hand side of the expression above,

$$v(a^j)\phi_j(a^j)\Big(2\phi_j(a^j) - 1\Big) = v(a^j),$$

since $\phi_i(a^j)=0$ if $i \neq j$. We take one mid-point to verify the theorem. On substituting a^{12} into the left expression, we have

$$\begin{aligned}
&v(a^1)\phi_1(a^{12})\Big(2\phi_1(a^{12}) - 1\Big) + v(a^2)\phi_2(a^{12})\Big(2\phi_2(a^{12}) - 1\Big)\\
&+ v(a^3)\phi_3(a^{12})\Big(2\phi_3(a^{12}) - 1\Big) + 4v(a^{12})\phi_1(a^{12})\phi_2(a^{12})\\
&+ 4v(a^{13})\phi_1(a^{12})\phi_3(a^{12}) + 4v(a^{23})\phi_2(a^{12})\phi_3(a^{12})\\
&= v(a^{12}),
\end{aligned}$$

since $2\phi_1(a^{12}) - 1 = 2 \times \frac{1}{2} - 1 = 0$, $2\phi_2(a^{12}) - 1 = 2 \times \frac{1}{2} - 1 = 0$, $\phi_3(a^{12}) = 0$ and $4\phi_1(a^{12})\phi_2(a^{12}) = 4 \times \frac{1}{2} \times \frac{1}{2} = 1$. Note that the local stiffness matrix is 6×6 when quadratic basis functions are used.

We have included a Matlab code of the finite element method using the quadratic finite element space over a uniform triangular mesh for solving a Poisson equation with a homogeneous (zero) Dirichlet boundary condition.

9.7.2 Cubic Basis Functions in $H^1 \cap C^0$

There are several ways to construct cubic basis functions in $H^1 \cap C^0$ over a triangulation, but a key consideration is to keep the continuity of the basis functions along the edges of neighboring triangles. We recall that the degree of freedom of a cubic function in 2D is ten, and one way is to add two auxiliary points along each side and one auxiliary point inside the triangle. Thus, together with the three vertices, we have ten points on a triangle to match the

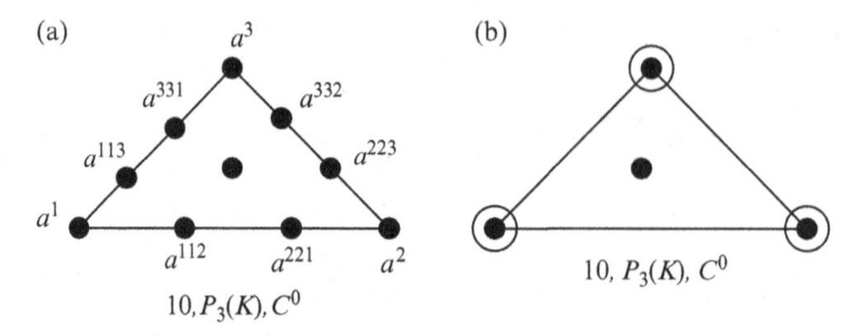

Figure 9.14. A diagram of the freedom used to determine two different cubic basis functions in $H^1 \cap C^0$. We use the following notation: • for function values and ∘ for values of the first derivatives.

degree of freedom (*cf.* Figure 9.14(a)). Existence and uniqueness conditions for such a cubic function are stated in the following theorem.

Theorem 9.19. *A cubic function $v \in P_3(K)$ is uniquely determined by the values of*

$$v(a^i), \quad v(a^{iij}),\ i,j=1,2,3,\ i \neq j \quad \text{and} \quad v(a^{123}), \tag{9.41}$$

where

$$a^{123} = \frac{1}{3}\left(a^1 + a^2 + a^3\right), \quad a^{iij} = \frac{1}{3}\left(2a^i + a^j\right),\ i,j=1,2,3,\ i \neq j. \tag{9.42}$$

Sketch of the proof: Similar to the quadratic case, we just need to prove that the cubic function is identically zero if $v(a^i) = v(a^{iij}) = v(a^{123}) = 0$. Again using the local coordinates where one of the sides of the triangle T is on an axis, we can write

$$v(\mathbf{x}) = C\phi_1(\mathbf{x})\phi_2(\mathbf{x})\phi_3(\mathbf{x}),$$

where C is a constant. Since $v(a^{123}) = C\phi_1(a^{123})\phi_2(a^{123})\phi_3(a^{123}) = 0$, we conclude that $C=0$ since $\phi_i(a^{123}) \neq 0$, $i=1,2,3$; and hence $v(\mathbf{x}) \equiv 0$.

With reference to the continuity along the common side of two adjacent triangles, we note that the polynomial of two variables again becomes a polynomial of one variable there, since we can substitute either x for y, or y for x from the line equations $l_0 + l_{10}x + l_{01}y = 0$. Furthermore, a cubic function of one variable is uniquely determined by the values of four distinct points.

There is another choice of cubic basis functions, using the first-order derivatives at the vertices (*cf.* Figure 9.14(b)). This alternative is stated in the following theorem.

Theorem 9.20. *A cubic function $v \in P_3(K)$ is uniquely determined by the values of*

$$v(a^i), \quad \frac{\partial v}{\partial x_j}(a^i),\ i=1,2,3,\ j=1,2 \text{ and } i \neq j, \quad v(a^{123}), \tag{9.43}$$

where $\partial v/\partial x_j(a^i)$ represents $\partial v/\partial x(a^i)$ when $j=1$ and $\partial v/\partial y(a^i)$ when $j=2$, at the nodal point a^i.

At each vertex of the triangle, there are three degrees of freedom, namely, the function value and two first-order partial derivatives; so in total there are nine degrees of freedom. An additional degree of freedom is the value at the centroid of the triangle. For the proof of the continuity, we note that on a common side of two adjacent triangles a cubic polynomial of one variable is uniquely determined by its function values at two distinct points plus the first-order derivatives in the Hermite interpolation theory. The first-order derivative is the tangential derivative along the common side defined as $\partial v/\partial t = \partial v/\partial x\, t_1 + \partial v/\partial y\, t_2$, where $\mathbf{t} = (t_1, t_2)$ such that $t_1^2 + t_2^2 = 1$ is the unit direction of the common side.

9.7.3 Basis Functions in $H^2 \cap C^1$

To solve fourth-order PDEs such as a 2D biharmonic equation

$$\Delta\left(u_{xx} + u_{yy}\right) = u_{xxxx} + 2u_{xxyy} + u_{yyyy} = 0\,, \tag{9.44}$$

using the finite element method, we need to construct basis functions in $H^2(\Omega) \cap C^1(\Omega)$. Since second-order partial derivatives are involved in the weak form, we need to use polynomials with degree more than three. On a triangle, if the function values and partial derivatives up to second order are specified at the three vertices, the degree of freedom would be at least 18. The closest polynomial would be of degree five, as a polynomial $v(\mathbf{x}) \in P_5$ has degree of freedom 21 (*cf.* Figure 9.15(a)).

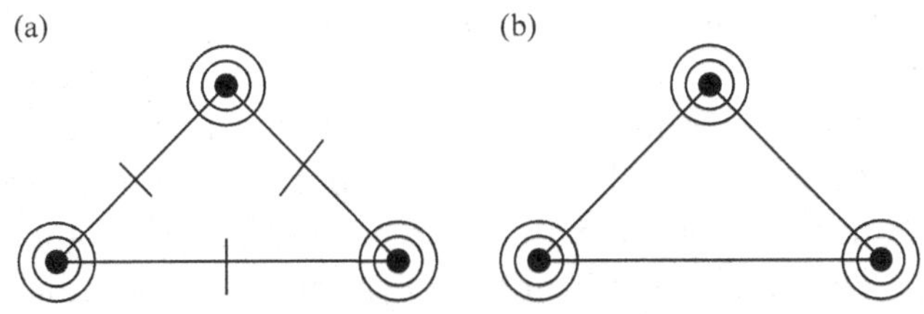

Figure 9.15. A diagram of the freedom used to determine two different fifth-order polynomial basis functions in $H^2 \cap C^1$. (a) We specify $D^\alpha v(a^i), 0 \le \alpha \le 2$ at each vertex ($3 \times 6 = 18$) plus three normal derivatives $\partial v/\partial n(a^{ij})$ at the mid-point of the three edges. (b) We can specify three independent constraints to reduce the degree of freedom, *e.g.*, $\partial v/\partial n(a^{ij}) = 0$ at the mid-point of the three edges.

Theorem 9.21. *A quintic function $v(x,y) \in P_5(K)$ is uniquely determined by the values of*

$$D^\alpha v(a^i),\ i=1,2,3,\ |\alpha| \le 2, \quad \frac{\partial v}{\partial n}(a^{ij}),\ i,j=1,2,3,\ i<j, \tag{9.45}$$

where $\partial v/\partial n(a^i) = n_1 \partial v/\partial x(a^i) + n_2 \partial v/\partial y(a^i)$ represents the normal derivative of $v(x)$ at a^i and $n=(n_1,n_2)$ $(n_1^2+n_2^2=1)$ is the outward unit normal at the boundary of the triangle.

Sketch of the proof: We just need to show that $v(\mathbf{x})=0$ if $D^\alpha v(a^i)=0$, $i=1,2,3$, $|\alpha|\le 2$ and $\partial v/\partial n(a^{ij})=0$, $i,j=1,2,3$, $i<j$. A fifth-order polynomial $v(s)$ of one variable s is uniquely determined by the values of v and its derivatives $v'(s)$ and $v''(s)$ at two distinct points, so along a^2a^3, $v(\mathbf{x})$ must be zero for the given homogeneous conditions. We note that $\frac{\partial v}{\partial n}(\mathbf{x})$ is a fourth-order polynomial of one variable along a^2a^3. Since all of the first- and second-order partial derivatives are zero at a^2 and a^3,

$$\frac{\partial v}{\partial n}(a^i)=0\,, \qquad \frac{\partial}{\partial t}\left(\frac{\partial v}{\partial n}\right)(a^i)=0,\ i=2,3\,,$$

and $\frac{\partial v}{\partial n}(a^{23})=0$. Here again, $\frac{\partial}{\partial t}$ is the tangential directional derivative. From the five conditions, we have $\frac{\partial v}{\partial n}(\mathbf{x})=0$ along a^2a^3, so we can factor $\phi_1^2(\mathbf{x})$ out of $v(\mathbf{x})$ to get

$$v(\mathbf{x}) = \phi_1^2(\mathbf{x})\, p_3(\mathbf{x}), \tag{9.46}$$

where $p_3(\mathbf{x}) \in P_3$. Similarly, we can factor out $\phi_2^2(\mathbf{x})$ and $\phi_3^2(\mathbf{x})$ to get

$$v(\mathbf{x}) = \phi_1^2(\mathbf{x})\, \phi_2^2(\mathbf{x})\, \phi_3^2(\mathbf{x})\, C, \tag{9.47}$$

where C is a constant. Consequently $C=0$, otherwise $v(\mathbf{x})$ would be a polynomial of degree six, which contradicts that $v(\mathbf{x}) \in P_5$.

The continuity condition along a common side of two adjacent triangles in C^1 has two parts, namely, both the function and the normal derivative must be continuous. Along a common side of two adjacent triangles, a fifth-order polynomial of $v(x,y)$ is actually a fifth-order polynomial of one variable $v(s)$, which can be uniquely determined by the values $v(s)$, $v'(s)$ and $v''(s)$ at two distinct points. Thus the two fifth-order polynomials on two adjacent triangles are identical along the common side if they have the same values of $v(s)$, $v'(s)$, and $v''(s)$ at the two shared vertices. Similarly, for the normal derivative along a common side of two adjacent triangles, we have a fourth-order polynomial of one variable $\partial v/\partial n(s)$. The polynomials can be uniquely determined by the values $\partial v/\partial n(s)$ and $(d/ds)\,(\partial v/\partial n)\,(s)$ at two distinct points plus the value of

$\partial v/\partial n(s)$ at the mid-point. Thus the continuity of the normal derivative is also guaranteed.

An alternative approach is to replace the values of $\frac{\partial v}{\partial n}(a^{ij})$ at the three mid-points of the three sides by imposing another three conditions. For example, assuming that along a^2a^3 the normal derivative of the fifth-order polynomial has the form

$$\frac{\partial v}{\partial n} = \widetilde{a_{00}} + \widetilde{a_{10}}\eta + \widetilde{a_{20}}\eta^2 + \widetilde{a_{30}}\eta^3 + \widetilde{a_{40}}\eta^4,$$

we can impose $\widetilde{a_{40}} = 0$. In other words, along the side of a^2a^3 the normal derivative $\partial v/\partial n$ becomes a cubic polynomial of one variable. The continuity can again be guaranteed by the Hermite interpolation theory. Using this approach, the degree of freedom is reduced to 18 from the original 21 (*cf.* Figure 9.15(b) for an illustration).

9.7.4 Finite Element Spaces on Rectangular Meshes

While triangular meshes are intensively used, particularly for arbitrary domains, meshes using rectangles are also popular for rectangular regions. Bilinear functions are often used as basis functions. Let us first consider a bilinear function space in $H^1 \cap C^0$. A bilinear function space over a rectangle K in 2D, as illustrated in Figure 9.16, is defined as

$$Q_1(K) = \left\{ v(x,y), \quad v(x,y) = a_{00} + a_{10}x + a_{01}y + a_{11}xy \right\}, \tag{9.48}$$

where $v(x,y)$ is linear in both x and y. The degree of freedom of a bilinear function in $Q_1(K)$ is four.

Theorem 9.22. *A bilinear function $v(x,y) \in Q_1(K)$ is uniquely determined by its values at four corners.*

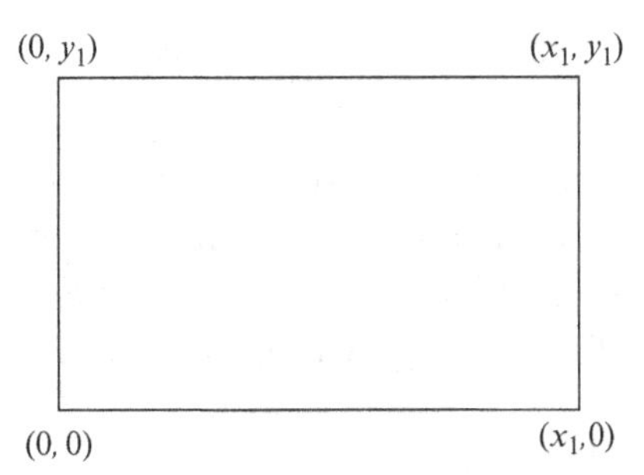

Figure 9.16. A standard rectangle on which four bilinear basis functions can be defined.

Proof Without loss of the generality, assume that the rectangle is determined by the four corners a^i: $(0,\,0)$, $(x_1,\,0)$, $(x_1,\,y_1)$, and $(0,\,y_1)$. The coefficient matrix of the linear system of algebraic equations that determines the coefficients a_{ij}, $i,j=0,1$ is

$$A=\begin{pmatrix} 1 & 0 & 0 & 0 \\ 1 & x_1 & 0 & 0 \\ 1 & 0 & y_1 & 0 \\ 1 & x_1 & y_1 & x_1y_1 \end{pmatrix},$$

with determinant $det(A)=x_1^2y_1^2\neq 0$ since $x_1y_1\neq 0$. Indeed, we have analytic expressions for the four nonzero basis functions over the rectangle, namely,

$$\phi_1(x,y)=1-\frac{x}{x_1}-\frac{y}{y_1}+\frac{xy}{x_1y_1}, \tag{9.49}$$

$$\phi_2(x,y)=\frac{x}{x_1}-\frac{xy}{x_1y_1}, \tag{9.50}$$

$$\phi_3(x,y)=\frac{xy}{x_1y_1}, \tag{9.51}$$

$$\phi_4(x,y)=\frac{y}{y_1}-\frac{xy}{x_1y_1}. \tag{9.52}$$

On each side of the rectangle, $v(x,y)$ is a linear function of one variable (either x or y) and uniquely determined by the values at the two corners. Thus any two basis functions along one common side of two adjacent rectangles are identical if they have the same values at the two corners, although it is hard to match the continuity condition if quadrilaterals are used instead of rectangles or cubic boxes.

A biquadratic function space over a rectangle is defined by

$$\begin{aligned} Q_2(K)=\Big\{v(x,y),\quad & v(x,y)=a_{00}+a_{10}x+a_{01}y+a_{11}xy \\ & +a_{20}x^2+a_{20}y^2+a_{21}x^2y+a_{12}xy^2+a_{22}x^2y^2\Big\}. \end{aligned} \tag{9.53}$$

The degree of freedom is nine. To construct basis functions in $H^1\cap C^0$, as for the quadratic functions over triangles, we can add four auxiliary points at the mid-points of the four sides plus a point, often in the center of the rectangle.

In general, a bilinear function space of order k over a rectangle is defined by

$$Q_k(K)=\left\{v(x,y),\quad v(x,y)=\sum_{i,j=0,i\leq k,j\leq k} a_{ij}x^iy^j\right\}. \tag{9.54}$$

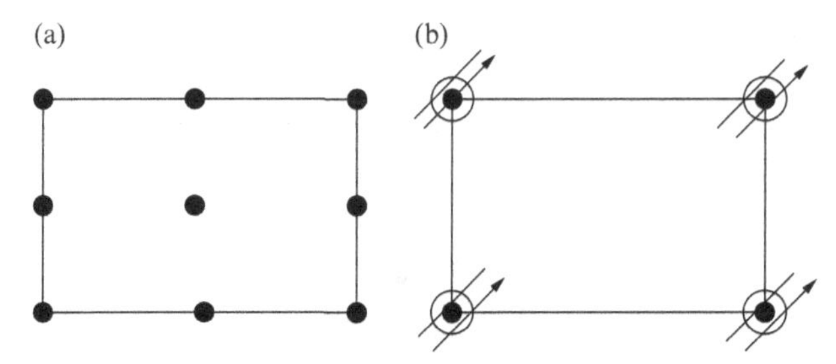

Figure 9.17. (a) $Q_2(K)$ (biquadratic) in $H^1 \cap C^0$ whose degree of freedom is 9 which can be uniquely determined by the values at the marked points and (b) $Q_3(K)$ (bicubic) in $H^2 \cap C^1$ whose degree of freedom is 16, which can be determined by its values, first-order partial derivatives marked as /, and mixed derivative marked as ↗, at the four corners.

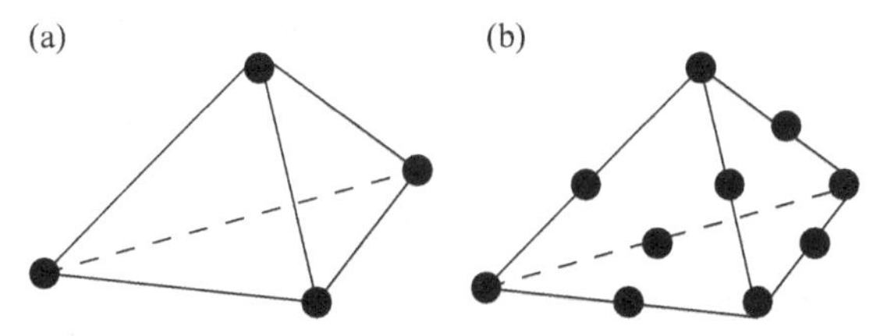

Figure 9.18. Finite element spaces in 3D. (a) $T_1(K)$ (linear) in $H^1 \cap C^0$ whose degree of freedom is 4 and (b) $T_2(K)$ (quadratic) in $H^1 \cap C^0$ whose degree of freedom is 10.

In Figure 9.17, we show two diagrams of finite element spaces defined on the rectangles and their degree of freedom. Figure 9.17(a) is the biquadratic $Q_2(K)$ finite element in $H^1 \cap C^0$ whose degree of freedom is 9 and can be determined by the values at the marked points. Figure 9.17(b) is the bicubic $Q_3(K)$ finite element in $H^2 \cap C^1$ whose degree of freedom is 16 and can be determined by the values at the marked points. The bicubic polynomial is the lowest bipolynomial in $H^2 \cap C^1$ space. The bicubic function can be determined by its values, its partial derivatives $\left(\frac{\partial}{\partial x}, \frac{\partial}{\partial y}\right)$, and its mixed partial derivative $\frac{\partial^2}{\partial x \partial y}$ at four vertices.

9.7.5 Some Finite Element Spaces in 3D

In three dimensions, the most commonly used meshes are tetrahedrons and cubics. In Figure 9.18, we show two diagrams of finite element spaces defined on the tetrahedrons and their degree of freedom. Figure 9.18(a) is the linear $T_1(K)$ finite element in $H^1 \cap C^0$ whose degree of freedom is 4 and can be determined by the values at the four vertices. Figure 9.18(b) is the quadratic $T_2(K)$ finite element in $H^1 \cap C^0$ whose degree of freedom is 10 and can be determined by the values at the four vertices and the mid points of the six edges.

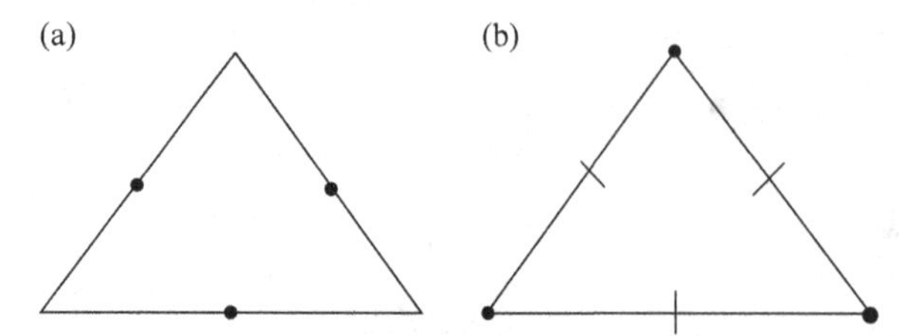

Figure 9.19. Diagram of non-conforming finite element spaces. (a) Crouzeix–Raviart (C–R) linear non-conforming element that is determined by the values at the middle points of the edges. (b) Morley quadratic non-conforming element that is determined by the values at the vertices and the normal derivatives at the middle points of the edges.

9.7.6 *Non-conforming Finite Element Spaces

For high-order PDEs, such as biharmonic equations ($\Delta^2 u = f$, where Δ is the Laplacian operator $\Delta = \frac{\partial^2}{\partial x^2} + \frac{\partial^2}{\partial y^2}$) in two or three dimensions, or systems of PDEs with certain constraints, such as a divergence free condition, it is difficult to construct and verify conforming finite element spaces. Even if it is possible, the degree of polynomial of the basis functions is relatively high, for example, we need fifth order polynomials for biharmonic equations in two space dimensions, which may lead to Gibbs oscillations near the edges. Other types of applications include non-fitted meshes or interface conditions for which it is difficult or impossible to construct finite elements that meet the conforming constraints. To overcome these difficulties, various approaches have been developed such as non-conforming finite element methods, discontinuous and weak Galerkin finite element methods. Here we mention some non-conforming finite element spaces that are developed in the framework of Galerkin finite element methods.

For triangle meshes, a non-conforming P_1 finite element space called Cronzeix–Raviart (C–R) finite element space is defined as a set of linear functions over all triangles that are continuous at the mid-points of all the edges. The basis functions can be determined by taking either unity at one middle point and zeros at other middle points of a triangle; see Figure 9.19(a) for an illustration. The theoretical analysis can be found in Brenner and Scott (2002) and Shi (2002), for example. A non-conforming Q_1 finite element space on rectangles called the Wilson element is defined in a similar way but with a basis $\{1, x, y, xy\}$ of degree four. A rotated non-conforming Q_1 is defined in the similar way but using $\{1, x, y, x^2 - y^2\}$ as the basis. Note that, for the conforming biquadratic finite element space, those bases are equivalent, but it is not true for non-conforming finite element spaces anymore.

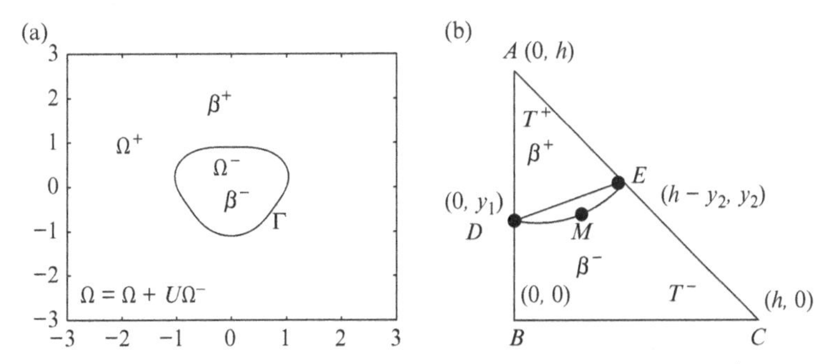

Figure 9.20. (a) A configuration of a rectangular domain $\Omega = \Omega^+ \cup \Omega^-$ with an interface Γ from an IFEM test. The coefficient $p(\mathbf{x})$ may have a finite jump across the interface Γ. (b) An interface triangle and the geometry after transformed to the standard right triangle.

For biharmonic equations, a nonconforming finite element space defined on triangle meshes called the Morley finite element (Shi, 2002) has been developed. A Morley finite element on a triangle is defined as quadratic functions that are determined by the values at the three vertices, and the normal derivative at the middle points of the three edges, see Figure 9.19(b) for an illustration. An alternative definition is to use the line integrals along the edges instead of the values at the middle points.

9.7.7 *The Immersed Finite Element Method for Discontinuous Coefficients

Following the idea of the immersed finite element method (IFEM) for 1D problems, we explain the IFEM for 2D interface problems when the coefficient $p(x, y)$ has a discontinuity across a closed smooth interface Γ. The interface Γ can be expressed as a parametric form $(X(s), Y(s)) \in C^2$, where s is a parameter, say the arc-length. The interface cuts the domain Ω into two subdomains Ω^+ and Ω^-; see the diagram in Figure 9.20(a).

For simplicity, we assume that the coefficient $p(x)$ is a piecewise constant

$$p(x, y) = \begin{cases} \beta^+ \text{ if } (x, y) \in \Omega^+, \\ \beta^- \text{ if } (x, y) \in \Omega^-. \end{cases}$$

Again, across the interface Γ where the discontinuity occurs, the natural jump conditions hold

$$[u]_\Gamma = 0, \qquad \left[\beta \frac{\partial u}{\partial n}\right]_\Gamma = 0, \tag{9.55}$$

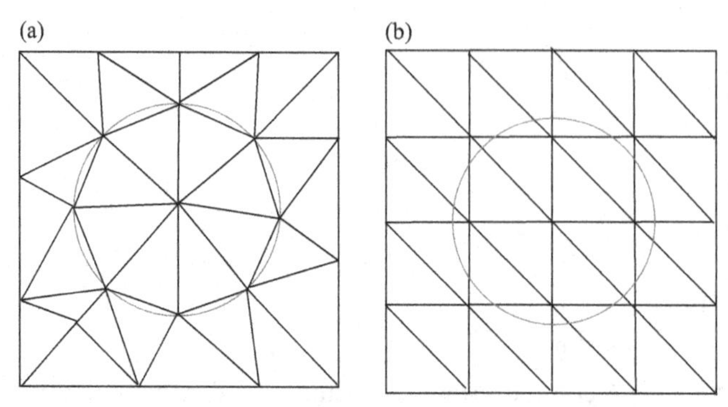

Figure 9.21. (a) A diagram of a fitted mesh (unstructured) and (b) an unfitted Cartesian mesh (structured).

where the jump at a point $\mathbf{X} = (X, Y) \in \Gamma$ on the interface is defined as

$$[u]_{\mathbf{X}} = u\Big|_{\mathbf{X}}^{+} - u\Big|_{\mathbf{X}}^{-} = \lim_{\mathbf{x}\to\mathbf{X},\mathbf{x}\in\Omega^{+}} u(\mathbf{x}) - \lim_{\mathbf{x}\to\mathbf{X},\mathbf{x}\in\Omega^{-}} u(\mathbf{x}),$$

and so on, where $\mathbf{x} = (x, y)$ is an interior point in the domain. Due to the discontinuity in the coefficient, the partial derivatives across the interface Γ are discontinuous although the solution and the flux (the second jump condition) are continuous. Such a problem is referred to as a 2D interface problem.

To solve such an interface problem using a finite element method, first a mesh needs to be chosen. One way is to use a fitted mesh as illustrated in Figure 9.21(a). A fitted mesh can be generated by many existing academic or commercial software packages, *e.g.*, Matlab PDE Toolbox, Freefem, Comsol, PLTMG, Triangle, Gmesh, *etc.* Usually there is no fixed pattern between the indexing of nodal points and elements, thus such a mesh is called an unstructured mesh. For such a mesh, the finite element method and most theoretical analysis is still valid for the interface problem.

However, it may be difficult and time-consuming to generate a body fitted mesh. Such a difficulty may become even more severe for moving interface problems because a new mesh has to be generated at each time step, or every other time step. A number of efficient software packages and methods that are based on Cartesian meshes, such as the FFT, the level set method, and others, may not be applicable with a body fitted mesh.

Another way to solve the interface problem is to use an unfitted mesh, *e.g.*, a uniform Cartesian mesh as illustrated in Figure 9.21(b). There is rich literature on unfitted meshes and related finite element methods. The nonconforming IFEM (Li, 1998) is one of the early works in this direction. The idea is to enforce the natural jump conditions in triangles that the interface cuts through, which we call an interface triangle. Without loss of generality, we consider a reference interface element T whose geometric configuration is given in Figure 9.20(b) in which the curve between points D and E is a part of the interface. We assume that the coordinates at A, B, C, D, and E are

$$(0,h), \quad (0,0), \quad (h,0), \quad (0,y_1), \quad (h-y_2,y_2), \tag{9.56}$$

with the restriction

$$0 \le y_1 < h, \quad 0 \le y_2 < h. \tag{9.57}$$

Given the values at the three vertices we explain how to determine a piecewise linear function in the triangle that satisfies the natural jump conditions. Assuming that the values at vertices A, B, and C of the element T are specified, we construct the following piecewise linear function:

$$u(\mathbf{x}) = \begin{cases} u^+(\mathbf{x}) = a_0 + a_1 x + a_2(y-h), & \text{if } \mathbf{x}=(x,y) \in T^+, \\ u^-(\mathbf{x}) = b_0 + b_1 x + b_2 y, & \text{if } \mathbf{x}=(x,y) \in T^-, \end{cases} \tag{9.58a}$$

$$u^+(D) = u^-(D), \quad u^+(E) = u^-(E), \quad \beta^+ \frac{\partial u^+}{\partial n} = \beta^- \frac{\partial u^-}{\partial n}, \tag{9.58b}$$

where $\mathbf{n}$ is the unit normal direction of the line segment $\overline{DE}$. Intuitively, there are six constraints and six parameters, so we can expect the solution exists and is unique as confirmed in Theorem 8.4 in Li and Ito (2006).

The dimension of the non-conforming IFE space is the number of interior points for a homogeneous Dirichlet boundary condition ($u|_{\partial\Omega} = 0$) as if there was no interface. The basis function centered at a node is defined as:

$$\phi_i(\mathbf{x}_j) = \begin{cases} 1 & \text{if } i=j \\ 0 & \text{otherwise}, \end{cases} \qquad [\phi_i]_{\bar{\Gamma}} = 0, \quad \left[\beta \frac{\partial \phi_i}{\partial n}\right]_{\bar{\Gamma}} = 0, \quad \phi_i|_{\partial\Omega} = 0. \tag{9.59}$$

A basis function $\phi_i(\mathbf{x})$ is continuous in each element T except along some edges if $\mathbf{x}_i$ is a vertex of one or several interface triangles (see Figure 9.22). We use $\bar{\Gamma}$ to denote the union of the line segment that is used to approximate the interface.

The basis functions in an interface triangle are continuous piecewise linear. However, it is likely discontinuous across the edges of neighboring interface triangles. Thus, it is a non-conforming finite element space. Nevertheless, the corresponding non-conforming finite element method performs much better than the standard finite element method without any changes. Theoretically, a second-order approximation property has been proved for the interpolation function in the L^∞ norm; and first-order approximation for the partial derivatives except for the small mismatched region depicted as bounded by the points D, E, and M. It has been shown that the non-conforming IFEM is second-order

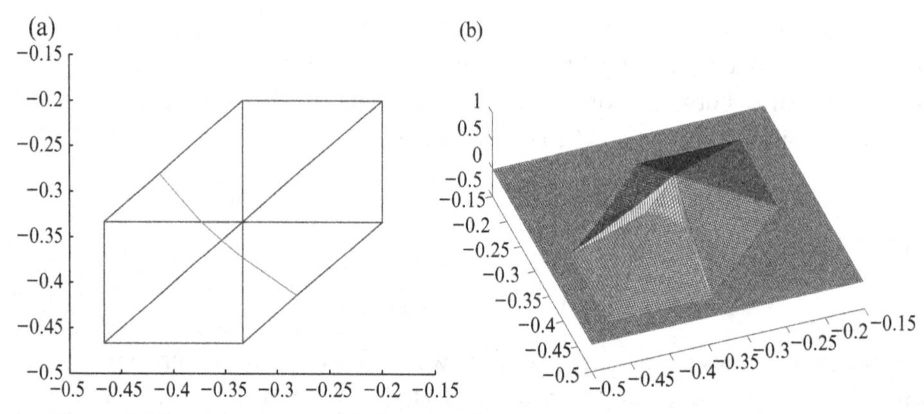

Figure 9.22. (a) A standard domain of six triangles with an interface cutting through. (b) A global basis function on its support of the non-conforming immersed finite element space. The basis function has small jump across some edges.

accurate in the L^2 norm. But its convergence order in the L^∞ norm is not so clear. Various variations, improvements, extensions, and applications can be found in the literature, particularly the symmetric and consistent IFEM that takes mismatched edge contributions into account in the variational form (Ji et al. (2014)), and various penalty methods. Note that, a conforming IFEM can also be found in Li et al. (2003) although its implementation is not so straightforward.

9.8 The FE Method for Parabolic Problems

We can apply the finite element method to solve time-dependent problems using two different approaches. One approach is to discretize the space variables using the finite element method while discretizing the time variable using a finite difference method. This is possible if the PDE is separable. Another way is to discretize both the space and time variables using the finite element method. In this section, we briefly explain the first approach, since it is simple and easy to implement.

Let us consider the following parabolic problem in 2D,

$$\frac{\partial u}{\partial t} = \nabla \cdot (p \nabla u) + qu + f(x,y,t)\,, \; (x,y) \in \Omega, \; 0 < t \le T, \tag{9.60}$$

$$u(x,y,0) = 0\,, \; (x,y) \in \Omega\,, \quad \text{the initial condition,} \tag{9.61}$$

$$u(x,y,t)\Big|_{\partial\Omega} = g(x,y,t), \quad \text{the boundary condition,} \tag{9.62}$$

where p, q, f, and g are given functions with usual regularity assumptions. Multiplying the PDE by a test function $v(x,y) \in H^1(\Omega)$ on both sides, and then integrating over the domain, once again we obtain the weak form below,

$$\iint_\Omega u_t v \, dxdy = \iint_\Omega (quv - p\nabla u \cdot \nabla v) \, dxdy + \iint_\Omega fv \, dxdy, \tag{9.63}$$

where $u_t = \partial u/\partial t$. The weak form above can be simplified as

$$(u_t, v) = -a(u,v) + (f,v) \qquad \forall v \in H^1(\Omega), \tag{9.64}$$

where $a(u,v) = \iint_\Omega (p\nabla u \cdot \nabla v - quv) \, dxdy$.

Given a triangulation T_h and finite element space $V_h \in H^1(\Omega) \cap C^0(\Omega)$, with $\phi_i(x,y)$, $i = 1, 2, \ldots, M$ denoting a set of basis functions for V_h, we seek the finite element solution of form

$$u_h(x,y,t) = \sum_{j=1}^{M} \alpha_j(t) \, \phi_j(x,y). \tag{9.65}$$

Substituting this expression into (9.64), we obtain

$$\left(\sum_{j=1}^{M} \alpha_i'(t)\phi_i(x,y) \,,\; v_h \right) = -a\left(\sum_{j=1}^{M} \alpha_i(t)\phi_i(x,y),\; v_h \right) + (f, v_h), \tag{9.66}$$

and then take $v_h(x,y) = \phi_i(x,y)$ for $i = 1, 2, \ldots, M$ to get the linear system of ODEs in the $\alpha_j(t)$:

$$\begin{bmatrix} (\phi_1,\phi_1) & (\phi_1,\phi_2) & \cdots & (\phi_1,\phi_M) \\ (\phi_2,\phi_1) & (\phi_2,\phi_2) & \cdots & (\phi_2,\phi_M) \\ \vdots & \vdots & \vdots & \vdots \\ (\phi_M,\phi_1) & (\phi_M,\phi_2) & \cdots & (\phi_M,\phi_M) \end{bmatrix} \begin{bmatrix} \alpha_1'(t) \\ \alpha_2'(t) \\ \vdots \\ \alpha_M'(t) \end{bmatrix}$$

$$= \begin{bmatrix} (f,\phi_1) \\ (f,\phi_2) \\ \vdots \\ (f,\phi_M) \end{bmatrix} - \begin{bmatrix} a(\phi_1,\phi_1) & a(\phi_1,\phi_2) & \cdots & a(\phi_1,\phi_M) \\ a(\phi_2,\phi_1) & a(\phi_2,\phi_2) & \cdots & a(\phi_2,\phi_M) \\ \vdots & \vdots & \vdots & \vdots \\ a(\phi_M,\phi_1) & a(\phi_M,\phi_2) & \cdots & a(\phi_M,\phi_M) \end{bmatrix} \begin{bmatrix} \alpha_1(t) \\ \alpha_2(t) \\ \vdots \\ \alpha_M(t) \end{bmatrix}.$$

The corresponding problem can therefore be expressed as

$$B\frac{d\vec{\alpha}}{dt} + A\vec{\alpha} = F, \qquad \alpha_i(0) = u(N_i, 0), \; i = 1, 2, \ldots, M. \tag{9.67}$$

There are many methods to solve the above problem involving the system of first-order ODEs. We can use the ODE Suite in Matlab, but note that the ODE system is known to be very stiff. We can also use finite difference methods that march in time, since we know the initial condition on $\vec{\alpha}(0)$. Thus, with the solution $\vec{\alpha}^k$ at time t^k, we compute the solution $\vec{\alpha}^{k+1}$ at the time $t^{k+1} = t^k + \Delta t$ for $k = 0, 1, 2, \ldots$

9.8.1 Explicit Euler Method

If the forward finite difference approximation is invoked, we have

$$B\frac{\vec{\alpha}^{k+1} - \vec{\alpha}^k}{\Delta t} + A\vec{\alpha}^k = F^k, \tag{9.68}$$

$$\text{or} \quad \vec{\alpha}^{k+1} = \vec{\alpha}^k + \Delta t B^{-1}\left(F^k - A\vec{\alpha}^k\right). \tag{9.69}$$

Since B is a nonsingular tridiagonal matrix, its inverse and hence $B^{-1}\left(F^k - A\vec{\alpha}^k\right)$ can be computed. However, the CFL (Courant–Friedrichs–Lewy) condition

$$\Delta t \le Ch^2, \tag{9.70}$$

must be satisfied to ensure the numerical stability. Thus we need to use a rather small time step.

9.8.2 Implicit Euler Method

If we invoke the backward finite difference approximation, we get

$$B\frac{\vec{\alpha}^{k+1} - \vec{\alpha}^k}{\Delta t} + A\vec{\alpha}^{k+1} = F^{k+1}, \tag{9.71}$$

$$\text{or} \quad (B + \Delta t A)\,\vec{\alpha}^{k+1} = B\vec{\alpha}^k + \Delta t F^{k+1}, \tag{9.72}$$

then there is no constraint on the time step and thus the method is called unconditionally stable. However, we need to solve a linear system of equations similar to that for an elliptic PDE at each time step.

9.8.3 The Crank–Nicolson Method

Both of the above Euler methods are first-order accurate in time and second order in space, *i.e.*, the error in computing $\vec{\alpha}$ is $O(\Delta t + h^2)$. We obtain a second-order scheme in time as well as in space if we use the central finite difference approximation at $t^{k+\frac{1}{2}}$:

$$B\frac{\vec{\alpha}^{k+1}-\vec{\alpha}^{k}}{\Delta t}+\frac{1}{2}A\left(\vec{\alpha}^{k+1}+\vec{\alpha}^{k}\right)=\frac{1}{2}\left(F^{k+1}+F^{k}\right), \tag{9.73}$$

$$\text{or}\quad \left(B+\frac{1}{2}\Delta tA\right)\vec{\alpha}^{k+1}=\left(B-\frac{1}{2}\Delta tA\right)\vec{\alpha}^{k}+\frac{1}{2}\Delta t\left(F^{k+1}+F^{k}\right). \tag{9.74}$$

This Crank–Nicolson method is second-order accurate in both time and space, and it is unconditionally stable for linear parabolic PDEs. The challenge is to solve the resulting linear system of equations efficiently.

Exercises

1. Derive the weak form for the following problem:

$$-\nabla\cdot(p(x,y)\nabla u(x,y))+q(x,y)u(x,y)=f(x,y)\,,\ (x,y)\in\Omega,$$
$$u(x,y)=0\,,\ (x,y)\in\partial\Omega_1\,,\qquad \frac{\partial u}{\partial n}=g(x,y),\ (x,y)\in\partial\Omega_2,$$
$$a(x,y)u(x,y)+\frac{\partial u}{\partial n}=c(x,y),\ (x,y)\in\partial\Omega_3,$$

   where $q(x,y)\geq q_{min}>0$, $\partial\Omega_1\cup\partial\Omega_1\cup\partial\Omega_3=\partial\Omega$ and $\partial\Omega_i\cap\partial\Omega_j=\phi$. Provide necessary conditions so that the weak form has a unique solution. Show your proof using the Lax–Milgram Lemma but without using the Poincaré inequality.

2. Derive the weak form and appropriate space for the following problem involving the biharmonic equation:

$$\Delta\Delta u(x,y)=f(x,y)\,,\ (x,y)\in\Omega\,,$$
$$u(x,y)|_{\partial\Omega}=0\,,\quad u_n(x,y)|_{\partial\Omega}=0\,.$$

   What kind of basis function do you suggest, to solve this problem numerically?
   Hint: Use Green's theorem twice.

3. Consider the problem involving the Poisson equation:

$$-\Delta u(x,y)=1\,,\ (x,y)\in\Omega,$$
$$u(x,y)|_{\partial\Omega}=0,$$

   where Ω is the unit square. Using a uniform triangulation, derive the stiffness matrix and the load vector for $N=2$; in particular, take $h=1/3$ and consider

   (a) the nodal points ordered as $(1/3,1/3)$, $(2/3,1/3)$; $(1/3,2/3)$, and $(2/3,2/3)$; and
   (b) the nodal points ordered as $((1/3,2/3)$, $(2/3,1/3)$; $(1/3,1/3)$, and $(2/3,2/3)$.

   Write down each basis function explicitly.

4. Use the Matlab PDE toolbox to solve the following problem involving a parabolic equation for $u(x,y,t)$, and make relevant plots:

$$u_t=u_{xx}+u_{yy},\quad (x,y)\in(-1,\ 1)\times(-1,\ 1),$$
$$u(x,y,0)=0.$$

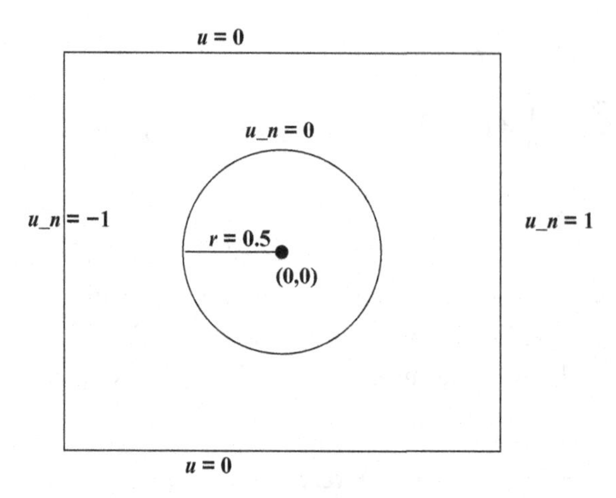

Figure 9.23. Diagram for Exercise 2.

The geometry and the BC are defined in Figure 9.23. Show some plots of the solution (mesh, contour, *etc.*).

5. Download the Matlab source code *f.m, my_assemb.m, uexact.m* from

   `www4.ncsu.edu/~zhilin/FD_FEM_Book.`

   Use the exported mesh of the geometry generated from Matlab, see Figure 9.23 to solve the Poisson equation

$$-(u_{xx} + u_{yy}) = f(x, y),$$

   subject to the Dirichlet BC corresponding to the exact solution

$$u(x, y) = \frac{1}{4}\left(x^2 + y^4\right) \sin \pi x \cos 4\pi y.$$

   Plot the solution and the error.

6. Modify the Matlab code to consider the generalized Helmholtz equation

$$-(u_{xx} + u_{yy}) + \lambda u = f(x, y).$$

   Test your code with $\lambda = 1$, with reference to the same solution, geometry and BC as in Problem 5. Adjust $f(x, y)$ to check the errors.

7. Modify the Matlab code to consider the Poisson equation

$$-\nabla \left(p(x, y) \cdot \nabla u(x, y)\right) = f(x, y), \tag{9.75}$$

   using a third-order quadrature formula. Choose two examples with nonlinear $p(x, y)$ and $u(x, y)$ to show that your code is bug-free. Plot the solutions and the errors.

9.8.4 Matlab PDE-Toolbox Lab Exercises

Purpose: to learn the Matlab *PDE* toolbox.

Use the Matlab PDE toolbox to solve some typical second-order PDE on some regions with various BC. Visualize the mesh triangulation and the solutions, and export the triangulation.

Reference: PDE Toolbox, MathWorks.

Test Problems

1. Poisson equation on a unit circle:

$$\begin{aligned} -\Delta u &= 1, \quad x^2+y^2<1, \\ u|_{\partial\Omega} &= 0, \quad x\le 0, \\ u_n|_{\partial\Omega} &= 1, \quad x>0. \end{aligned}$$

2. Wave equation on a unit square $x \in [-1,1] \times y \in [-1,1]$:

$$\begin{aligned} \frac{\partial^2 u}{\partial t^2} &= \Delta u, \\ u(x,y,0) &= \arctan\left(\cos\frac{\pi x}{2}\right), \\ u_t(x,y,0) &= 3\sin(\pi x)e^{\sin(\pi y/2)}, \\ u=0 \text{ at } x=-1 \text{ and } x=1, &\quad u_n=0 \text{ at } y=-1 \text{ and } y=1. \end{aligned}$$

3. Eigenvalue problem on an L-shape:

$$-\Delta u = \lambda u, \quad u=0 \text{ on } \partial\Omega.$$

   The domain is the L-shape with corners (0,0) , $(-1,0)$, $(-1,-1)$, $(1,-1)$, (1,1) , and (0,1).

4. The heat equation:

$$\frac{\partial u}{\partial t} = \Delta u\,.$$

   The domain is the rectangle $[-0.5,\ 0.5]\times[-0.8,\ 0.8]$, with a rectangular cavity $[-0.05,\ 0.05]\times[-0.4,\ 0.4]$; and the BC are:

   - $u=100$ on the left-hand side;
   - $u=-10$ on the right-hand side; and
   - $u_n=0$ on all other boundaries.

5. Download the Matlab source code 2D.rar from

   `www4.ncsu.edu/~zhilin/FD_FEM_Book`

   Export the mesh of the last test problem from Matlab and run assemb.m to solve the example.

General Procedure

- Draw the geometry;
- define the BC;
- define the PDE;
- define the initial conditions if necessary;
- solve the PDE;
- plot the solution;
- refine the mesh if necessary; and
- save and quit.

Appendix: Numerical Solutions of Initial Value Problems

This textbook is about finite difference and finite element methods for BVPs of differential equations assuming that the readers have knowledge of numerical analysis which includes numerical methods for IVPs of differential equations. The purpose of the appendix is to provide a necessary supplement for those readers who have not been exposed to numerical methods for IVP.

A.1 System of First-Order ODEs of IVPs

We briefly explain finite difference methods for IVPs, particularly how to use the Matlab ODE suite to solve such problems. The purpose of this appendix is to make the book more complete and self-contained. Usually, the more advanced material in the appendix can be found in the later stage of a numerical analysis class. The readers who have not been exposed to the materials will find the appendix useful. The methods described here can be used as time discretization techniques for various applications.

Consider an IVP of the following,

$$\begin{aligned} \frac{d\mathbf{y}}{dt} &= \mathbf{f}(t, \mathbf{y}), \\ \mathbf{y}(t_0) &= \mathbf{v}, \end{aligned} \tag{A.1}$$

where $\mathbf{f}(t, \mathbf{y})$ is a given vector (or scalar) function, and $\mathbf{v}$ is a known vector (or scalar). The following is the component form of the problem,

$$\begin{aligned} \frac{dy_1}{dt} &= f_1(t, y_1(t), y_2(t), \ldots, y_m(t)), \\ \frac{dy_2}{dt} &= f_2(t, y_1(t), y_2(t), \ldots, y_m(t)), \\ &\vdots \\ \frac{dy_m}{dt} &= f_m(t, y_1(t), y_2(t), \ldots, y_m(t)), \end{aligned} \tag{A.2}$$

with a known initial condition $y_1(t_0) = v_1$, $y_2(t_0) = v_2$, ..., $y_m(t_0) = v_m$. Such an IVP is a quasilinear system since the highest derivative terms (here is the first order) are linear. Often we have $t_0 = 0$. We also use the notation $\frac{dy}{dt} = \mathbf{y}'(t)$. Below are two such examples. The first one is a scale IVP while the second one is a system of ODEs of an IVP.

Example A.1.

$$y'(t) = \sin(y(t)), \qquad y(0) = 1.$$

Example A.2.

$$y_1'(t) = y_2(t), \tag{A.3}$$

$$y_2'(t) = -y_1(t) + y_1(t)\,y_2(t), \tag{A.4}$$

$y_1(0) = 1$, $y_2(0) = 0$.

Note that a high-order quasilinear ODE IVP can be converted to a system of first-order ODEs of an IVP as shown in the following example.

Example A.3. Convert the IVP,

$$y'' + \cos^2 t\, y' + \sin y = 0, \qquad y(0) = v_1, \quad y'(0) = v_2,$$

to a first-order system of ODEs of an IVP, where v_0 and v_1 are two given constants.

Solution: We set $y_1(t) = y(t)$, and $y_2(t) = y'(t)$. Thus we have $y_1'(t) = y'(t) = y_2(t)$; $y_2'(t) = y''(t) = -\cos^2 t\, y' - \sin y = -(\cos^2 t)\, y_2 - \sin y_1$. The original second-order IVP has been transformed to the following first-order system of ODEs of the IVP,

$$\begin{aligned} y_1'(t) &= y_2, \qquad y_1(0) = v_1 \\ y_2'(t) &= -\sin y_1 - (\cos^2 t)\, y_2, \qquad y_2(0) = v_2. \end{aligned}$$

A.2 Well-posedness of an IVP

Most of the numerical methods are designed for well-posed problems. Ill-posed problems need special algorithms and analysis. Here we discuss only well-posed first-order IVPs (A.1). A well-posed problem means that the solution exists, and is unique, and is not sensitive to perturbations of the data, such as the initial condition. For an IVP (A.1), the Lipschitz continuity can guarantee the well-posedness.

Theorem A.4. *Assume that* $\mathbf{f}(t,\mathbf{y})$ *is Lipschitz continuous in a neighborhood* $\mathcal{D}$ *of* $(t_0,\mathbf{y}_0)$, *that is, there is a constant L such that*

$$\|\mathbf{f}(t,\mathbf{y}_1)-\mathbf{f}(t,\mathbf{y}_2)\| \le L\,\|\mathbf{y}_1-\mathbf{y}_2\| \tag{A.5}$$

for all $(t,\mathbf{y}_1)$ *and* $(t,\mathbf{y}_2)$ *in* $\mathcal{D}$. *Then in the neighborhood* $\mathcal{D}$ *there is a unique solution* $\mathbf{y}(t)$ *to (A.1) that is not sensitive to perturbation of the data in the original problem.*

For a well-posed problem, the solution should be insensitive to the data such as the initial condition and coefficients involved. For the IVP (A.1), this implies the dynamical stability. If all the eigenvalues of the following matrix

$$\frac{\mathbf{Df}}{\mathbf{Dy}}\Big(t,\mathbf{y}(t)\Big)=\begin{bmatrix} \frac{\partial f_1}{\partial y_1} & \frac{\partial f_1}{\partial y_2} & \cdots & \frac{\partial f_1}{\partial y_m} \\ \frac{\partial f_2}{\partial y_1} & \frac{\partial f_2}{\partial y_2} & \cdots & \frac{\partial f_2}{\partial y_m} \\ \vdots & \vdots & \vdots & \vdots \\ \frac{\partial f_m}{\partial y_1} & \frac{\partial f_m}{\partial y_2} & \cdots & \frac{\partial f_m}{\partial y_m} \end{bmatrix} \tag{A.6}$$

at $(t_0,\mathbf{y}(t_0))$ are negative, that is, $\lambda_i\left(\frac{\mathbf{Df}}{\mathbf{Dy}}(t_0,\mathbf{y}(t_0))\right)\le 0$, $i=1,2,\ldots,m$, then the IVP is dynamically stable in a neighborhood of $(t_0,\mathbf{y}(t_0))$.

A.3 Some Finite Difference Methods for Solving IVPs

Assume that the IVP (A.1) is well-posed, and we wish to find the solution $\mathbf{y}(t)$ at some final time $T>t_0$. For the sake of simplicity, we set $t_0=0$ and discuss time marching methods. We start with a uniform time discretization. Given a parameter N, let $\Delta t=int(T/N)$, $t^0=0$, $t^n=n\Delta t$ with Δt be the uniform time step size. We wish to find an approximate solution $\mathbf{y}^n$ to the IVP problem at t^n, $\mathbf{y}^n\approx\mathbf{y}(t^n)$, $n=1,2,\ldots,N$.

A.3.1 The Forward Euler Scheme

The simplest method is the forward Euler method

$$\frac{\mathbf{y}^{n+1}-\mathbf{y}^n}{\Delta t}=\mathbf{f}(t^n,\mathbf{y}^n), \qquad n=0,1,\ldots \tag{A.7}$$

The method is first-order accurate, that is, $\|\mathbf{y}^n-\mathbf{y}(t^n)\|\le C\Delta t$. The scheme is conditionally stable meaning that we cannot take very large Δt.

We can give a reasonable guess of $\Delta t < 1$. Theoretically, the method is stable if we choose Δt such that $|\Delta t\,\lambda_i\left(\frac{D\mathbf{f}}{D\mathbf{y}}(t_0, \mathbf{y}(t_0))\right)| \leq 1$ for all i's.

A.3.2 The Backward Euler Scheme

The backward Euler method is

$$\frac{\mathbf{y}^{n+1} - \mathbf{y}^n}{\Delta t} = \mathbf{f}\left(t^{n+1}, \mathbf{y}^{n+1}\right), \qquad n = 0, 1, \ldots. \tag{A.8}$$

The method is also first-order accurate, that is, $\|\mathbf{y}^n - \mathbf{y}(t^n, \mathbf{y}(t^n))\| \leq C\Delta t$. The scheme is unconditionally stable, meaning that we can take any Δt. The difficulty of the backward Euler method is that we need to solve a nonlinear system of equations in order to get the approximate solution at the next time step. The advantage of this method is the strongest stability sometimes it is called a metastable method.

A.3.3 The Crank–Nicolson Scheme

The Crank–Nicolson scheme is the following

$$\frac{\mathbf{y}^{n+1} - \mathbf{y}^n}{\Delta t} = \frac{1}{2}\left(\mathbf{f}(t^n, \mathbf{y}^n) + \mathbf{f}(t^{n+1}, \mathbf{y}^{n+1})\right), \qquad n = 0, 1, \ldots. \tag{A.9}$$

The method is second-order accurate, that is, $\|\mathbf{y}^n - \mathbf{y}(t^n, \mathbf{y}(t^n))\| \leq C(\Delta t)^2$. The scheme is unconditionally stable for linear problems. The difficulty of the Crank–Nicolson scheme is that we need to solve a nonlinear system of equations in order to get the approximate solution at the next time step. The stability is better than the forward Euler method but not as good as the backward Euler method.

The methods that we discussed above are all one step methods. We refer the readers to a general textbook on *Numerical Analysis*, for example, the one by Burden and Faires (2010) about other methods including multistep methods for IVPs and analysis.

A.3.4 Runge–Kutta ($RK(k)$) Methods

In numerical analysis, the Runge–Kutta methods are a family of implicit and explicit finite difference methods, which includes various Euler Methods, used in temporal discretization for the approximate solutions of ODEs. In this subsection, we explain explicit Runge–Kutta methods for scalar IVPs ($m = 1$). Below are some examples.

Example A.5. The Crank–Nicolson scheme is an implicit scheme. We can use a predictor and a corrector scheme to make the scheme explicit,

$$\begin{aligned}\text{predictor:}\quad & \tilde{y}^{n+1} = y^n + \Delta t f(t^n, y^n) \\ \text{corrector:}\quad & y^{n+1} = y^n + \frac{\Delta t}{2}\left(f(t^n, y^n) + f(t^{n+1}, \tilde{y}^{n+1})\right).\end{aligned} \tag{A.10}$$

The scheme is called Heun's scheme which is second-order accurate and it is conditionally stable. The scheme above can be written as a one-step scheme

$$y^{n+1} = y^n + \frac{\Delta t}{2}\left(f(t^n, y^n) + f(t^{n+1}, y^n + \Delta t f(t^n, y^n))\right). \tag{A.11}$$

A class of Runge–Kutta (2) methods can be written as

$$y^{n+1} = y^n + \Delta t\left((1 - \frac{1}{2\alpha})f(t^n, y^n) + \frac{1}{2\alpha}f(\, t^n + \alpha\Delta t, y^n + \alpha\Delta t f(t^n, y^n))\right),$$

for some $\alpha > 0$. If $\alpha = 1$, we get the above Heun's scheme. If $\alpha = 1/2$, we get the middle point method. The middle point scheme is also second-order accurate.

The general k-th Runge–Kutta methods (RK(k)) can be written as

$$y^{n+1} = y^n + \Delta t \sum_{i=1}^{k} b_i f_i, \tag{A.12}$$

where

$$\begin{aligned} f_1 &= f(t^n,\, y^n), \\ f_2 &= f\left(\, t^n + c_2\Delta t,\, y^n + \Delta t\,(a_{21} f_1)\right), \\ f_3 &= f\left(\, t^n + c_3\Delta t,\, y^n + \Delta t\,(a_{31} f_1 + a_{32} f_2)\right), \\ &\vdots \\ f_k &= f\left(\, t^n + c_k\Delta t,\, y^n + \Delta t\,(a_{k1} f_1 + a_{k2} f_2 + \cdots a_{k,k-1} f_{k-1})\right). \end{aligned}$$

To specify a particular method, one needs to provide the integer k, and the coefficients a_{ij}, (for $1 \le j < i \le k$), b_i (for $i = 1, 2, \ldots, k$), and c_i (for $i = 2, 3, \ldots, k$). The matrix a_{ij} is called the Runge–Kutta matrix, while the b_i and c_i are known as the weights and the nodes. Those coefficients can be arranged as in Table A.1.

The Runge–Kutta (RK(k)) method is consistent if

$$\sum_{j=1}^{i-1} a_{ij} = c_i \quad \text{for } i = 2, \ldots, k.$$

Table A.1. *The table of the coefficients of an RK(k) method*

$$
\begin{array}{c|ccccc}
0 & & & & & \\
c_2 & a_{21} & & & & \\
c_3 & a_{31} & a_{32} & & & \\
\vdots & \vdots & \cdots & & & \\
c_k & a_{k1} & a_{k2} & \cdots & a_{k,k-1} & \\
\hline
 & b_1 & b_2 & \cdots & b_{k-1} & b_k
\end{array}
$$

A well-known RK(4) has the following form. Given a step size $h > 0$ and define

$$y^{n+1} = y^n + \frac{\Delta t}{6}\left(f_1 + 2f_2 + 2f_3 + f_4\right),$$

where

$$
\begin{aligned}
f_1 &= f(t^n, y^n), \\
f_2 &= f\left(t^n + \Delta t/2, y^n + \Delta t f_1/2\right), \\
f_3 &= f\left(t^n + \Delta t/2, y^n + \Delta t f_2/2\right), \\
f_4 &= f\left(t^n + \Delta t, y^n + \Delta t f_3\right).
\end{aligned}
$$

The coefficients table is

Table A.2. *The table of the coefficients of an RK(4) method*

$$
\begin{array}{c|cccc}
0 & & & & \\
\frac{1}{2} & \frac{1}{2} & & & \\
\frac{1}{2} & 0 & \frac{1}{2} & & \\
1 & 0 & 0 & 1 & \\
\hline
 & \frac{1}{6} & \frac{1}{3} & \frac{1}{3} & \frac{1}{6}
\end{array}
$$

A.4 Solving IVPs Using Matlab ODE Suite

It is quite easy to use the Matlab ODE Suite to solve a system of first-order ODEs of an IVP and visualize the results. The Matlab ODE Suite is a collection of five user-friendly finite difference codes for solving IVPs given by first-order systems of ODEs and plotting their numerical solutions. The three codes ode23, ode45, and ode113 are designed to solve nonstiff problems and the two codes ode23s and ode15s are designed to solve both stiff and nonstiff problems. The mathematical and software developments and analysis are given in Lawrence (1997). The Matlab ODE Suite is based on Runge–Kutta methods and can choose time step size adaptively.

To call one of five methods in the Matlab ODE Suite, for example, ode23, we can simply type in the Matlab command window the following

```
[t,y] = ode23('yp',[0,5], [1,0]');
```

assume that we have already defined the ODE system in a Matlab file called *yp.m*. For Example A.2, the Matlab script file *yp.m* can be written as

```
function yp = yp(t,y)
k = length(y); yp = zeros(k,1); % detect the dimension,
                                % using column vector.
yp(1) = y(2);              % Definition of f(t,y): y(1), y(2),
                           % ..., y(k).
yp(2) = -y(1) + y(1)*y(2);
```

Running the Matlab command will return outputs t, an array of different time from $t=0$ to $t=5$; the matrix $y \in R^{N,2}$, the approximate solution of $y(t)$ corresponding to the time. The column vectors $y(:,1)$ and $y(:,2)$ are the components of the approximation solution of $y_1(t)$ and $y_2(t)$, respectively. In Figure A.1, we plot the solution of $y_1(t)$ against time (the top plot); $y_2(t)$ against time (the middle plot); and the phase plot $y_2(t)$ against $y_1(t)$ (the bottom plot).

We can put everything into a Matlab script file called *ivp_ex2.m*, below, which we can modify later.

```
[t,y] = ode23('yp' ,[0,5], [1,0]');
y1=y(:,1); y2=y(:,2);
subplot(3,1,1); plot(t,y1)
xlabel('t'); ylabel('y_1(t)')
subplot(3,1,2); plot(t,y2)
xlabel('t'); ylabel('y_2(t)')
subplot(3,1,3); plot(y1,y2)
xlabel('y_1'); ylabel('y_2')
```

In the Matlab command window, we just need to type "*ivp_ex2*" and then we will see the plot.

We can replace ode23 with four other subroutines in the Matlab ODE Suite, ode45 and ode113, which are designed to solve nonstiff problems, and the two codes ode23s and ode15s, which are designed to solve both stiff and nonstiff problems. If we are not sure whether a problem is stiff or not, we can always use the codes for stiff problems.

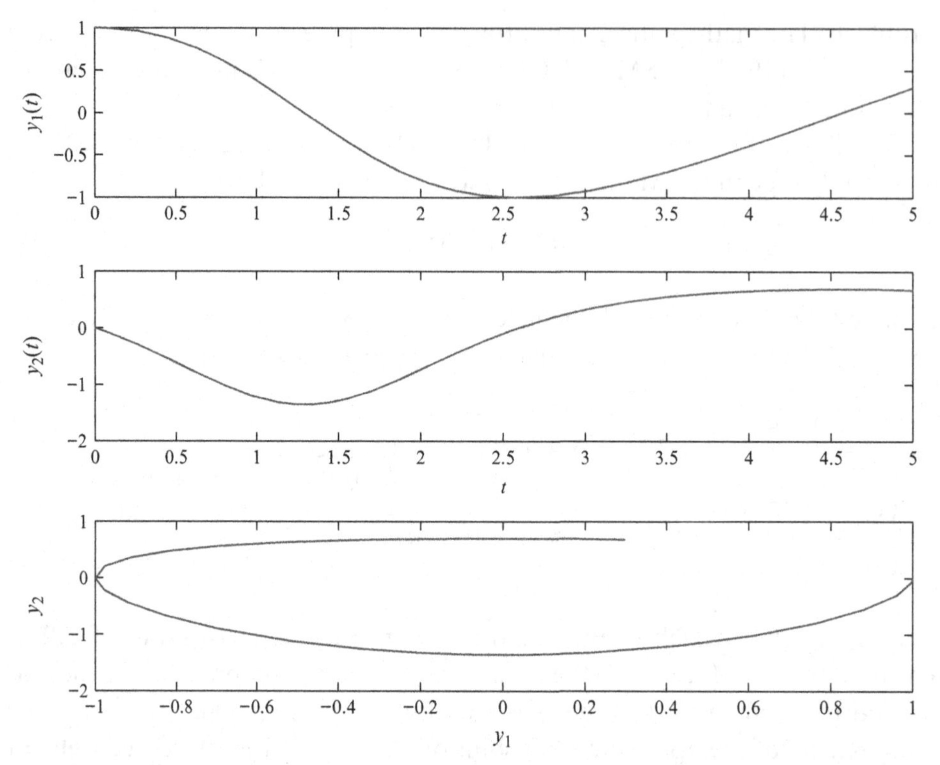

Figure A.1. Plots of the approximate solution. (a) $y_1(t)$ against time; (b) $y_2(t)$ against time; and (c) the phase plot $y_2(t)$ against $y_1(t)$.

Example A.6. As another demonstration, we solve the nondimensionalized Lotka–Volterra predator–prey model of the following system,

$$\begin{aligned} y_1' &= y_1 - y_1 y_2, \\ y_2' &= -a y_2 + y_1 y_2, \\ y_1(0) &= p_1, \qquad y_2(0) = p_2, \end{aligned} \tag{A.13}$$

where p_1 and p_2 are two constants, $y_1(t)$ is the population of a prey, while $y_1(t)$ is the population of a predator. Under certain conditions, predator and prey can coexist.

This problem is potentially stiff as we can see from some of plots of the solutions. We define the system in a Matlab function called *prey_prd.m* whose contents are

```
function yp = prey_prd(t,y)
global a
k = length(y); yp = zeros(k,1);
```

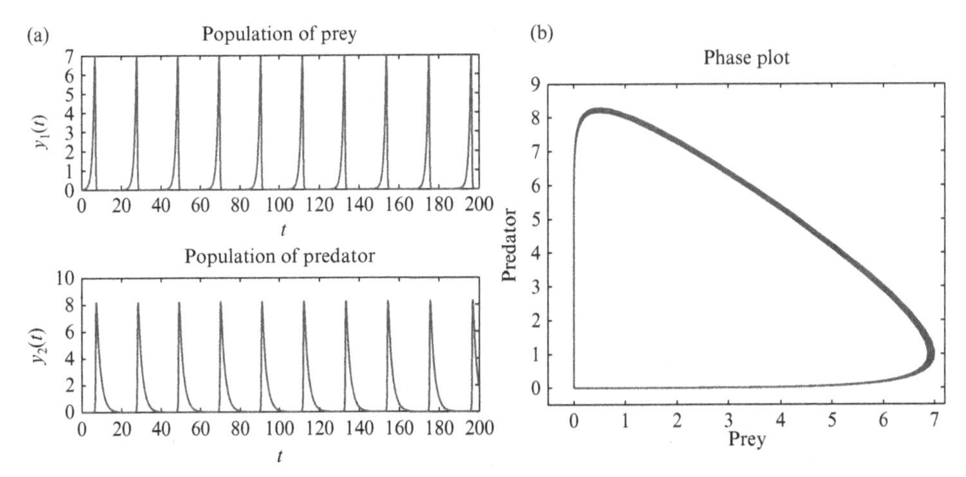

Figure A.2. Plots of the solution of the prey–predator model from $t = 0$ to $t = 200$ in which we can see the prey and predator coexist. (a) solution plots against time and (b) the phase plot in which the limit cycle can be seen.

```
yp(1)  =  y(1)  -  y(1)*y(2)  ;
yp(2)  =  -a*y(2)  +  y(1)*y(2);
```

To solve the problem, we write a Matlab script file called *prey_prd_drive.m* containing the following.

```
global a
a = 0.5; t0 = 0; y0 = [0.01 0.01];  tfinal=200;
[t y] = ode23s('prey_prd',[t0,tfinal],y0);
y1 = y(:,1); y2=y(:,2);                % Extract solution components.
figure(1); subplot(211); plot(t,y1);  title('Population of prey')
subplot(212); plot(t,y2);             title('Population of predator')
figure(2); plot(y1,y2)                % Phase plot
xlabel('prey'); ylabel('predator'); title('phase plot')
```

In Figure A.2, we plot the computed solution for the parameters $a = 0.5$, the initial data is $y_1(0) = y_2(0) = 0.01$. The final time is $T = 200$. The left plot is the solution of each component against time. We can observe that the solution changes rapidly in some regions indicating the stiffness of the problem. The right plot is the phase plot, that is, the plot of one component against the other. The phase plot is more like a closed curve in the long run indicating the existence of the limiting cycle of the model.

Exercises

1. Consider a model for chemical reaction,

$$y_1' = a - (b+1)y_1 + y_1^2 y_2,$$
$$y_2' = by_1 - y_1^2 y(2).$$

   Use the Matlab ODE Suite to solve the problem with various parameters, initial data, and final time.

2. Use the Matlab ODE Suite to solve the Lorenz equations,

$$y_1' = \sigma (y_2 - y_1)$$
$$y_2' = y_1 (\rho - y_3) - y(2),$$
$$y_3' = y_1^2 y(2) - \beta y_3,$$

   where σ, ρ, and β are constants. Try to solve the problem with various parameters, initial data, and final time. In particular, try to get the Lorenz attractor.

References

Adams, J., Swarztrauber, P., and Sweet, R., Fishpack: Efficient Fortran subprograms for the solution of separable elliptic partial differential equations. www.netlib.org/fishpack/.

Braess, D., *Finite Elements: Theory, Fast Solvers, and Applications in Solid Mechanics*. Cambridge University Press, Cambridge, 3rd edition, 2007.

Brenner, S. C., and Scott, L. R., *The Mathematical Theory of Finite Element Methods*. Springer, New York, 2002.

Burden, R. L., and Faires, J. D., *Numerical Analysis*. PWS-Kent Publ. Co., Brooks/Cole Cengage Learning, Boston, MA, 9th edition, 2010.

Calhoun, D., A Cartesian grid method for solving the streamfunction-vorticity equation in irregular regions. *J. Comput. Phys.*, 176:231–75, 2002.

Carey, G. F., and Oden, J. T., *Finite Element, I–V*. Prentice-Hall, Inc., Englewood Cliffs, NJ, 1983.

Chorin, A. J., Numerical solution of the Navier–Stokes equations. *Math. Comp.*, 22:745–62, 1968.

Ciarlet, P. G., *The Finite Element Method for Elliptic Problems*. North Holland, 1978. Reprinted in Classics in Applied Mathematics 40. SIAM, Philadelphia, 2002.

De Zeeuw, D., Matrix-dependent prolongations and restrictions in a blackbox multigrid solver. *J. Comput. Appl. Math.*, 33:1–27, 1990.

Dennis, J. E., Jr. and Schnabel, R. B., *Numerical Methods for Unconstrained Optimization and Nonlinear Equations*. SIAM, Philadelphia, PA, 1996.

Evans, L. C., *Partial Differential Equations*. AMS, Providence, RI, 1998.

Golub, G., and Van Loan, C., *Matrix Computations*. Johns Hopkins University Press, Baltimore, MD, 2nd edition, 1989.

Huang, H., and Li, Z., Convergence analysis of the immersed interface method. *IMA J. Numer. Anal.*, 19:583–608, 1999.

Iserles, A., *A First Course in the Numerical Analysis of Differential Equations*. Cambridge University Press, Cambridge, 2008.

Ji, H., Chen, J., and Li, Z., A symmetric and consistent immersed finite element method for interface problems. *J. Sci. Comput.*, 61(3):533–57, 2014.

Johnson, C., *Numerical Solution of Partial Differential Equations by the Finite Element Method*. Cambridge University Press, Cambridge, 1987.

LeVeque, R. J., Clawpack and AMRClaw – Software for high-resolution Godunov methods. *4th International Conference on Wave Propagation*, Golden, CO, 1998.

LeVeque, R. J., *Finite Difference Methods for Ordinary and Partial Differential Equations, Steady State and Time Dependent Problems.* SIAM, Philadelphia, PA, 2007.

Li, Z., *The Immersed Interface Method – A Numerical Approach for Partial Differential Equations with Interfaces.* PhD thesis, University of Washington, Seattle, WA, 1994.

Li, Z., The immersed interface method using a finite element formulation. *Appl. Numer. Math.*, 27:253–67, 1998.

Li, Z., and Ito, K., The immersed interface method – Numerical solutions of PDEs involving interfaces and irregular domains. *SIAM Front. Ser. Appl. Math.*, FR33, 2006.

Li, Z., and Lai, M.-C., The immersed interface method for the Navier–Stokes equations with singular forces. *J. Comput. Phys.*, 171:822–42, 2001.

Li, Z., Lin, T., and Wu, X., New Cartesian grid methods for interface problems using the finite element formulation. *Numer. Math.*, 96:61–98, 2003.

Li, Z., and Wang, C., A fast finite difference method for solving Navier–Stokes equations on irregular domains. *J. Commun. Math. Sci.*, 1:180–96, 2003.

Minion, M., A projection method for locally refined grids. *J. Comput. Phys.*, 127: 158–78, 1996.

Morton, K. W., and Mayers, D. F., *Numerical Solution of Partial Differential Equations.* Cambridge University Press, Cambridge, 1995.

Ruge, J. W., and Stuben, K., Algebraic multigrid. In S. F. McCormick, ed., *Multigrid Method.* SIAM, Philadelphia, PA, 1987, 73–130.

Saad, Y., GMRES: A generalized minimal residual algorithm for solving nonsymmetric linear systems. *SIAM J. Sci. Stat. Comput.*, 7:856–69, 1986.

Shampine, L. F., and Reichelt, M. W., The MATLAB ODE suite. *SIAM J. Sci. Comput.*, 18(1):1–22, 1997.

Shi, Z.-C., Nonconforming finite element methods. *J. Comput. Appl. Math.*, 149: 221–5, 2002.

Strang, G., and Fix, G. J. *An Analysis of the Finite Element Method.* Prentice-Hall, Upper Saddle River, NJ, 1973.

Strikwerda, J. C., *Finite Difference Scheme and Partial Differential Equations.* Wadsworth & Brooks, Belmont, CA, 1989.

Stüben, K., Algebraic multigrid (AMG): An introduction with applications. *Gesellschaft für Mathematik und Datenveranbeitung*, Nr. 70, 1999.

Thomas, J. W., *Numerical Partial Differential Equations: Finite Difference Methods.* Springer, New York, 1995.

White, R. E., *An Introduction to the FE Method with Applications to Non-linear Problems.* John Wiley & Sons, Charlottesville, VA, 1985.

Index

1D IFEM, 221
1D IIM, 39
1D Sturm–Liouville problem, 27, 169
1D interface problem, 39
1D interpolation function, 175
2D IFEM, 269
2D interface problem, 270
2D second order self-adjoint elliptic PDE, 230

abstract FE method, 219
ADI method, 99
  consistency, 103
  stability, 103
algebraic precision of Gaussian quadrature, 196
assembling element by element, 203

backward Euler method
  1D parabolic, 84
  2D parabolic, 98
  stability, 96
Beam–Warming method, 118
biharmonic equation, 263
bilinear basis on a rectangle, 265
bilinear form, 137
  1D elliptic, 169
  2D elliptic PDE, 231
boundary value problem (BVP), 3

Cauchy–Schwarz inequality, 163
CFL condition, 83, 111
characteristics, 109
compatibility condition, 50, 230
conforming FE methods, 168
conforming FEM, 143
Courant–Friedrichs–Lewy, 83
Crank–Nicolson scheme
  for 1D advection equation, 119
  1D parabolic equation, 87
  2D, 99
  in FEM for 2D parabolic, 274

cubic basis function
  in 1D H^1, 195
  in 1D H^2, 210
  in 2D $H^1 \cap C^0$, 261
  in 2D $H^2 \cap C^1$, 263

D'Alembert's formula, 122
degree of freedom, 190, 210
discrete Fourier transform, 92
discrete inverse Fourier transform, 92
discrete maximum principle in 2D, 57
distance in a space, 159
domain of dependence, 122
domain of influence, 122
double node, 212
dynamical stability, 281

eigenvalue problem, 26, 225
energy norm, 170, 216
essential BC, 209
essential boundary condition, 181, 209
explicit method, 81

fast Fourier transform (FFT), 71
FD approximation for $u'(x)$, 14
  backward finite difference, 14
  central finite difference, 14
  forward finite difference, 14
FD in polar coordinates, 69
FE method for parabolic problems, 272
FE space in 1D, 144
finite difference grid, 9
finite difference (FD) method, 5
  1D grid points, 9
  1D uniform Cartesian grid, 9
  consistency, 23
  convergence, 24
  discretization, 23
  finite difference stencil, 10
  five-point stencil in 2D, 52

ghost point method in 2D, 60
local truncation error, 10, 22
master grid point, 23, 52
stability, 24, 25
step size, 15
finite element method (FEM), 5, 135
a 1D element, 143
a 1D node, 143
assembling element by element, 147
hat functions, 136
piecewise linear function, 136
weak form in 1D, 136
finite element solution, 136
first order accurate, 15
FM spaces on rectangles, 265
forward Euler method
1D parabolic, 81
2D heat equation, 97
in FEM for 2D parabolic PDE, 274
stability, 94
Fourier transform (FT), 89
fourth order BVP in 1D, 209
fourth order compact scheme in 2D, 67
fourth-order compact scheme, 69
functional spaces, 158

Galerkin method, 143
Gauss–Seidel iterative method, 63
Gaussian points and weights, 196
Gaussian quadrature formulas, 195
global basis functions, 238
grid refinement analysis, 16
growth factor, 95

hat functions, 145

immersed finite element method (IFEM), 269
implicit Euler method, 274
initial value problem (IVP), 2, 279
inner product in H^m, 166
inner product in L^2, 162
interpolation function in 2D, 241
inverse Fourier transform, 72, 89

Jacobi iterative method, 62

$L^2(\Omega)$ space, 160
Lax–Friedrichs method, 110
Lax–Milgram Lemma, 214, 215, 231
Lax–Wendroff scheme, 117
leap-frog scheme, 114
for heat equation, 96
linear form, 137
linear transform in 2D, 246
local load vector, 149
local stiffness matrix, 149
local stiffness matrix and load vector, 204
local truncation error
1D parabolic, 82

maximum principle in 2D elliptic PDE, 55
mesh generation, 233
mesh parameters, 233
mesh size in 1D, 143
method of line (MOL), 85
method of undetermined coefficients, 19
minimization form, 220
mixed (Robin) boundary condition, 34
modified PDE, 115
Lax–Wendroff method, 118
multi-index notation, 159

natural boundary condition, 182, 210
natural jump conditions
in 1D, 40, 221
in 2D, 269
natural ordering, 52
neutral stability, 114
nine-point discrete Laplacian, 69
nonconforming IFE space, 271
numerical boundary condition, 120
numerical dissipation, 116
numerical solutions, 2

one-sided finite difference, 19
one-way wave equations, 108
ordinary differential equation (ODE), 1

partial differential equation (PDE), 1
piecewise linear basis function
in 2D H^1, 234
piecewise quadratic function in 2D H^1, 258
Poincaré inequality, 217, 231
pole singularity, 71
p-th order accurate, 15

quadratic basis function in 1D H^1, 190
quadrature formula in 2D, 247
quintic function, 264

red–black ordering, 52
Ritz method, 143, 146
round-off errors, 26, 55
Runge–Kutta methods RK(k), 283

second Green's theorem, 228
shape function, 199, 212
simplified FE algorithm in 2D, 251
Sobolev embedding theorem, 167
Sobolev space, 164
SOR(ω) method, 64
stability
Lax–Wendroff scheme, 117
staggered grid in polar coordinates, 71
steady state solution, 79, 105

Sturm–Liouville problem in 1D, 182
symmetric positive definite (SPD), 54

Taylor expansion, 14
time marching method, 81, 281
triangulation, 233
truncated Fourier series, 71

unconditional stability, 96
upwind scheme for 1D advection equation, 111
upwinding discretization, 29

von Neumann stability analysis, 89, 94

wave equation, 108
weak derivative, 164
weak form
  1D Sturm–Liouville BVP, 183
  2D elliptic PDE, 231